Modern Humanism

Science, Ethics, and the Stewardship of Earth

Kai Taraporevala

Wisdom and compassion for planetary flourishing

© **Kaikhushru Vicaji Taraporevala,** 2025

Published by MH Eudaimon Editions

Limited reproduction of excerpts up to 500 words is permitted without prior permission, provided full and clear attribution to the author and title of this work is included. Any reproduction exceeding this limit, or any use for commercial purposes, adaptation, translation, or distribution, requires prior written permission from the author. Case studies and accompanying teaching notes are available for educational use by contacting the author. The author reserves all other rights, including but not limited to rights of reproduction, distribution, adaptation, translation, performance, and derivative works. Correspondence: modernhumanism@gmail.com

ISBN:
979-8-99377 19-0-8 eBook
979-8-99377 19-1-5 Paperback
979-8-99377 19-2-2 Hardcover

Cover image and cartoons: © Kaikhushru Vicaji Taraporevala

The cover image symbolizes the interconnectedness of life and the universe, reflecting the spirit of science, the creative arts, service, and modern humanism. A strand of DNA forms the bark of a tree whose branches and roots represent Earth, evoking evolution by natural selection. To the left, an atom suggests the foundations of matter; to the right, a diagram alludes to life-forms. Stars and a distant galaxy in the background suggest the vastness and mystery of the cosmos. The hands represent human stewardship. What's missing in the image is chance. That is what comes into play when the reader sees the image and opens this book.

Select quotations are included under *fair use* for scholarly commentary and critical analysis.

The extract in *"A universe of atoms, an atom in the universe"* is from *The Pleasure of Finding Things Out* by Richard P. Feynman, copyright © 1999. Reprinted by permission of Basic Books, an imprint of Hachette Book Group, Inc.

The extract in Chapter 17 from *The Life You Can Save: How to Do Your Part to End World Poverty* by Peter Singer, copyright © 2009 is reprinted by kind permission of Peter Singer. The website at https://www.thelifeyoucansave.org provides details of charities through which you can make effective donations in the fight against extreme poverty.

Figure 7.3 in Chapter 7: 'Average blood lead in the U.S. and lead used in gasoline' ©1983, John Wiley & Sons, Ltd. is reprinted with permission from John Wiley & Sons, Ltd. The figure is from J.L Annest, "Trends in the blood lead levels of the U.S. population: The Second National Health and Nutrition Examination Survey (NHANES II) 1976-1980," in *Lead Versus Health*, M. Rutter and R.R. Jones, eds., New York: John Wiley & Sons.

The author is grateful to Leb (lebanonraingam.com) whose thoughtful design and meticulous formatting have enhanced the book's readability and aesthetic appeal.

The author thanks the Xavier's Resource Centre for the Visually Challenged (XRCVC), St. Xavier's College, Mumbai, for kind assistance in composing the ALT-TEXT that is heard when the images are read out from the eBook while utilising a screen reader.
www.xrcvc.org

This book was edited with limited assistance from AI tools functioning in an assistive capacity, used only for clarity and grammar. All intellectual content remains solely the author's own.

Loving gratitude to:

My parents whose encouragement allowed me to explore widely, make my own mistakes, and learn, and who always made sure I knew I had a home to return to.

Kamal, for her kind partnership in nurturing our wonderful children.

My children, Kayan, Kahan, and Katayun, for their love (storge and philia) and unwavering support: cheerful and foundational to my creative work. My discussions with them have clarified my thoughts and improved my work.

Katayun lent her gentle touch to refine the cartoons included throughout the book. These serve as an aide-mémoire to the ideas explored and, I hope, give rise to a few smiles.

N1 Does he know anything?

Immense indebtedness

Those hundreds of thousands of creative artists, scientists, mathematicians, and humanists who over millennia have contributed to human knowledge embody a spirit of agape—selfless devotion to truth, beauty, and the betterment of humankind. To be a small part of this community is to celebrate the privileges of our humanity, giving meaning to life.

Deep thanks

Rustom Antia, Merwan Engineer, and Ruggero Huesler. You opened so many new avenues of thought. You reasoned and helped me reason and change my mind. You read drafts, offered kind suggestions and comments, and made this book immeasurably better. Our philia has infused my spirit and given me courage and hope.

Warm acknowledgment

The many participants of my Science and Humanism workshops, whose generous engagement with evolving case studies and arguments enriched the dialogue and offered thoughtful insights across a wide spectrum of themes.

Sincere appreciation

Yoshihiko Abe, Arthur Cornell, Tony Hambro, Ludo Van der Heyden, Jay Sanklecha, Manoj Sanklecha, Anand Shanbhag, and Alan Thomson for reviewing the antepenultimate draft and for lending their insight to the book's final polish.

This book reflects my own views and interpretations. Any errors are mine alone.

How to Read this Book

Read this section first!

This is an ambitious book, a manifesto of modern humanism. I recommend beginning by reading the following sections first:
1. How to read this book
2. What's in a Word – 'Modern Humanism'
3. Outline of the book
4. Objectives and hope
5. Chapters 1 through 4

These selections will help readers understand why I wrote this book, what it aims to explore, how it is structured, and what kinds of case study-stories and commentaries it contains.

Reading guide

<u>From beginning to end</u>

To follow the book's linked dialogues and multiple streams of thought, the most coherent path is to read it from beginning to end.

<u>Stories first – commentaries later</u>

Begin with the case studies and return to the commentaries afterward. This approach allows the narratives to speak for themselves before engaging with analysis and reflection.

<u>Read by interest</u>

Although the book builds its arguments progressively, each Part can be read independently. Cross-references are provided where needed, allowing readers to move between sections based on their interests.

What's in a Word – 'Modern Humanism'

This book uses 'modern humanism' with intent, qualifying and distinguishing it from narrower or conflicting meanings. While rooted in traditions of reason, empathy, and freethought, it moves beyond human-centred inwardness. It embraces science, creative arts, ethical responsibilities, and the animating role of emotions. Drawing insights from evolution and cosmology, it expands care to all life on Earth, ecology and future generations, envisioning a flourishing planet with humans as reflective, compassionate, responsible, and accountable co-creators.

At its core this book is motivated by timeless questions that fuelled the quest of humanists: What does it mean to live well? To be a responsible and dignified human? It then extends this to the modern era. What kind of world shall we shape together now that we have acquired immense and potentially destructive powers? Where do we and shall we fit in the cosmos? These vital inquiries ought to shape how we aspire, relate, and live.

This is a hopeful, grounded vision shaped by the context and insights of modernity, and with a willingness to learn from our inevitable errors. As Goethe wrote, "By seeking and blundering we learn."[1]

Outline of the Book

While the book is formally divided into nine Parts, these cohere into five conceptual domains each reflecting a distinct mode of exploration. These domains arise from an expanded reading of Hume's 'is–ought' distinction, which frames the relationship between empirical knowledge (how the world IS) and normative reflection (how it OUGHT to be) as a method for navigating life's central questions.

Domain I:
The central **QUESTIONS OF LIFE**
Part 1

Domain II Science - How the world **IS** How we know and what we know Parts 2, 3, 4	**Domain III** Humanism - How the world **OUGHT** to be Humanism's foundational building blocks Part 5

Domain IV
CONSEQUENCES of is and ought

Needs, purposes, responsibilities, frameworks, and institutions

Parts 6, 7, 8

Domain V
VISIONS OF LIFE in a better world

Part 9

Objectives and Hope

My aim is to engender a serious dialogue and persuade. In 1946 Albert Camus wrote:

> "Throughout the coming years an endless struggle is going to be pursued between violence and friendly persuasion, a struggle in which, granted, the former has a thousand times the chances of success than that of the latter. But I have always held that, if he who bases his hopes on human nature is a fool, he who gives up in the face of circumstances is a coward. And henceforth, the only honourable course will be to stake everything on a formidable gamble: that words are more powerful than munitions."[1]
>
> CAMUS

I hope this book finds many readers, though it may speak most deeply to a few who carry the conversation forward. As Clair Patterson envisioned, a scattered band of emerging scientists, creative artists, and humanists might connect to explore the illness afflicting our culture (PATTERSON)[2]. From such exchanges, momentum may build toward fresh insight and humanist leadership. Ideologues have led us to the brink. It is our responsibility to shape a better world.

Contents

i. Loving gratitude to: ii
ii. Immense indebtedness iii
iii. Deep thanks iii
iv. How to Read this Book iv
v. What's in a Word – 'Modern Humanism' v
vi. Outline of the Book vi
vii. Objectives and Hope vii
viii. Figures, Tables, and Cartoons 1
ix. Key to References 5
x. A Libation to the Universe 6

1. Preface 8
   Modern Humanism 8
   The challenges we face 11
   Formative epochs of thought 12
   From pre-modern to humanist ethics 13
   A matter of urgency and survival 15
   Transformations 16
   Wagering everything 18

2. Narrative Foundations of This Book 22

I. Central QUESTIONS OF LIFE

Part 1. Central Questions of Life 24

3. Tetraethyl Lead in Gasoline: History and Politics 25
   Motor gasoline, knocking, and solutions 25
   Leaded gasoline and Ethyl Corporation 28
   Lead poisoning in 2025 36

4. Humanity's Defining Questions and a Roadmap 37
   I. Central QUESTIONS OF LIFE 37

II. How the world IS 37
III. Humanism—How the world OUGHT to be 39
IV. CONSEQUENCES of is and ought 40
V. VISIONS OF LIFE in a better world 42

II. Science - How the world IS

Part 2. Self-knowledge 44

5. Two Ascents of Everest 45
Two expeditions to Everest 45
The southeast ridge route up Everest 46
1996 - Adventure Consultants and Mountain Madness 48
John Hunt and the 1953 team 55
The first ascent in perspective 61

6. Origins of Human Judgment and Actions 63
An approach to self-knowledge 63
Cognitive biases 63
Genetic drivers of cognition 67
Cognitive neuroscience 71
Memes 72
The priming environment 74
Evolutionary perspectives 76
Strategies for cognitive integrity 77
Emotions as judgments 79

Part 3. A Test of knowledge 83

7. Tetraethyl Lead in Gasoline – The Science 84
Lead poisoning 84
Robert Kehoe 86
Clair Patterson 88
The effects of lead in children 94
No natural level of lead in humans 95

    The effects of lead at the molecular level 95
    The use of probabilities and statistics 96

8. The Cold Fusion Fiasco 98

9. Science as a Test of Knowledge 104
    An empirical basis and falsification 104
    Bayesian plausible reasoning 106
    Feedback loops and living with uncertainties 109
    Instruments 111
    The design of experiments and the use of controls 112
    Targeted observations 114
    Replication, universal findings, and no authorities 116
    Theories, models, and stories 117
    The reductionism – emergence spectrum 120
    Questions science addresses and those it cannot 123
    Non-specialists - the challenge of scientific understanding 123

III. Humanism - How the world OUGHT to be

Part 4. Evolution and Chance 126

10. A Brief History of Life 127
    The nature of life 127
    The Cambrian explosion 132
    Evolution by natural selection 135
    Mendel's experiments and findings 142
    The modern synthesis 143
    Mass extinction events 144
    Homo sapiens 146
    The historical record 150

11. Evolution as Chance and Necessity 151
    Evolution without purpose 151
    Evolution without direction 152

Insights into adaptations 154
Evolution shaped by contingency 155
The nature of chance 158

Part 5. Humanist Principles and Processes 161

12. Eradication of Smallpox and the COVID-19 Crisis 162
Smallpox and COVID-19 – postmortem questions 162
Infectious diseases and major pandemics 162
Modelling the spread of pandemics 164
Smallpox 165
COVID-19 171

13. The Building Blocks of Modern Humanism 180
Characterizing the building blocks 180
The basics of knowledge and belief 181
Ethical processes and decision-making 186
Philosophical boundaries and moral humility 189
Human-centred values and systems 192

IV. CONSEQUENCES of is and ought

Part 6. Needs, Purposes, and Activities 196

14. Tim Berners-Lee and Mark Zuckerberg 197
Questions on needs, purposes, and activities for flourishing 197
Tim Berners-Lee 198
Mark Zuckerberg 202

15. Needs, Purposes and a Humanist Worldview 210
Evaluating the actions of Berners-Lee and tech billionaires 210
Worldviews 214
Needs and capabilities 217
Deprivations and emergent worldviews 226
A humanist worldview 232

16. Activities in the Sciences, Arts, and Humanism 237
- Cooperation and monopoly seeking competition 237
- Science and engineering 241
- Humanism, the creative arts, and science 245
- Humanist flourishing 251
- Professional scientists – practices, pressures, and challenges ... 258

Part 7. Planetary Flourishing 261

17. The Shallow Pond 262

18. Collapse of the Aral Sea 265

19. The UN Durban Conference 2001 269

20. Poverty, conservation, and historical responsibility 272
- An approach to global concerns 272
- Moral choices - poverty and precarity 273
- Conservation 275
- Humanism and history 280
- Planetary flourishing 288

Part 8. Structures and Frameworks 289
- Clarifying the Scope and Intent of Part 8 290

21. Keynes' Vision for the Future 291

22. Diagnosing the Crisis and Evaluating Possibilities 293
- Humanism and economics 293
- Revolutions vs. dialogic justice 294
- Status of Keynes' vision 2025 296
- Ortega y Gasset's 'Mass-Man' 297
- A chakra - meaningless work to mindless time-wasting 298
- The fallacy of rankings and a question on change 299
- Essential embodied well-being 300

23. Approaches to Economic and Social Structures 303
Summary of humanist-economic themes 303
Economic institutional designs 309
Cautionary tales - India 316
Cautionary tales - China 319
Experimentation in the social sciences 320
Libertarian economics vs. humanist counterpoints 322
Social Choice theory and Arrow's Theorem 323
Examples of change 325

24. Approaches to Democracy - Power and Politics 327
Mapping humanist thinkers 327
The social contract and boundaries of power 327
Historical and alternative models 333
M. N. Roy – Radical Humanism 335
Civic foundations and democratic renewal 338

25. A Framework for a Humanist Education 341
Education – A vital lever 341
A modern humanist framework for education 342
Education, technology, and AI 349

V. VISIONS OF LIFE in a better world

Part 9. Directions Towards a Better World 351

26. An Ape Looks at a Pale Blue Dot 352
The Pale Blue Dot and Carl Sagan's vision for the future 352
The Voyager missions 356

27. Augmenting Reasoned Discourse 359
A landscape of knowledge 359
Humanism vs Idols of the Mind 361
From persecution to exclusion to tolerance to humanism 363
Distinctions between intelligence and rationality 365

xiii

Humanist discourse as a layered process 367
Flourishing without fixed moral truths 372

28. From Cleverness to Wisdom 373
Homo sapiens – A clever animal 373
The S Curve 373
Tipping points 376
Elixirs of death - Testaments of hope 377
A humanist scale of civilizations 378
A civilizational rite of passage 381
Supreme existential challenges –reflections of a humanist 382

Supplementary Dimensions

Appendices 387

29. Aspects from the Philosophy of Science 388

30. Plausible Reasoning and A Probability Calculation 392
Understanding risks 392
Plausible reasoning 394
The normal distribution 396
A binomial distribution calculation 396
Kardashev scale – energy estimates and calculations 397

31. The Allure of Mathematics and Moral Certainty 399

32. Afterword - In Defence of Depth 403

33. Select Bibliography 404

Endnotes and Citations 410

Index 444

Figures, Tables, and Cartoons

Cartoons

N1 Does he know anything?	ii
N2 Nero fiddling – Mobile scrolling	21
N3 Two faces of TEL	43
N4 The Orange King	82
N5 Galileo fatwaed	125
N6 Octopus - Cogito, Arm Sum	160
N7 Trade-offs and Precautionary principle - Boiling Alive	195
N8 Last man on Earth	260
N9 The Earth and Mars	268
N10 A manual for febrile triviality	350
N11 Last wager	386

Economic / political frameworks

T23.1 Inequality and Redistribution	304
T23.2 Role of institutions	304
T23.3 Justice and fairness	305
T23.4 Growth and development	306
T23.5 Inequality and Redistribution	307
T23.6 Measurement of well-being	308
F24.1 Coercion/Consent vs Political Orientation	328

Everest

F5.1 The Southeast Ridge route up Everest	47

Evolution and chance

F6.1 The evolutionary relationship between apes	68
T10.1 Taxonomy for humans	131
T10.2 Biomass of life on Earth	132
F10.1 Geological ages of Earth	134
F10.2 Bell-curve - effects of mutations and selection	136
F10.3 Evolution in terms of change and adaptation	137
F10.4 The 3-D Bell-curve collection of traits	138
F10.5 Adaptation traits showing several local optimizations	139
F10.6 Evolution is net-like not branch like	142
T10.3 Big five extinctions over the last 500 million years	145
T10.4 Examples of human vestigial organs / features	149
F11.1 Evolutionary Complexity	151

Human behaviour

T6.1 Results of money priming experiments	75
T22.1 Average daily hours spent on digital content	297

Humanism

F1.1 Conjoined triplets: science, the creative arts, and humanism	8
T13.1 Humanism's building block	181
F15.2 The close compact between moral decisions and science	233
T31.1 Utilitarianism vs. Parfit vs. Modern Humanism	402

Leaded gasoline

F3.1. U.S. Car Production and Motor Gasoline	25
F3.2. Consumption of White lead and Tetraethyl lead	33
T3.1 Maximum acceptable Blood Lead Levels in children	35

F7.1 Lead Concentrations in Polar Ice of Northern Greenland	91
F7.2 Estimated lead accumulations across human populations	92
F7.3 Average blood lead in the U.S. and lead used in gasoline	95
F7.4 Violent crime and lead in the U.S.	97

Needs, capabilities, worldviews

F15.1 Needs, capabilities, the environment and worldviews	211
T15.1 Shanahan's AI model used to interpret human behaviour	213
T15.2 Maslow's hierarchy of needs	219
T15.3 Nussbaum's list of capabilities	221
T15.4 Emergent worldviews due to deprivations	227
T15.5 Emergent worldviews – historic and contemporary	229
T15.6 Global status of human needs / capabilities	231
T15.7 Comparison of worldviews	236
T17.1 Global extreme poverty	264
T27.2 Characteristics of dogmatic and humanist cultures	366

Sagan's vision

F26.1 First photograph of Earth from space	353
F26.2 First Hubble Deep Field Image	354
F26.3 The Pale Blue Dot	355
T30.2 Kardashev scale-energy estimates	397

Science

F9.1 Conceptual framework for the scientific learning process	110
F9.2 Ruler showing centimetres and millimetres	112
F9.3 Data on O-Rings from previous flights in cold weather	115
F9.4 Data on O-Ring incidents from all flights	115

F9.5 Kehoe and Patterson's abstractions	118
F9.6 Top Down and Bottom-Up effects	122
T16.1 Professional Scientists Worldwide	258
T27.1 The Johari Window Landscape of knowledge	359
T30.1 Illustration of Type I and Type II errors	393

Smallpox, Covid-19, pandemics

T12.1 Most harmful pandemics/ epidemics in human history	163
T12.2 Reported and excess deaths due to COVID-19	175
T12.3 Vaccine type, roll out and efficacy	176

Technology and growth

F28.1 Increased rate of technological inventions	374
F28.2 The fallacy of non-stop compounding	375
F28.3 Limits to compound growth – the logistics function	376

Key to References

The exemplar-protagonists of this book, whose seminal ideas I have happily wrestled and often resonated with, are acknowledged using a special typographic attribution format. For example:

- "To understand the actual world as it is, not as we should wish it to be, is the beginning of wisdom" (RUSSELL)[1].

Or:

- "The best we can do is a compromise: learn to recognize situations in which mistakes are likely and try harder to avoid significant mistakes when the stakes are high."[2]
KAHNEMAN

As eloquently expressed by William Harvey, the many references are not intended "to make a parade of the strength of my memory, the extent of my reading, or the amount of my pains."[3] I have included these references so that readers may enter directly into the adventure of conversations with these guiding minds.

Syntax

I occasionally use the ungrammatical (colon + list within a sentence that continues):

"Sen's five instrumental freedoms: political, economic, social, transparency, and protective security, were unevenly realized."

I hope pedants will forgive the indulgence where rhythm trumps grammar.

A Libation to the Universe

A universe of atoms, an atom in the universe
Richard P. Feynman

From *The Pleasure of Finding Things Out* by Richard P Feynman, copyright © 1999. Reprinted by permission of Basic Books, an imprint of Hachette Book Group, Inc.[1]

I stand at the seashore, alone, and start to think
There are the rushing of waves…
mountains of molecules, each stupidly minding its own business…
trillions apart…
yet forming white surf in unison.

Ages on ages…
before any eyes could see…
year after year…
thunderously pounding the shore as now.
For whom, for what?…
on a dead planet,
with no life to entertain.

Never at rest…
tortured by energy…
wasted prodigiously by the sun…
poured into space.
A mite makes the sea roar.

Deep in the sea,
all molecules repeat the patterns of one another
till complex new ones are formed.
They make others like themselves…
and a new dance starts.

Growing in size and complexity…
living things,
masses of atoms, DNA, protein…
dancing a pattern ever more intricate.

Out of the cradle
onto the dry land…
here it is standing…
atoms with consciousness…
matter with curiosity.

Stands at the sea…
wonders at wondering…
I…
a universe of atoms…
an atom in the universe.

1

Preface

Modern Humanism

Science, the creative arts, and modern humanism are best understood as conjoined triplets, inseparable human endeavours forming a larger unity where none can exist for long without the others (F1.1).[1] Motivated by this framework, especially the role of modern humanism, I wrote this book to explore how we might think and act, both individually and collectively, in ways that lead to a better world.

The creative arts and emotion provide the spark and animate scientific and humanist endeavours. They engender compassion and kindness, while illuminating and celebrating the human condition.	Humanism provides the essential environment within which science and the creative arts thrive, self-reflection cultivated, emotions tempered, service pursued, and moral decisions taken towards planetary flourishing.

Science embodies rational inquiry – a test of knowledge in understanding the universe. The findings of science enable emotions and service to be based on true beliefs and provide perspectives and inputs to the creative arts and humanist discourse.

F 1.1 Conjoined triplets: science, the creative arts, and humanism

Humanism makes the moral choice to apply science's test of knowledge about all aspects of the universe. 'Science' is a learning process that self-corrects through a feedback loop of replicable observations and experiments, hypothesis and model making, plausible reasoning, falsification attempts, and revision.

Where scientific knowledge is uncertain or incomplete, humanism embraces living with uncertainty, but also prudence due to deep care for life on Earth. It bases ethical decisions on the connection between our moral choices, purposes, emotions, and the findings of science.

A narrow focus is necessary for detailed investigations into natural phenomena, artistic processes, or human needs. However, limiting attention to narrow disciplines does not help us learn how to live flourishing lives. "Knowledge," wrote Will Durant, "split into a thousand isolated fragments, no longer generates wisdom" (DURANT)[2]. Humanism is a systemic discourse embracing the full spectrum of human thought, emotion, and imagination, guiding us toward lives of kindness and compassion.

Modern humanism considers a widening circle of stakeholders and generational time scales. It is a way of journeying through life by sharing knowledge, collaborating through vigorous discourse, embracing dissent, and revising one's views in light of reasoned argument.

Importantly, humanism does not arrogate to itself delusions of a perfect or best of all possible worlds. Utopias are grounded in unfalsifiable theories that justify any means in pursuit of purportedly better ends. The result, as Stephen Kotkin writes, is that "The most terrible crimes became morally imperative acts in the name of creating paradise on Earth."[3]

Invigorated by this synthesis we become responsible, here and now, for answering Carl Sagan's question: "What kind of species do we want to be in this vast cosmos?" (SAGAN)[4].

The humanist answer provides a direction towards a world where:
- Basic needs and capabilities are not just secured but precarity alleviated, and pleasures embraced as celebrations of our human condition.
- Love, kindness, and compassion are transformed from nouns into living verbs — embodied through deliberate action and the agency to shape one's environment.
- Activities promoting individual and planetary flourishing include:
  - Scientific inquiry as a human drive to understand and discover, and a means of grounding emotion in true belief. It reveals our embeddedness within living systems and illuminates the patterns of the cosmos.
  - The creative arts convey beauty with emotional resonance and encompass all forms of imagination and artistic expression.
  - Self-reflection, which integrates philosophical reasoning with ethical and historical awareness, guiding moral conscience.
  - Service activities—education, caregiving, and civic participation—enhance dignity, equity, and shared purpose. Grounded in empathy, they extend across political, commercial, and engineering domains, not for power or profit, but to foster resource sharing, ecological responsibility, and social concern that shape humane systems and sustainable futures.

These activities are not solitary pursuits, collective endeavours, or forms of dehumanized homogeneity. They are expressions of human creativity enriched by community and dialogue and aim at planetary flourishing.

The challenges we face

Many of our problems are long-standing: poverty and precarity persist; pandemics spill from animals to humans; autocrats wield malignant power; wars and pogroms remain a grim constant. But in our interconnected world, these risks have multiplied and new ones have emerged: climate change, engineered pathogens, nuclear catastrophe, pervasive surveillance, and AI, each threatening life itself.

In "The Dawn of Everything," David Graeber and David Wengrow argue that we have lost three fundamental freedoms to:[5]

- Move away or relocate. Geographic mobility, once a hallmark of freedom and dignity, is now constrained by tightening borders, surveillance regimes, and nationalist policies. The humanist ideal of welcoming those seeking a better life has eroded and under threat. Meanwhile, global externalities — climate change, nuclear risks, pandemics — make true escape impossible. We have only one Earth.
- Disobey. The right to dissent or ignore commands without punishment is eroding — not just in authoritarian regimes but across all nations.
- Create new social realities. The ability to imagine, experiment with, and construct alternative forms of social organization is increasingly suppressed by economic and political powers.

Dogmas promise salvation but demand cognitive rigidity, blind to contradiction and impervious to evidence. Many defer responsibility to divine intervention or demagogues, thereby abdicating the urgent need for action. This mindset becomes an obstacle to collective survival. Modern Humanism seeks to reclaim these lost freedoms as the foundation for ethical, participatory societies.

Formative epochs of thought

<u>Before the Enlightenment</u>

Across cultures and centuries, human societies have been part of exchanges based on rational inquiry, artistic innovation, and civic life.

- In classical India, philosophical schools such as Nyaya and Saṃkhya formalized logic, epistemology, and metaphysics, while yoga codified empirical approaches to anatomy and health.
- In Ancient Greece, thinkers from the Presocratic era through the classical period laid foundational principles of rational inquiry, civic ethics, and artistic expression. Epicurus advanced a materialist ethics centred on friendship and freedom from fear, linking atomism with moral philosophy.
- During the Islamic Golden Age (8th–13th centuries), scholars in Baghdad, Cordoba, and Cairo translated and expanded Greek, Persian, and Indian texts, developing algebra (al-Khwarizmi), optics (Ibn al-Haytham), and clinical medicine (Avicenna). The House of Wisdom fostered interdisciplinary research across astronomy, philosophy, and engineering.

- In Song Dynasty China, polymaths such as Shen Kuo and Su Song advanced astronomy, geology, and mechanical design, while Confucian and Daoist traditions debated ethics, governance, and education.
- In Renaissance Italy, figures such as Galileo, Leonardo da Vinci and Girolamo Cardano pursued anatomical and planetary studies, hydrodynamics, and probability theory.

These traditions, shaped through centuries of influence and exchange, required courage, perseverance, and often led to reprimand and punishment. They reveal the futility of dividing human intellectual, artistic, and ethical achievements into 'Western,' 'Eastern,' or any other compartments. They all built the foundations of humanism into what it is today.

The European Enlightenment

Its central ideas sought happiness in the present over preparation for an afterlife, treated reason as a shared faculty refined through dialogue, embraced emotion alongside logic, and championed the spread of knowledge to empower individuals and challenge authority.

These ideals helped shape modern humanism. Yet, as Richie Robertson warns, Enlightenment thought must be understood in its own cultural context.[6] Many thinkers focused on the interests of property-owning men and often excluded women, the poor, colonized peoples, and slaves.

From pre-modern to humanist ethics

Earlier epochs, including the Enlightenment, were shaped by faith in a universal moral code, fuelling visions of utopia and emancipation. But they were untouched by Darwin, Hutton, Einstein, quantum mechanics, ecological systems, and cognitive science. They lacked concepts of evolution, deep time, cosmic immensity, ecological interdependence, neural complexity, and scientific uncertainty. Static categories of 'man' and 'nature' prevailed; the Earth was divine, the universe small, nature passive, and the cosmos predictable.

Modern humanism rejects dogma and proof as moral anchors. It draws from the conjoined triplets of humanism, science, and the creative arts, respecting the gap between 'is' and 'ought', and treating moral issues as choices, not certainties.

The idea of progress is now tempered by history. Though millions live longer and healthier lives, the twentieth century saw 238 million killed in wars and pogroms, and the twenty-first has added millions more. Survival is not success, and neither is it assured.

Yet Enlightenment and earlier epoch thinkers grappled with justice, liberty, education, and human rights. Today, many knowledge-makers avoid these imperatives, while artists often overlook the role of science and humanism. To critique thinkers of earlier eras is not to diminish them, but to honour their legacy by deepening their ideals through a modern lens.

Before continuing, the following is an acknowledgement of a profoundly reflective work on the Enlightenment.

Ritchie Robertson's "The Enlightenment: The Pursuit of Happiness 1680–1790" is a work of immense scholarship and insight (ROBERTSON)[7]. Robertson offers a sweeping account of Enlightenment thought, illuminating its pursuit of happiness across philosophy, politics, and culture.

While Robertson's focus remains historical and literary, this book builds on that legacy by incorporating frameworks for moral decisions and practical actions in economics, social structures, political institutions and education as they confront us today.

A matter of urgency and survival

Many embrace the fruits of science, evident in the near-universal electrification of the world, but discard science as a way of thinking when confronting doctrine. This split reflects a failure to distinguish science from technology. The latter, wielded by engineers and their masters, including governments, corporations, and billionaires, reshapes our world, often without understanding or care.

On 26 September 1983, we came close to nuclear war. One of the Soviet Union's monitoring systems indicated that there were multiple missiles launched by the U.S. On that day Stanislav Petrov was the Soviet Union officer in charge of monitoring such launches. Petrov's computer system said that the reliability of the missile launches was 'highest'. Yet Petrov was "not quite sure it was possible."[8]

The Soviet Union's early warning system had mistaken sunlight reflecting off clouds for the launch of five U.S. Minuteman missiles. Thirty years later, Stanislav Petrov recalled putting the odds of nuclear war at 50-50. He was the only officer on duty with a civilian education and believed that if anyone else had been there, the alarm would have been raised. For saving the world, Petrov was relentlessly interrogated and never rewarded by Russia.[9]

Such near misses are not rare. As of 2023, 3,720 nuclear warheads remain on high alert. Even a 0.001% chance of accidental launch per warhead yields a 4% probability of global catastrophe (see C30). Is this not existentially terrifying?

Yet most remain comforted by belief systems and blind to science and chance. We have already used atomic weapons on Hiroshima and Nagasaki. We have not changed. COVID-19 showed there are no islands of safety.

Earth has survived five mass extinctions. But now, with unprecedented power, global interconnection, and ecological damage, we may be driving ourselves toward a sixth. The time for fundamental action is now.

Transformations

<u>I, Robot</u>

This is a brief account of my transformation, offered for context, that my journey towards humanism may inspire wider renewal.

As a teenager, I was electrified by feedback mechanisms—a self-correcting learning process. Revising beliefs with evidence felt honest and subversive. I admired Charles Darwin's temper of mind: "I have steadily endeavoured to keep my mind free so as to give up any hypothesis, however much beloved (and I cannot resist forming one on every subject), as soon as facts are shown to be opposed to it" (DARWIN)[10].

Though I earned two Master's degrees in Physics, I left science for business, completing an MBA. Instead of escaping Plato's cave, I walked into it. Yet the commercial world taught me much about human nature. Facing my own vulnerabilities, living amid poverty and ecological destruction in India, and caring for aging parents strengthened my compassion and awareness.

I invoke the Jain sentiment of *Michhami Dukkadam*—seeking forgiveness from any sentient life I may have harmed, including through extractive consumption.

At fifty, I reengaged with science, the arts, and philosophy, exploring what it means to live a well-thought life. This book began over a decade ago, shaped by workshops, case studies, and conversations that opened new paths of thought.

The journey continues. There is still much to learn. But as Laozi wrote, "Fill your bowl to the brim and it will spill. Keep sharpening your knife and it will blunt."[11] There comes a moment to pause, reflect, and share. That moment has arrived.

Awakening the robot

My transformation is just one example of a possibility, that humans, far from being passive products of biology and culture, can awaken to new ways of living and learning. It is not, as Keith Stanovich describes, a robot's rebellion.[12] Stanovich was referring to the perspective, suggested by Richard Dawkins, that humans are "survival machines—robot vehicles blindly programmed to preserve the selfish molecules known as genes."[13] The possibility of change is better understood as humans awakening to how they can learn about the world, how they are influenced by their genes, mammalian and primate instincts, and social memes, and how they can begin to live with new purposes as they continue a voyage of discovery.

We have evolved, as Jacques Monod said, through "Chance and Necessity."[14] Chance through random mutations shaped by physical laws, without purpose or design, the emergence of the Solar System, Earth, and our improbable evolutionary path. 'Necessity' governed by both natural selection and the physics by which molecules interact, and chemical processes occur. The emergence of *Homo sapiens* carried no progress or certainty.

Our desires and thoughts are rooted in ancient environments. We cannot deny these impulses, but we can live with awareness, guided by science, kindness, compassion, and creative pursuits. In doing so, we may cultivate flourishing lives and become stewards of life on Earth.

Wagering everything

There is much cause to be a resigned pessimist. Stephen Emmott's "Ten Billion," concludes that Earth's system will go into a "runaway" process, a tipping point like that of a fever over 42°C. He says, "I hope I'm wrong, but the science points in my not being wrong."[15]

The brave Daniel Ellsberg, responded to the questions "Do you find yourself being cynical about our future? Is it possible to change the course of history?" with: "I can't bring myself to say that the odds are any more than low. But it's not impossible. My own case shows that. The challenge will be with us all the way. But is it worth trying? Yes" (ELLSBERG)[16].

Albert Camus believed that "there is only one honourable choice: to wager everything on the belief that in the end words will prove stronger than bullets."[17]

Preface

There are two additional reasons that make Camus' wager worth taking.

Most of us are filled with awe and wonder when looking at a star-filled night sky. This triggers our imagination and curiosity and gives rise to emotions that bring awareness of our being part of a spectacular universe. There is a "grandeur in this view of life,"[18] and this provides a powerful spark for change.

Second, humanism provides a positive and hopeful vision that we can use to go forward. There are innumerable examples of individual and team achievements. A recent exemplary project is the building and successful deployment of the James Webb Space Telescope (JWST).

Amber Straughn, Astrophysicist at NASA explains, that the JWST is "the biggest, most complex observatory we [have] ever sent to space. [It enables us] to see the first light, the first galaxies in our universe. JWST's images will help us get closer to answering those questions of; 'Where do we come from?' 'How did we get here?,' and 'Are we alone?'"[19]

Astonishingly, observations suggest that there are roughly one septillion stars (1 followed by 24 zeros)[20] in the universe making astronomy "a humbling and character-building experience" (SAGAN)[21].

After three decades of planning and work JWST was launched into space on 25 December 2021 and is continuing the exploration of our vast cosmos from its home 1.5 million kilometres from Earth. Thomas Zurbuchen, Head of Science at NASA during the development and launch of JWST, said:

"JWST is the toughest and biggest leap of any NASA mission ever done. Making and deploying JWST was a super hard thing that [was] almost impossible…JWST is a demonstration of what is possible when we come together and do something hard. 10,000 people, all of them with strengths and weaknesses, all of them with many reasons why JWST should not work [came] together as an excellent team. If we can do that imagine all the other problems we can solve. JWST is humanity at its best, a selfless pursuit of what is really out there, learning more about ourselves and our history. It is incredibly amazing."[22]

What a shame it would be, as we continue to learn about our incredible universe, including being on the cusp of evidence of life (existing, extant or possible in the future) on other planets and moons inside and outside our solar system, if we were to destroy life on Earth. As Scarlin Hernandez, Spacecraft Engineer NASA says, "There is so much more to explore in our universe and how we are connected. This is only the beginning."[23]

N2 Nero fiddling – mobile scrolling

"But man is a part of nature, and his war against nature is inevitably a war against himself" (Rachel Carlson)[24]

2

Narrative Foundations of This Book

<u>The use of stories</u>

I began writing parts of this book in September 2012. What emerged was a series of essays centred on 'facts' and 'rational' arguments, which elicited only lukewarm responses from the friends who read portions of the draft. In 2015, I started writing business-related case studies, which sparked ideas for similar stories to include in the book. By 2020, I had written most of the case study narratives featured here and began using them in workshops on 'Science and Humanism.' The largely enthusiastic reactions to these stories stood in stark contrast to the earlier responses to my more austere drafts.

I learned that while facts alone rarely change minds, meaningful conversations can. But how can an author effectively share ideas and engage a broader audience without the benefit of close personal connection?

Studies have shown that "individual cases have a powerful impact and are a more effective tool for teaching psychology [and other subjects] because the incongruity must be resolved and embedded in a causal story" (KAHNEMAN)[1].

Stories allow readers to identify with characters and feel their perspectives acknowledged. This can make them more open to reconsidering views than they might be through the mere presentation of evidence.

Narrative Foundations of This Book

To harness the power of stories, each of the book's nine parts are centred on case studies. The styles vary—some are descriptive, others explanatory; two draw from Peter Singer and John Maynard Keynes, and one distils current research on evolution.

Addressing narrative biases

Stories carry risks. They can mislead and feed the narrative fallacy, which explains events through imagined cause and effect without evidence. To help readers navigate these dangers, the book offers tools to avoid such errors, grounded in science's self-correcting process. Part 3 introduces this approach for analysing and learning from case studies. Each story is followed by commentary and analysis that engages what Daniel Kahneman calls 'System 2' thinking — slow and deliberate — rather than intuitive 'System 1' judgments. References are included for enhanced insight, and readers are encouraged to discuss ideas with thought and humility.

Alfred North Whitehead wrote, "The child should experience first the stage of romance, then the stage of precision, and finally the stage of generalization" (Whitehead)[2]. We are all children in the journey of learning.

Part 1

Central Questions of Life

What really matters

Diagnosis of our challenges and a framework for solutions

Tetraethyl Lead in Gasoline – History and Politics[a]

"No toxic substance has been more widely distributed throughout man's environment than the lead additive Tetraethyl Lead ('TEL') in gasoline. For over seven decades, millions of autos of all descriptions have successfully dispersed this toxic substance to all corners of the world. How did such a toxic substance ever gain approval to expose hundreds of millions of people?"[1]

HERBERT L. NEEDLEMAN

"You will observe with Concern how long a useful Truth may be known and exist, before it is generally received and [acted] on."[2]

Benjamin Franklin, July 31, 1786

Writing to a friend on the poisonous effects of lead

---

a Gasoline (used in North America) and petrol (used in most other English-speaking regions)

3

Tetraethyl Lead in Gasoline: History and Politics

Motor gasoline, knocking, and solutions

Car and motor gasoline sales

The invention of the car radically changed the world. In the early 1900s, several engine types competed, but the internal combustion engine eventually became dominant.

The adoption of motor gasoline as the standard fuel was not guaranteed. Though crude oil drilling and refining predated the car, the demand for gasoline was vital to the oil industry's growth (F3.1).

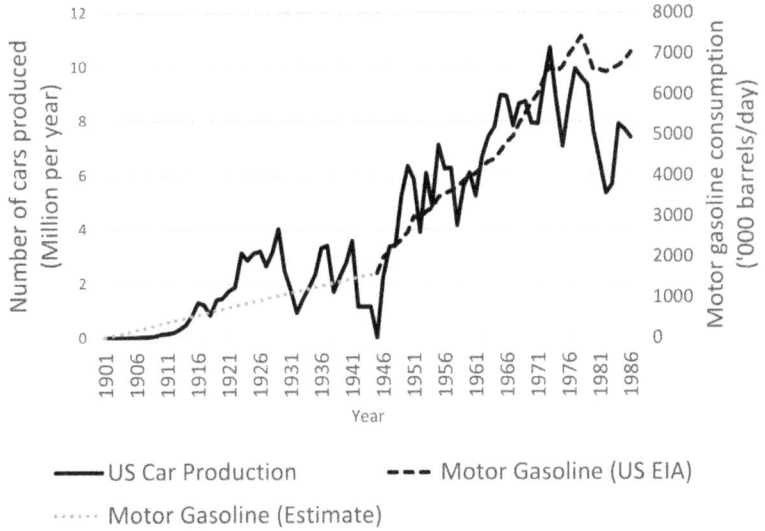

——— US Car Production   --- Motor Gasoline (US EIA)
······ Motor Gasoline (Estimate)

F3.1. Car production and motor gasoline consumption 1901 to 1986[1,2]

From under 5 million barrels in 1900, motor gasoline consumption rose to over 403 million barrels by 1931, a gain of 8,000% in just three decades.[3] Between 1945 and 1986, motor gasoline accounted for 40% to 50% of all petroleum products supplied in the US.[4]

The problem of knocking

Knocking occurs when gasoline ignites prematurely in an engine, producing a loud sound and reducing efficiency.

Ford's Model T, launched in 1908, used a modest engine that avoided knocking. GM's Cadillac, aiming for power, required fuel that could handle higher pressures. Solving knocking became a priority for GM.

Charles Kettering

Kettering, an electrical engineer, made his name with the 1911 self-starter for Cadillac. His company, Delco, was acquired by GM in 1916, and he became vice-president of GM Research.

Between 1918 and 1921, Kettering focused on air-cooled engines, but the project failed. By 1923, under pressure to deliver, he considered resigning. GM President Alfred Sloan refused his resignation.

Thomas Midgley Jr.

Midgley, a mechanical engineer, joined GM in 1916 and was tasked with solving engine knocking. Jamie Kitman, who first publicly exposed the leaded gasoline story, noted that Midgley tested every substance he could find, with little success.[5] Yet, as William Kovarik observed, Kettering and Midgley presented their work as scientific rather than trial-and-error engineering.[6]

Alcohols as solutions to knocking

In 1917, Midgley found that ethanol had anti-knock properties, and tests for the U.S. Army Corps confirmed this. In 1920 "Scientific American" summarized consensus: "Alcohol in some form will be a constituent of the motor fuel of the future."[7]

Ethanol was efficient and compatible with gasoline. But two problems loomed:
- GM couldn't patent ethanol.
- Ethanol was a potential alternative to gasoline and threatened oil companies, including GM's powerful shareholders like the du Pont family.

In the late 1800s, kerosene demand had made Standard Oil a giant. After its 1911 breakup, companies like Standard Oil of New Jersey (SONJ), later Exxon and now ExxonMobil remained highly profitable and influential. Ensuring gasoline's dominance in cars would protect their profits as electrification reduced kerosene demand.

The discovery of tetraethyl lead

On December 9, 1921, Midgley identified tetraethyl lead (TEL) as a highly effective anti-knock additive. First synthesized in 1853 by Carl Jacob Loewig, TEL had previously no practical use. TEL offered two key advantages:
- Its complex manufacturing process could be patented.
- TEL wouldn't replace gasoline—only one part TEL per 1,000 parts gasoline was needed—preserving the oil industry's dominance.

In a 1925 interview, Midgley made no mention of ethanol and GM and oil companies stopped referring to alcohols as viable alternatives. TEL became the standard, setting the stage for decades of environmental harm.

Leaded gasoline and Ethyl Corporation

The Ethyl Corporation

In 1923 Du Pont, tied to the du Pont family, began production in a new plant. SONJ independently developed a more efficient TEL process and began production, paying royalties to GM. This likely prompted GM to form a joint venture, Ethyl Gasoline Corporation, with SONJ. Though GM had to share profits, the partnership allowed it to focus on automobiles while incentivizing the largest U.S. oil company to push TEL into the market.

As Peter Drucker noted, "GM, in effect, made money on almost every gallon of gasoline sold by anyone."[8] SONJ also profited from TEL and reinforced gasoline's dominance.

Tetraethyl lead — a known poison

Lead was known to be toxic. To avoid public alarm, Kettering named the TEL-based product "Ethyl," omitting any mention of lead. Ethyl alcohol was used to suspend lead compound.

Evidence shows that GM and Du Pont leadership were aware of TEL's dangers:
- In March 1922, Pierre du Pont warned his brother Irenee that TEL was highly toxic and could cause rapid lead poisoning through skin contact.[9]
- In January 1923, Midgley and three lab workers took medical leave due to lead exposure.[10]

- Charles Kraus, a leading chemist, warned GM that TEL was deadly and possibly linked to a colleague's death.[11]
- In 1924, the U.S. Army Chemical Warfare Service found that 100 drops of TEL on the skin could be fatal within 24 hours.[12]

Tetraethyl lead in gasoline

Despite these warnings production proceeded. Surgeon General Hugh Cumming raised safety concerns. Midgley responded it had received attention but admitted no experimental data had been collected.[13]

Midgley had consulted Yandell Henderson of Yale, who warned that widespread lead poisoning was likely and offered to conduct an independent study on the condition it remain free from influence. Midgley declined.[14]

Instead, GM turned to the Bureau of Mines, then considered the authority on metals. The Bureau agreed to first submit all reports of its study to GM for approval.

On October 31, 1924, the Bureau released a GM-approved report claiming minimal risk from leaded exhaust. But Harvard public health professor Cecil K. Drinker criticized the study's methodology, test animals were kept in well-ventilated cages, and lead dust was deliberately prevented from accumulating.[15]

As these discussions concluded, deaths began at TEL production sites. The first fatalities were at GM's pilot plant followed by more at Du Pont's facility.

Public outcry

Between September 26 and October 30, 1924, several workers died at SONJ's TEL plant in Bayway, New Jersey. Survivors were institutionalized.[16] The press responded with intense criticism.

On December 24, 1924, Kettering and Du Pont's technical director met with the U.S. Surgeon General, who announced a public hearing for May 1925. In the months leading up to it:
- GM, Du Pont, and SONJ engineers worked to make TEL manufacturing safer. Though more deaths occurred, immediate hazards were eventually reduced.
- Ethyl Corporation launched a public relations campaign. Midgley led the defence, claiming TEL was the only anti-knock compound. At a press conference, he immersed his hands in TEL to demonstrate its safety.[17]
- On May 4, 1925, Ethyl Corporation temporarily suspended TEL sales.

The Surgeon General's conference

Held on May 20, 1925, the conference was shortened to one day and limited to TEL manufacturing risks. No discussion of alternative anti-knock solutions was allowed.

The event exposed a deep divide. Industrial representatives emphasized technological progress and economic benefit, downplaying lead's toxicity. Public health experts focused on long-term risks to health. Henderson described the gathering as a clash of incompatible priorities.[18]

Hamilton and Henderson

Alice Hamilton of Harvard Medical School and Henderson warned that leaded gasoline posed serious public and environmental health risks. They argued that emissions from widespread use would cause gradual, insidious poisoning. Henderson predicted that leaded gasoline would become universal before the public grasped its dangers.[19]

Other public health experts and industry response

Dr. Henry F. Vaughan of the American Public Health Association pointed out that while lead and TEL were toxins, there was no proof of harm from gasoline emissions.[20,21]

GM, Du Pont, SONJ, and Ethyl positioned Kettering as an impartial authority, a government role.[22] Frank Howard of SONJ called TEL a "Gift of God."[23] The companies focused on occupational safety and framed TEL as an anti-knock vital for the U.S. economy.

The outcome

The conference ended without resolution. The Surgeon General announced a committee to investigate TEL and leaded gasoline. Ethyl Corporation was praised for suspending sales pending the findings.

The Committee's report

The committee moved quickly. It studied 252 workers in Dayton and Cincinnati, some exposed to leaded gasoline, others not, and 61 men in industrial plants with known lead exposure. Released on January 17, 1926, the report concluded:[24]

1. "There are at present no good grounds for prohibiting the use of ethyl gasoline.. provided that its distribution and use are controlled by proper regulations."
2. "It remains possible that if the use of leaded gasoline becomes widespread, conditions may arise very different from those studied by us, which would render its use more of a hazard.... [This] may lead to chronic degenerative diseases of a less obvious character."
3. "In view of such possibilities, the investigation begun must not be allowed to lapse... It should be possible to determine whether or not it may constitute a menace to the health of the general public after prolonged use or other conditions not now foreseen. The committee urges strongly... for the continuation of these investigations."

Despite these warnings, no further independent studies were pursued.

The use of leaded gasoline

With improved TEL manufacturing and no further factory deaths, public pressure for testing faded. Ethyl Corporation resumed TEL sales in May 1926 with signs reading "Ethyl is back." The company aggressively marketed TEL which became a universal additive to gasoline.

In 1924, Cadillac held 9% of the U.S. car market compared to Ford's 60%. By 1927, with a powerful engine running on leaded gasoline, Cadillac overtook the Model T—38% to 14%.

Just as white lead was being phased out from paints and cosmetics, lead in gasoline surged (F3.2).

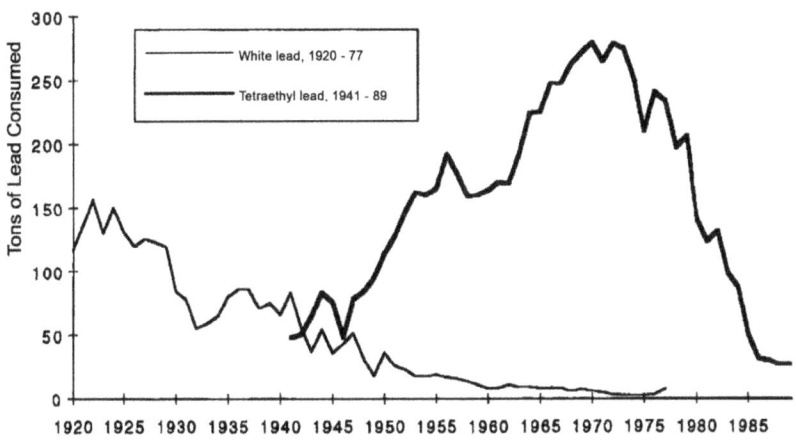

F3.2. Consumption of White lead and Tetraethyl lead (1920-1989)[25]

In 1940, Midgley contracted polio. On November 2, 1944, he was found strangled in his home. The coroner ruled it suicide. In addition to TEL he developed chlorofluorocarbons (CFCs), two of the most environmentally harmful chemical inventions of the 20th century.[26]

1926–1963: Kehoe and industry control

In 1930, GM, Du Pont, and Ethyl funded the University of Cincinnati's Kettering Laboratory with USD130,000. Kehoe became director and professor of industrial medicine. For nearly 50 years, most research on leaded gasoline was funnelled through Kehoe, who consistently defended TEL's safety.

In 1962, Albemarle Paper bought Ethyl Corporation in a USD200 million leveraged buyout. GM and SONJ, though no longer financially involved, continued supporting TEL.

1963–1970: Patterson's challenge

In 1963 Caltech scientist Clair Patterson showed that environmental lead levels were a phenomenon caused by leaded gasoline. Petroleum executives offered him funding to change research direction. When Patterson refused the American Petroleum Institute unsuccessfully tried to have him fired from Caltech.[27]

In 1966, Kehoe and Patterson testified before a Senate subcommittee. Kehoe began testimony by "informing the chairman that he knew so much of the subject that he was forced to leave a great deal of information out."[28] Kehoe said that the amount of TEL could be increased without risk and claimed that "no other hygienic problem in the field of air pollution has been investigated so intensively, over such a prolonged period of time, and with such definitive results."[29]

Patterson countered with experimental data that showed otherwise, but political resistance and industry influence stalled reform.[30]

1970–1980: mounting evidence

In the 1970s, Drs. Herbert Needleman and Philip Landrigan showed that lead exposure harmed children's IQ and behaviour. Molecular studies later confirmed lead's toxicity even in small doses.

In 1971, the EPA began funding studies on airborne lead. By 1972, it linked lead pollution to blood lead levels in children and proposed reducing lead in gasoline, but no action was taken.

In 1975, smog concerns led to regulations requiring catalytic converters in new cars. TEL couldn't be used with these devices, prompting research into substitutes. Christian Warren noted that "it was lead's threat to pollution-control devices, not humans, that first encouraged unleaded gasoline."[31] In 1976, the EPA began phasing out leaded gasoline, targeting a full ban by 1986.

1980–2020: final phase-out

In 1981, the Reagan administration appointed Ann Gorsuch as EPA Administrator. She cut the agency's budget by 22% and proposed relaxing the lead phase-out. Public opposition forced a reversal in 1982.

Leaded gasoline was finally banned in the U.S. in 1996, but it continued to be used abroad, where by 1979, foreign TEL sales had surpassed those in the U.S.

No acceptable lead levels

According to the United Nations Environment Programme, 82 countries were still using leaded gasoline as recently as 2002. It took nearly two more decades, until September 2021, for Algeria to become the last country to end its sale.

The U.S. CDC's guidance on the maximum acceptable levels of blood lead in children has been revised downwards (T3.1).

Year	1930-1970	1970	1991	2012
Max. Blood Lead Level in Children	80μg/dL	40μg/dL	10μg/dL	5μg/dL
Recommended by	Kehoe	CDC	CDC	CDC

T3.1: CDC recommendation for maximum acceptable blood lead levels in children[32,33]

A July 2020 report by UNICEF states that around 1 in 3 children, up to 800 million, had blood lead levels at or above 5 µg/dL. Richard Fuller, a co-author of the UNICEF-Pure Earth report said, "We did a terrific job of taking lead out of gasoline, but the use of lead has [now] plateaued."[34]

Powering the car in the 21st century

The rise of electric vehicles (EVs) has now shifted attention away from motor gasoline. A 2025 survey projects that 64% of global consumers will opt for EVs.[35] Yet concerns remain: mining for battery materials like nickel and cobalt can increase sulphur dioxide pollution, and lithium extraction, especially from deep-sea sources, raises environmental alarms.[36]

Hydrogen fuel cells have lost momentum, and the global power grid may struggle to meet surging electricity demand. Meanwhile, public transport investment remains insufficient, despite its potential to cut emissions and congestion.

Lead poisoning in 2025

Despite TEL no longer used in gasoline and decades of regulation, lead exposure remains a public health crisis. According to the World Health Organization, in 2025 an estimated 1 in 3 children worldwide still have blood lead levels above the threshold, with 1.5 million deaths annually attributed to lead-related causes, primarily cardiovascular disease.[37]

A 2025 report by the Centre for Global Development reveals that 3,000 metric tons of lead chromate, a banned pigment in all developed countries, are still exported annually to poorer nations, potentially poisoning up to 277,000 children.[38]

4

Humanity's Defining Questions and a Roadmap Through the Book

I. Central QUESTIONS OF LIFE

<u>What really matters</u>

Building on the questions that arise from the story of leaded gasoline, this chapter outlines the book's structure. It introduces the central inquiries of the book: how we understand ourselves, what we accept as knowledge, and how choices shape the world, and maps how each of the book's case studies addresses these concerns.

II. How the world IS

<u>What do we know about ourselves</u>

Part 2 Origins of human judgment and actions

In 1917, Thomas Midgley confirmed ethanol's anti-knock properties. Yet he pursued TEL, a known toxin, and later claimed it was the only viable solution. He ignored warnings from Charles Kraus and Yandell Henderson, even after suffering lead poisoning himself. At a press event, he dipped his hands in TEL to demonstrate its "safety." Despite the Surgeon General's call for continued investigation, Midgley never pursued further research.

Why do we ignore evidence? What makes us cling to beliefs despite contradiction? The ancient Greek maxim "Know thyself" urges us to reflect, not on Midgley's choices, but our own.

Part 2 explores the roots of human behaviour through a case study comparing two Everest expeditions:

Case Study: Two Ascents of Everest – 1953 and 1996
- 1996: Rob Hall and Scott Fischer led commercial climbs; five people died.
- 1953: Edmund Hillary and Tenzing Norgay reached the summit; all returned safely.

How do we know? What test of knowledge should we rely on?

Part 3 Science as a test of knowledge

How do we know what we know? What makes knowledge reliable?

For decades, TEL was added to gasoline, causing lead pollution. Part 3 examines how science can uncover truth, sometimes tentatively. Questions include:
- Did the Bureau of Mines conduct meaningful risk assessments?
- On what grounds did Robert Kehoe claim TEL's safety?
- How did Clair Patterson trace lead from car emissions to oceans?
- Are Needleman's studies of children's teeth as valid as Patterson's experiments?

Richard Feynman wrote that the "principle of science, the definition, almost, is the following: 'The test of all knowledge is experiment. Experiment is the sole judge of scientific 'truth'" (FEYNMAN)[1]. But is it so simple? What more is there to this principle of science?

Part 3 contrasts science's self-correcting process with myth and error, through two case studies:

Case Study: Tetraethyl Lead in Gasoline – The Science
- The science behind TEL's environmental and health effects from the 1920s to 1970s.

Case Study: The Cold Fusion Fiasco
- A look at a failed scientific claim and what it reveals about the process of inquiry.

How did life evolve? Are we the result of progress? What role does chance play in life?

Part 4 Evolution – progress or chance?

The TEL story reminds us that however sophisticated our tools, we remain primates adapted for short-term survival, not reasoning.

Understanding human evolution through natural selection and random variation sheds light on our ethical and creative choices. Part 4 explores evolution as a contingent process shaped by environment, and chance.

Case Study: A Brief History of Life – Evolution and Chance
- Was *Homo sapiens* inevitable? Is evolution a march toward complexity or a product of randomness?

III. Humanism—How the world OUGHT to be

What humanist principles and processes might we use in moral decisions?

Part 5 The foundational building blocks of modern humanism

At the 1925 Surgeon General's Conference on TEL, experts clashed over its risks. Should they have focused only on factory safety, or considered broader public health? What principles should guide decisions when science offers incomplete answers or cannot address moral issues?

Part 5 examines the core moral standards that should guide humanists in personal and collective decision-making.

Case Study: The Eradication of Smallpox and Facing COVID-19
- Smallpox killed millions for centuries. A global campaign led by WHO eradicated it by 1980.
- COVID-19 caused at least 19 million deaths—far more than official counts. Why did so many die, and what principles should have guided the response?

IV. CONSEQUENCES of is and ought

What are our worldviews? What are we living for?

Part 6 Needs and purposes, a humanist worldview, and expressions of flourishing

Kehoe, Patterson, and Needleman studied lead exposure but with different motives. Kehoe was funded by TEL's backers; Patterson and Needleman pursued scientific and medical truth. Their contrasting worldviews shaped their research and its impact.

Part 6 explores how deprivation of needs and capabilities, along with our personal exemplars and environments, shape worldviews. It asks: What defines a well-lived life? How do science, engineering, and technology differ in purpose, and how do they influence approaches to cooperation and competition?

Case Study: Tim Berners-Lee and Mark Zuckerberg
- Berners-Lee built the Web to foster openness and collaboration.

- Zuckerberg built Facebook into a dominant platform, prioritizing profit, growth and control.

How wide are our circles of concern and influence?

Part 7 Poverty, conservation, and historical responsibility

Kettering, Midgley, and the leadership at SONJ, GM, and Du Pont focused narrowly on finding a patentable anti-knock agent. In contrast, Patterson and Needleman expanded their circles of concern, moving from personal inquiry to global advocacy. Their growing awareness of environmental damage led them to advocate for the banning of leaded gasoline, transforming their influence into meaningful change.

Part 7 highlights three global challenges often dismissed as 'externalities': poverty, environmental degradation, and historical injustice. These issues are too often treated as distant concerns rather than urgent priorities.

Case Study: The Shallow Pond
- Peter Singer's allegory urges action against global poverty. The case explores how humanists might respond to extreme poverty.

Case Study: Collapse of the Aral Sea
- Once the world's fourth-largest lake, the Aral Sea was destroyed by human activity. The case evaluates E.O. Wilson's "Half-Earth" proposal for conservation.

Case Study: The UN Durban Conference 2001
- Intended to address racism, the conference ended in discord. The case examines how humanists might approach history and pursue meaningful remedies.

<u>What economic, political, social, and educational frameworks might humanists select?</u>

Part 8 Institutions and education

Kettering and Midgley operated in a corporate system driven by profit and hierarchy. Patterson and Needleman resisted commercial pressures, finding purpose in education and service.

Part 8 explores how economic, political, and educational systems can support human flourishing. It emphasizes education as a key pathway to humanism.

Case Study: Keynes' "Letter to My Grandchildren"
- In 1930, Keynes imagined a future where automation meets human needs, freeing people to pursue meaningful lives. The case reflects on how we might build institutions that support well-being and the art of living.

V. VISIONS OF LIFE in a better world

<u>What humanist directions point towards a better world?</u>

Part 9 From cleverness to wisdom

Kettering insisted that the technology of leaded gasoline was essential for the U.S. economy. Patterson and Needleman asked what the longer-term effects of this technology might be. They acted from a commitment to evidence, justice, and human dignity.

Part 9 centres on the iconic Voyager 1 image of Earth—a distant "pale blue dot" that invites reflection on our place in the cosmos. What kind of civilizations should we strive to build? Is endless technological growth our only path, or can we pursue wiser futures grounded in humanist values? It introduces the Sagan Scale, a new framework for evaluating civilizations by humanist criteria, contrasting with the Kardashev Scale's focus on energy and technology.

Humanist thinkers urge for a shift from cleverness to wisdom, reframing our aspirations toward moral depth and collective well-being.

Case Study: An Ape Looks at a Blue Dot
- Through three space images of Earth, the case explores Carl Sagan's vision of unity, humility, and the insights gained from science and the arts.

N3 Two faces of TEL

"We are drowning in information while starving for wisdom" (E.O. Wilson)[2]

Part 2

How the world IS

Self-knowledge

<u>What do we know – Human behaviour</u>

Origins and drivers of human judgement and actions

Two Ascents of Everest – 1953 and 1996

Hillary and Tenzing's first ascent of Everest was "Comparable in its visceral impact to the first manned landing on the moon."[1]

Jon Krakauer, Climber and writer, quoting an "old friend."

"Your Sherpa will tell you, 'You're too slow, you have to turn around, or you'll die,' and some people don't. Mountains don't kill people; people kill themselves."[2]

Mark Jenkins, Adventurer and writer

5

Two Ascents of Everest

Two expeditions to Everest

May 1996

On May 10, despite warnings to turn back by 2 pm, Rob Hall and Scott Fischer remained in Everest's death zone. Both led expeditions, charging clients around USD 65,000. By May 11, five climbers had died, including Hall and Fischer, and six others barely survived. Hall's client, Dr. Beck Weathers, was twice abandoned and presumed dead before staggering to safety. Severely frostbitten, Weathers was rescued by pilot Madan Khatri Chhetri, who flew to 21,000 feet—well above the helicopter ceiling. Including a Sherpa from Fischer's team who later died of HAPE, seven lives were lost.

May 1953

On May 29, Edmund Hillary and Tenzing Norgay, working as a harmonious team, crossed narrow ridges with sheer drops, climbed steep rock faces and an ice spur, and became the first humans to reach Everest's summit (29,032 feet/8,849 m) at 11:30 am.

Tenzing wrote, "I was not thinking of 'first' and 'second'... We went on slowly, steadily. And then we were there."[1] Hillary recalled, "I put out my hand... but that wasn't enough for him. He threw his arms around my shoulders, and I threw my arms around his."[2]

The southeast ridge route up Everest

Seventeen routes have been used to climb Everest. Before 1953, most attempts came from the North Face via Tibet. In 1935, Eric Shipton identified the Southeast Ridge route from Nepal, the most popular path. F5.1 shows this route and the major camps used.

Base Camp (~17,600 feet / 5,365 m)

Base Camp is high enough to cause altitude sickness. It is now equipped with satellite phones and even courier services—a tourist-climber from Fischer's 1996 team had Vogue and Vanity Fair delivered to her there.

From Base Camp, climbers cross the Khumbu Glacier and its shifting Icefall, navigating crevasses and ice towers using fixed ropes and ladders, usually early in the morning before melting ice worsens conditions.

Camp I (19,685 feet / 6,000 m)

Just above the Icefall, Camp I leads into the Western Cwm, a glacial valley ending at the Lhotse Face. This stretch includes treacherous crevasses and ice walls.

Camp II (21,000 feet / 6,750 m)

Located at the base of Lhotse, Camp II serves as an advanced base. Climbers often sleep here during acclimatization before returning to Base Camp.

Between Camp II and III lies a crevasse before the steep Lhotse Face begins, an icy slope angled 40 to 50 degrees, with some pitches reaching 80 degrees. Ropes are fixed from this point onward.

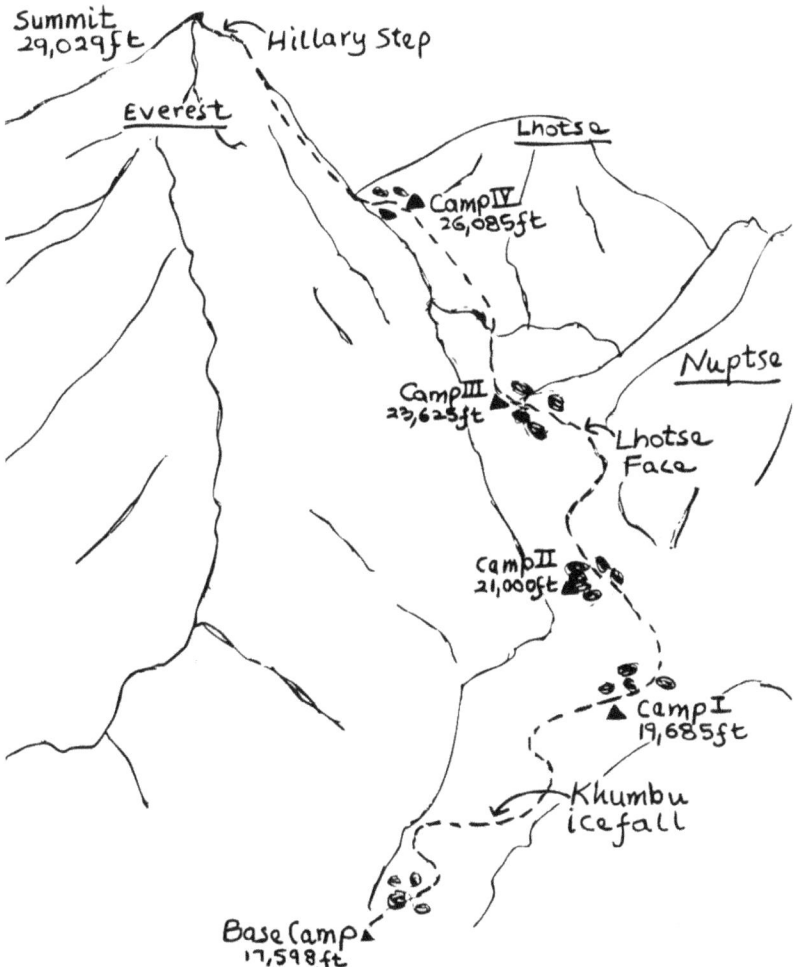

F5.1 The Southeast Ridge route up Everest

Camp III (23,625 feet / 7,100 m)

Perched on a ledge of the Lhotse Face, Camp III leads to the 'yellow band,' a narrow rock layer crossed single file—and then to the Geneva Spur, a series of rock benches ending in a steep climb to Camp IV.

Camp IV (26,085 feet / 7,920 m)

From here, almost all climbers use supplemental oxygen. This is the start of the 'death zone,' where winds are fierce and pitching tents is arduous.

Climbers ascend to the 'Balcony' for a brief rest, then continue along Everest's ridge with steep drops on either side. After reaching the South Summit, they cross the Hillary Step, once a vertical rock chimney, now a gentler slope after the 2015 Nepal earthquake, to reach the summit.

Additional camps in 1953

The 1953 team used extra camps above Camp IV. Later expeditions found it more efficient to summit directly from Camp IV, avoiding higher overnight stays and reducing logistical strain.

1996 - Adventure Consultants and Mountain Madness

In May 1996, 30 expeditions gathered at Everest Base Camp. Nepal's government charged USD 70,000 for up to seven climbers, plus USD 10,000 per additional climber. Each expedition was allowed one summit attempt unless they paid extra for a full-season permit.

Rob Hall and Scott Fischer

Rob Hall, from New Zealand, led Adventure Consultants. With four Everest summits and 39 clients successfully guided between 1990 and 1995, Hall was respected enough to act as coordinator among expeditions.

Scott Fischer, an American climber, led Mountain Madness. Though this was his first guided Everest expedition, he had summited Everest without supplemental oxygen and famously climbed K2 with a dislocated arm taped to his body. Charismatic and bold, Fischer aimed to match Hall's success.

Both men understood Everest's dangers. Yet, Hall feared not for his own team, but for others who might ignore the mountain's risks. Fischer, speaking to Jon Krakauer, emphasized mindset over experience and believed climbing Everest had become routine.[3] Unknown to his team, Fischer was under financial strain and battling a long-ignored liver illness.

John Hunt, leader of the 1953 expedition, had outlined six major risks warning of "factors of awe-inspiring magnitude."[4]

Risk 1: Altitude

Hunt observed that the thin air near Everest's summit makes simple movements arduous, and that the lack of oxygen impairs mental clarity. He emphasized that beyond a certain point, life itself is no longer possible (HUNT)[5].

Everest's summit has only one-third the oxygen at sea level. Blood oxygen saturation can drop from 98% to 40%. Above 26,000 feet—the "death zone"—the body deteriorates rapidly. Acclimatization helps, but only up to this point.

Supplemental oxygen is essential but tricky: valves freeze, tubes clog, and cylinders last 4–5 hours. Yet Fischer allowed Anatoli Boukreev, his top guide, and Lopsang Jangbu Sherpa, his sirdar (lead Sherpa guide), to climb without it.

Hall enforced a strict acclimatization regimen, confident in its success.[6] Fischer used the same schedule but let clients manage it themselves. This led to multiple exhausting trips for Fischer between Base Camp and Camp II, leaving him depleted before summit day. Prior to the summit day most climbers were already fatigued, poorly rested, and undernourished.

Risk 2: Climbing challenges and fixed ropes

Hunt challenged the notion that Everest is easy. While acknowledging that other climbs may present more complex manoeuvres, he insisted that climbing Everest demanded respect and was far from trivial.

In 1996, with so many teams on the mountain, Hall led coordination efforts. He and Fischer agreed their Sherpas would fix ropes on summit day. But by the morning of May 10, key ropes were still missing, causing delays and bottlenecks on the mountain.

One critical section, a 200-foot ice wall near Camp IV, had no fixed ropes. That evening, seven tourist climbers, two guides, and two Sherpas descended to the wall in worsening weather. Camp IV, just 200 feet below, was invisible. Disoriented, the group split. Two guides, two Sherpas, and two climbers found Camp IV, leaving five behind.

Boukreev, who had descended earlier without clients, was at Camp IV. He returned and rescued three climbers but left Beck Weathers and Yasuko Namba behind, believing they were dead.

Risk 3: The weather

Hunt underscored the lethal impact of bad weather on Everest, noting that windows of opportunity for summit attempts were brief and that Everest's scale amplified the consequences of rapid weather changes.

Jet streams reaching 240 to 320 km/h batter the summit most of the year. Climbers rely on two seasonal windows, Spring and Fall, with Spring offering the best chance before monsoons arrive. Even then, conditions remain unpredictable.

Hall scheduled his summit attempt for May 10, a date he considered significant and auspicious. He coordinated with other expeditions to secure it. Fischer also opted for the same date.[7]

Edmund Viesturs, part of an IMAX team aiming to summit on May 9, had observed worsening conditions: cloud buildup and high winds. On May 8, sensing the weather was wrong, he retreated to Camp II, yet Hall and Fischer pressed ahead.

By May 9, fifty climbers were at Camp IV. Though the afternoon brought high winds, the evening cleared, and the summit push was confirmed. But on May 10, with climbers still high on the mountain, the weather suddenly turned, unleashing a deadly storm.

Risk 4: The turn-around time

Hunt stressed the urgency of saving time in the final stages of the ascent, not only to counter physical decline, but also because the weather could deteriorate at any moment.

Hall insisted on a strict 2 pm turnaround, regardless of how close climbers were to the summit.

Hall and Fischer's teams left Camp IV around midnight. The goal was to reach the summit early, but progress was slow. Delays from missing ropes forced climbers to wait in freezing winds.

Despite paying USD 70,000, four of Hall's clients turned back on their own. Frank Fischbeck turned around early in the day. By 11:30 am, Drs. Hutchison and Taske realized they lacked enough oxygen and descended. Lou Kasischke, initially hesitant, joined them after a discussion.

Dr. Beck Weathers, suffering from an eye condition, was told by Hall's sirdar to turn around, but Hall allowed him to continue. No guide or Sherpa stayed with him.

At 3 pm, Doug Hansen was still climbing, out of oxygen. Hall's sirdar ordered him to turn back, but Hall overruled him, determined to help Hansen summit, perhaps to make up for turning him around the previous year.

Many of Fischer's clients summited after 2 pm and lingered there for an hour. At 3:10 pm, guide Neal Beidleman began descending with four clients. Fischer, still ascending, reached the summit at 3:45 pm. Soon thereafter the storm hit, engulfing the upper mountain.

Risk 5: Teams

Hunt attributed much of the 1953 expedition's success to "Our unity as a party. This was undoubtedly the biggest single factor in the final result" (HUNT)[8].

The 1953 expedition relied on support teams at high altitudes. In 1996, climbers were strangers. Hall and Fischer had only two guides each for 16 clients. Boukreev, Fischer's top guide, believed clients should be self-reliant. He descended alone early, although he later rescued several climbers.

Hall demanded obedience, emphasizing that on the mountain, he would "tolerate no dissension" and that his "word will be absolute law beyond appeal."[9] Clients were expected to follow instructions without question, a mindset that extended to junior guides.

Beidleman, paid less than Boukreev, didn't enforce the 2 pm rule. Seeing himself as subordinate, he hesitated to challenge clients, a mistake regretted.[10]

Andy Harris, a guide on Hall's team, reached the summit at 1 pm. On descent, his oxygen regulator froze. Likely suffering from hypoxia, he mistakenly claimed cylinders were empty. Krakauer and Groom later found six full ones, but Harris refused to believe it. Krakauer notes that a major reason he did not challenge Harris was the "guide-client protocol," explaining, "We had been specifically indoctrinated not to question our guides' judgement."[11]

Risk 6: Logistics and coordination

Hunt detailed the logistical complexity of operating above 21,000 feet, where multiple high camps must be established. These required substantial equipment, tents, food, fuel, and protective gear, all of which needed to be hauled upward despite the harsh conditions.

Establishing high camps requires carrying tents, food, fuel, and oxygen through brutal conditions, a task that falls largely to the climbing Sherpas and is logistically complex.

In 1996, Sherpas earned between USD 1,500 and USD 3,000 per expedition, about a tenth of what Western guides made. Sherpas, a Tibetan ethnic group have supported Himalayan climbs since the 1920s. Their genetic adaptations allow more efficient oxygen use, conserving energy at altitudes and contributing to their elite performance.

Despite their skill, Sherpas face disproportionate risk. About one-third of Everest deaths are Sherpas, due to their constant exposure to dangerous sections like the Khumbu Icefall.[12] They spend up to 16 hours a day fixing ropes and ladders, making 15 to 20 trips per season to ferry supplies. Still, many Sherpa families see high-altitude work as a path out of poverty and toward better futures for their children.

Hall and Fischer delegated logistics to their sirdars, but oversight was minimal. With so many teams on the mountain, coordination was essential, yet their sirdars weren't on good terms. Hall likely assumed instructions would be followed, including fixing summit ropes, a task left incomplete, causing critical delays.

Fischer's sirdar, Lopsang, was only twenty-three. Though a strong climber, his inexperience and eagerness to impress led to questionable choices like carrying Sandy Pittman, a socialite, during critical parts up the mountain.

Tourism to the top of Everest

Between 10 and 11 May, Hall and Fischer died high on Everest from exposure to extreme cold and lack of oxygen. Two of Hall's clients and one of his guides also perished. Three of Fischer's clients were fortunate to survive. Hall's client, Beck Weathers, endured severe frostbite and later underwent amputations—losing his nose, all the fingers on his left hand, and half of his right arm.[13]

However, as Krakauer writes, the percentage of people who died in 1996 while going higher than Base Camp was 3 percent as compared to the historical record, up to 1996, of 3.3 percent.[14]

Krakauer points out that the ratio of people who died to those who reached the summit between 1921 and May 1996 was one in four.

There are multiple dead bodies left on Everest. Categorizing these deaths as 'unavoidable' when due to avalanches and falling ice and 'avoidable' when due to altitude sickness, exhaustion, exposure to cold, falls, heart attacks, and pneumonia, an analysis of 310 recorded deaths of climbers on Everest from 1921 to May 2021 shows that:
- 62% were avoidable
- 30% were unavoidable
- 8% the cause of death was not certain

The deaths and injuries in the Hall and Fischer expeditions were all avoidable deaths.

Hall had been stung by Hillary's disdain for Everest's commercialization. Hillary had said, "I like to think of Everest as a great mountaineering challenge, and when you've got people just streaming up the mountain - well, many of them are just climbing it to get their name in the paper. They might as well [be] provided [with] a bus."[15]

A less visible dimension of today's outdoor pursuits is their dependence on synthetic, plastic-based apparel— prompting concern over microplastic pollution and ecological sustainability.

John Hunt and the 1953 team

The Joint Himalayan Committee, which organized the 1953 Everest expedition, was expected to appoint Eric Shipton, a legendary Himalayan mountaineer, as leader. But in a late decision, they chose John Hunt, a military officer with logistical skills and climbing experience. The goal was clear: ensure a British team reached the summit before the French (approved for 1954) or the Swiss (1955), as Nepal then allowed only one expedition per year.

Hunt's first challenge was winning over team members loyal to Shipton, who was revered for his pioneering Himalayan climbs. His contributions included:[16]

- Identifying the Southeast Ridge route used in 1953.
- Selecting a young Tenzing Norgay in 1935, launching his mountaineering career.
- Exploring the 1951 route with Edmund Hillary.

Hunt acknowledged the awkwardness of his appointment, telling the team it was "most unfortunate."[17] His humility and collaborative approach won over sceptics. Hillary later praised Hunt's "energy, organizing ability, and charm."[18,19] George Lowe noted that Hunt turned "a group of prima donnas" into a cohesive team.[20]

Bound by the Committee's requirement to select only Commonwealth climbers, Hunt added four criteria:

- Age. 25–40 (with exceptions for George Band, 23, and Hunt himself, 42).
- Technical climbing experience on snow, ice, and rock.
- Temperament. ambition balanced with teamwork.
- Selflessness. willingness to support others, even without personal glory.[21]

To build unity, Hunt arranged for UK-based members to travel together by ship to India. The full team met in Kathmandu and trekked 17 days to Thyangboche, near Base Camp. The journeys forged strong bonds.

Hunt selected six Sherpas for high-altitude work, including Tenzing's brothers. At high camps, he made no distinction between British, New Zealanders, and Sherpas, fostering mutual respect.

Training and equipment

Despite their experience, the team trained in winter 1952. They tested oxygen systems, new tent fabrics, lighter boots, dehydrated food, air mattresses, and aluminium ladders for the Khumbu Icefall.[22]

In December, Hunt and several climbers trained in the Alps and met with the 1952 Swiss Everest team. The Swiss generously shared their findings and locations of oxygen caches left high on the mountain.[23]

Acclimatization

The long trek from Kathmandu (4,000 ft) to Thyangboche (12,687 ft) included climbs of smaller peaks to test oxygen needs and altitude response. Base Camp was established on April 12; the summit was reached May 29. The gradual ascent ensured acclimatization.

Logistics and support

Hunt drafted a detailed logistics plan before departure, knowing it would need adapting on the mountain. Once in Nepal, he formed a core decision-making trio with Hillary and Charles Evans. On the mountain, Hunt was also assisted by Tenzing, who had participated in five previous Everest expeditions.

Over 350 porters carried supplies to Base Camp. From there, Hunt divided Everest into "three mountains," assigning teams to establish and stock higher camps. Backup systems ensured no critical gaps.

Hunt's meticulous attention to load management and camp provisioning was evident throughout. Nutrition was also prioritized. He enlisted Dr. Griffith Pugh, a pioneering physiologist and mountaineer whose later studies laid the foundation for sports medicine.[24]

Hunt's plan

Hunt planned for up to three summit attempts emphasizing that the final summit attempt should proceed only if the oxygen equipment was functioning reliably, the weather remained stable, and the climb allowed for a safe round trip within the available time. Each attempt would involve two climbers supported by teams at high camps:

- Tom Bourdillon and Charles Evans.
- Edmund Hillary and Tenzing Norgay.
- Wilf Noyce and Mike Ward.

Should all three teams fail to reach the summit, Hunt had contingency plans for a second expedition in the autumn.

The first attempt in 1953

On May 26, Bourdillon and Evans launched the first attempt from Camp VIII. Before setting out, Evans experienced problems with his oxygen equipment, which took an hour to fix. Once underway, they climbed rapidly, about 1,000 feet (300 m) per hour. At 28,000 feet (8,500 m), Evans' oxygen set failed again.

Unable to repair it, they chose to continue, with Evans climbing without supplemental oxygen.

At 1 pm, they reached the South Summit just 300 feet (100 m) below the final summit. But Evans was exhausted, and the time window for a safe ascent and descent had narrowed dangerously. After some discussion, a reluctant Bourdillon agreed to turn back.[25]

On their trip down they stumbled and fell down a gully but luckily without injury. Despite their good fortune in coming down safely, Bourdillon kept thinking about how he might have climbed to the summit himself. A few days prior to this attempt, Hillary and other team members had discussed:

"The possibility of the first assault party having to make this same decision. I felt quite happy about Charles…but I wasn't quite so sure about Tom. I had great respect for Tom's bulldog determination, but I didn't feel too sure that it mightn't at times influence his judgment."[26]

The day after the failed attempt, on the way down to lower camps, Bourdillon was very weak and fell full length onto the ice. Bourdillon was given supplemental oxygen and revived.

An avid climber, Bourdillon died in 1956 along with a climbing partner during an ascent of Jegihorn, a mountain in the Swiss Alps.

<u>The first successful ascent</u>

During the acclimatization period, Hillary and Tenzing spent time climbing together. They developed mutual confidence and respect, a bond that strengthened on April 26 while crossing the

Khumbu Icefall. Hillary nearly fell into a crevasse, but Tenzing, roped to him, held firm and prevented an injury — or worse. Hunt wrote that "These [were] two exceptional men."[27]

The day before their summit attempt Hillary and Tenzing were accompanied by three team members, Lowe, Gregory, and Sherpa Ann Nyima, who together carried most of the equipment and provisions up to Camp IX. They set up camp for Tenzing and Hillary before going back down, Lowe to the South Col and Gregory and Sherpa Ann Nyima to Camp VII.

On May 29, coincidentally Tenzing's thirty ninth birthday, Hillary and Tenzing started climbing from their high camp at 6:30 am and reached the summit at 11:30 am.

On their descent, Lowe, who was waiting for them at the South Col (Camp VIII), gave them hot soup and supplemental oxygen. At Camp VII, additional team members met them, and together they descended to Camp IV, where Hunt and the remaining team members were waiting. Hillary wrote, "As we greeted them all, I felt more than ever before that very strong feeling of friendship and co-operation that had been the decisive factor throughout the expedition."[28]

As important as the successful ascent was the fact that the only injury on the expedition was mild frostbite on Sherpa Namgyal's finger.

The press in 1953

George Lowe, a fellow climber, captured on film the camaraderie and joy with which the team greeted Hillary and Tenzing on their safe descent to Camp IV at 21,000 feet (6,400 m). It remains one of the most emotionally uplifting moments ever filmed, an enduring testament to shared purpose, humility, and the triumph of collective spirit.[29]

James Morris (later Jan Morris), the correspondent for The Times newspaper, was also part of the team and famously sent a coded message on the successful climb back to London, in time for the coronation of Queen Elizabeth II on June 2, 1953.

The first ascent in perspective

Despite the organizing Committee's nationalistic aim of having a British team 'conquer' Everest, Hunt saw the expedition in a very different light. For him, the motivation behind climbing Everest went far beyond personal enjoyment or a passion for mountaineering. He viewed it as a quest to confront and resolve a formidable challenge, one that had long resisted the skill and perseverance of earlier climbers. Hunt saw the climb as a chance to venture into the unknown and reach the highest point on Earth, a pursuit that mirrors the universal human drive to explore and overcome.[30]

Hunt also framed the expedition as a scientific and collaborative endeavour. He likened it to a relay, where each team built upon the insights and experiences of those who came before, contributing in turn to the cumulative understanding of high-altitude climbing. In his view, the struggle between climbers and the mountain reflects a broader human effort to engage with and adapt to the forces of nature, a continuous undertaking that forges connection among all who participate in it.[31]

"Ultimately, the justification for climbing Everest will lie in the seeking of … [others' own] 'Everests', stimulated by this event as we were inspired by others before us."[32]

HUNT

6

Origins of Human Judgment and Actions

An approach to self-knowledge

Self-knowledge is elusive, why chase it? For many the motivations stem from goals like:
- Financial gain
- Advancing oneself, accumulating power, and dominance
- Navigating major life choices—career shifts, relationships, ethical crossroads

But modern humanism's aim is not conquest or advantage, but the cultivation of a life that is thoughtfully constructed, ethically anchored, emotionally attuned, and capable of flourishing in the most meaningful sense.

This chapter draws on the Everest expeditions of 1953 and 1996 as mirrors for self-reflection through which we uncover truths about ourselves. The analysis traverses a wide terrain of disciplines: cognitive biases, neuroscience, social memes, priming, genetics, primatology, and emotional drivers. Within this framework, strategies for cultivating cognitive integrity are examined.

Cognitive biases

The proximate causes of the 1996 Everest disaster are well known: overcrowding, missing fixed ropes, and delays that pushed climbers past the 2 pm turnaround. Senior guides and Sherpas didn't accompany tourist climbers, who were left exposed when the weather turned. But what underlying forces shaped these decisions?

Rob Hall knew Everest's risks. He believed disasters were inevitable—but only for others. That early signal foreshadowed the unravelling. Studies show overconfidence is widespread. It distorts judgment under pressure, amplified by past success, social validation, or the illusion of control.

In the U.S., only one-third of small businesses survive five years. Yet over 80% of founders estimate their chances of success at 70% or higher—and one-third believe they won't fail at all (KAHNEMAN)[1].

Overconfidence reflects systematic errors—biases that recur predictably. So when we ask, "Why did Hall and Fischer fail?" or "What led to the 1953 success?" we should also ask, "Why do I make poor decisions?" and "How can I improve my judgment?"

Consequences of overconfidence in 1996
- Hall trusted his acclimatization strategy, but it failed some climbers. Doug Hansen struggled throughout.
- Hall knew the summit had to be reached by 2 pm. Yet at 3 pm, with Hansen far from the summit and out of oxygen, Hall allowed him to continue.
- Many of Fischer's clients summitted after 2 pm and lingered, well above the death zone.
- Fischer himself summitted at 3:45 pm, despite a liver condition.
- Tourist-climbers assumed paying a fee guaranteed a safe summit. Few had trained or prepared adequately.

Kahneman attributes overconfidence to the WYSIATI principle: "What You See Is All There Is." Here, "neither the quantity nor the quality of the evidence counts for much in subjective confidence."[2] Hall and Fischer ignored warnings from Ed Viesturs since the weather earlier in the day was benign.

Overconfidence is tied to excessive optimism and the narrative fallacy. Optimism fuels invention and art but misleads politicians and military leaders. It's the twin of overconfidence.

The narrative fallacy, the stories we tell ourselves, also played a role. Hall had summitted Everest four times and guided 39 climbers. Fischer had repeatedly moved between high camps. Both believed they were invincible.

A troubling study compared ICU autopsies with doctors' diagnoses. Physicians who were "certain" were wrong 40% of the time.[3,4]

Another particularly dangerous bias is Authority bias: the tendency to defer excessively to perceived authorities. Rob Hall told his team in 1996, "I will tolerate no dissension. My word will be absolute law beyond appeal."[5] This silenced junior guides and clients, not just from dissenting, but from voicing views.

Neal Beidleman, the third guide, later admitted he held back to avoid overstepping. In hindsight, he regretted not speaking. His silence mirrored that of experienced Sherpas, who deferred to Hall and Fischer. Only Hall's sirdar intervened, ordering Hansen to turn around at 3 pm. But when Hall insisted he'd assist Hansen, the sirdar relented.

Stanley Milgram's 1960s Yale experiments exposed the dangers of obedience to authority.[6] Participants, guided by a figure in a lab coat, were told to administer escalating electric shocks to an unseen 'learner' (an actor). As the shocks intensified, the learner feigned pain, pounded on the wall, and eventually fell silent. Psychiatrists predicted only 0.1% would deliver the maximum shock. In reality, 65% did. Just 14 of 40 stopped before the highest level.

Milgram initially believed Nazi atrocities stemmed from German culture. But across 18 variations, results stayed consistent: about two-thirds of participants succumbed to authority bias.

Equally disturbing was the 1971 Stanford Prison Experiment, led by Philip Zimbardo.[7] College students were randomly assigned roles as guards or prisoners. Zimbardo, acting as Prison Superintendent, reinforced power dynamics. Within hours, guards grew sadistic, prisoners submissive. The prisoners began accepting mistreatment, exhibiting learned helplessness. In 1996, Hall and Fischer's tourist-climbers similarly accepted violations of the strict turnaround time without protest.

The Stanford study showed how authority bias can be amplified. Beyond Zimbardo's role, the prison setting reinforced hierarchy. As some guards grew aggressive, others conformed. The study was halted after six days due to extreme psychological distress. Even Zimbardo, immersed in his role, delayed intervention. The study showed that the presence of authority, shaped by context, can magnify harmful behaviour often followed by regret.[8]

Historian Christopher Browning found similar patterns in WWII. Ordinary men in the German Reserve Police Battalion 101, middle-aged conscripts from Hamburg, mostly working-class and not Nazis, succumbed to authority and peer pressure. From 1941 to 1942, this 500-member unit was responsible for an estimated 83,000 Jewish deaths.[9]

Milgram summarized his findings starkly: "A substantial proportion of people do what they are told to do, irrespective of the content of the act and without limitations of conscience, so long as they perceive that the command comes from a legitimate authority."[10]

Other biases surfaced in 1996. Belief bias occurs when beliefs persist despite contradictory evidence (Stanovich). In the Tetraethyl Lead case, Thomas Midgley ignored the dangers of TEL, even after discovering ethanol as a viable alternative and suffering lead poisoning himself. At an Ethyl Corporation PR event, he dipped his hands in TEL, claiming he could do so daily without harm. Midgley knew TEL was harmful and ethanol was safer. Yet he clung to TEL, possibly due to belief bias.

The stories reveal something unsettling but vital: our judgments are shaped not just by evidence, but by unconscious biases. Confronting these biases is a first step toward self-knowledge.

Genetic drivers of cognition

How can we tell if behavioural traits like overconfidence are innate, part of our genetic heritage? "Behaviour," writes primatologist Frans de Waal, "doesn't fossilise. This is why speculations about human prehistory are often based on what we know about other primates. Their behaviour indicates the range of behaviour our ancestors may have shown" (FRANS DE WAAL)[11].

The fossil record and genetic sequencing show that orangutans, gorillas, chimpanzees, bonobos, and humans shared a common ancestor 13 million years ago. Chimpanzees and bonobos are our closest cousins, with a shared ancestor 5.5 to 7 million years ago. Humans share around 98.7% of their genome with both, and chimpanzees and bonobos share 99.6% with each other.[12,13,14]

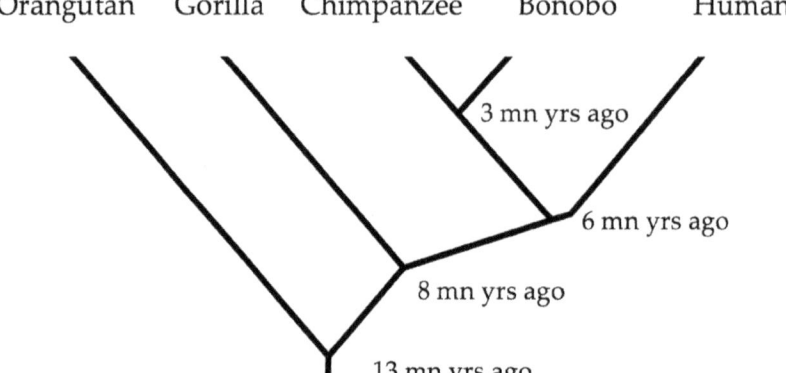

F6.1 The evolutionary relationship between apes. Adapted from NOVA Teachers[15] with modifications.

A 'pecking order' with 'authorities' in power exists in both chimpanzees and bonobos, though their hierarchies and conflict resolutions differ. Chimpanzees live in male-dominated groups, led by politically astute alpha males who maintain dominance through aggression, intimidation, and alliances. The alpha male enacts Hall's dictum "I will tolerate no dissension,"[16] through "impressive displays, going around with his hair on end, hitting anyone who doesn't move out of the way in time" (DE WAAL)[17].

While empathy exists, aggression is embedded in their social structure and strategic advantage. The Ngogo chimpanzees in

Uganda's Kibale National Park have been studied for over two decades. In the documentary "Rise of The Warrior Apes,"[18] gangs from one group expand territory by invading and exterminating neighbouring groups. These chimpanzees patrol borders and engage in lethal clashes.

Bonobos, equally close cousins, have a female-led hierarchy where violence is rare. An older female leads group movement and foraging. Though males are stronger, their status depends on their mother's rank. Sex is the preferred method of conflict resolution. In over thirty documented intergroup encounters, individuals of opposite sexes greeted each other with sexual and affiliative behaviour. While males often avoided unfamiliar males, intergroup mating was common.[19]

Frans de Waal documented bonobos' prosocial behaviour. They comfort distressed individuals by embracing, kissing, or grooming. Chimpanzees show similar behaviour but less often and are more competitive. Bonobos share food willingly, even with outsiders; chimpanzees do not. Bonobos have cared for injured birds and mammals, showing concern beyond their species.

Sexual drive, like aggression, is a primal force shaping behaviour across species. In chimpanzees, it often fuels dominance and conflict—males assert control over mating through intimidation and strategic alliances. Bonobos, by contrast, use sexual contact to diffuse tension and foster group cohesion. These divergent strategies reflect evolutionary adaptations to reproductive competition and social stability. In humans, echoes of both persist: sexual behaviour intersects with power, status, and emotional regulation. Where female choice is high, male aggression may evolve as a reproductive strategy. Yet cultural stigma around sex distorts these impulses, punishing expression and silencing trauma. Far from being isolated, sex influences broader domains—leadership, creativity, mental health—making it central to understanding both primate and human social structures.

Our ancestry is equally distant from chimpanzees and bonobos, but we are not half of each. Genomic studies show some segments are closer to bonobos, others to chimpanzees. We reflect traits of both: aggression and empathy.

Our genetic heritage is revealed not only through primates but across mammals. In Marco and Golden's experiment, rats showed empathy.[20] One rat was trapped in a restrainer tube; the other was free. After repeated attempts, the free rat learned to release its companion. Of 30 pairs, all showed distress; 23 succeeded in freeing the trapped rat. Some did so before eating chocolates offered as a reward.

Frans de Waal highlights other examples of mammalian empathy:[21]
- Elephants comfort and mourn herd members
- Dolphins support injured companions
- Dogs mirror their owners' distress and try to console them

De Waal argues empathy evolved from maternal care—mammals sense their offspring's needs to ensure survival.

Paul Bloom's research found that babies, even before speaking or walking, show empathy and fairness.[22] In one study, babies chose between puppets, one helpful, one selfish.

They consistently preferred the helper. Bloom concludes babies:
- Judge others' actions
- Show compassion to the distressed
- Prefer fair behaviour and reward helpfulness

However, this early morality is often limited to familiar individuals and does not extend to strangers. We are born with elements of empathy, which are shaped by, and shape, the cultures we grow up in.

Cognitive neuroscience

Kahneman describes two modes of decision-making: 'System 1,' fast and intuitive; and 'System 2,' slower, deliberate, and mentally demanding. System 1 is our default, requiring less energy, while System 2 consumes more brain resources (KAHNEMAN)[23].

Evidence for this model comes from cognitive neuroscience, which shows how external stimuli and internal brain processes shape behaviour. The brain relies on glucose for energy, during self-control and problem-solving. When glucose is low, these functions weaken. Nourishment, hydration, rest, and sleep also affect thinking, making us impulsive, fatigued, or focused depending on our physical state.

In an experiment on 100 participants, Roy Baumeister and colleagues found that acts of self-control reduced blood glucose and predicted poor performance on a subsequent task.[24] A glass of glucose eliminated these impairments. As participants became skilled, energy demands dropped.

Decision-making is rooted in specific brain regions. Physical stimuli, electrical, chemical, or sensory, can alter neural activity and behaviour. Two studies show how targeted interventions influence how we decide and respond:

- In research led by Daria Knoch, low-frequency rTMS disrupted the right dorsolateral prefrontal cortex (DLPFC).[25] Participants became more likely to accept unfair offers, though they still judged them as unfair.
- In work by MIT neuroscientists led by Rebecca Saxe, magnetic stimulation disrupted the right temporo-parietal junction (TPJ).[26] This impaired participants' ability to make moral judgments requiring an understanding of others' intentions, such as a failed murder attempt.[27]

These studies reveal the brain's vulnerability to external manipulation. Saxe remarked, "You think of morality as being a really high-level behaviour. To be able to apply (a magnetic field) to a specific brain region and change people's moral judgments is really astonishing."[28]

Such findings help explain why decision-making at Everest's extreme altitudes is so impaired. Above 26,000 feet, oxygen drops, pulse rate soars, blood thickens, and stroke risk rises. This disrupts nourishment, rest, and sleep. Under such physiological depletion and emotional strain, life-and-death decisions become erratic.

It is worth remembering: "You are not always the same person, and you are less consistent over time than you think" (KAHNEMAN)[29].

Memes

Our environments shape our beliefs and biases, beginning with the earliest one—our families. In the 1960s and '70s, James W.

Prescott conducted research on the role of physical affection, touch, and movement in early childhood development. He proposed that tactile and kinaesthetic stimulation during infancy is essential for brain development and emotional regulation. Societies that provide infants with high levels of physical affection, holding, carrying, cuddling, tend to be less violent than those that do not. Prescott supported this with experiments on infant monkeys and cross-cultural data showing an inverse correlation between early tactile nurturing and adult violence.[30]

But not all environmental effects stem from early childhood or family dynamics. Beliefs can be seen as a form of possession that we believe are grounded in 'rational' arguments. Yet Stanovich points out that biases like myside bias persist even in those considered 'intelligent.'[31] He argues that beliefs are memes, 'cultural replicators,' that spread from person to person. Like genes, memes evolve over time. He outlines four reasons why memes survive and spread:[32]

- They benefit those who believe in them. Kettering and Midgley's careers were funded by organisations promoting that brought them financial gain.
- They reinforce genetic predispositions.
- They help replicate genes in those who hold the meme— such as religious beliefs encouraging large families.
- They "survive and spread because of the self-perpetuating properties of the memes themselves" (STANOVICH)[33]. Even if a meme doesn't serve the believer, it can be hard to dislodge. This helps explain persistent biases like belief bias and myside bias.

Memes compete for survival in the cultural landscape, evolving through variation, selection, and replication within the "memosphere" (DENNETT)[34]. Memes can thus propagate themselves, even at the expense of human well-being. Dennett argues that our thoughts, beliefs, and even consciousness are shaped by the memes we absorb.

The 1996 Everest disaster reflects the 'shortcut to success' meme. Many climbers believed money could compensate for lack of experience. Hiring experts was seen as a guarantee of achievement. But Everest is unforgiving. The tragedy shows that memes cannot replace hard work and the time it takes to master a craft.

The priming environment

A priming environment refers to external stimuli that influence our behaviour and decision-making at an unconscious level. Priming occurs when certain cues, such as words, images, sounds, or even social interactions, activate mental associations that then shape our perceptions and reactions.

Priming is psychological in nature, whereas external factors like drugs or magnetic stimulation physically alter brain function. Priming environments operate subtly, and changing our surroundings can substantially enhance or diminish decision-making.

Kathleen Vohs and colleagues conducted nine experiments to explore how money influences perseverance, helpfulness, and social interaction.[35] In these experiments, participants were primed with money-related stimuli (such as images, words, or currency), and then tested on tasks measuring perseverance, helpfulness, and social engagement. The results are summarized in T6.1.

Behaviour	Money primed	Not money primed
Self-sufficiency and perseverance	Greater	Less
Independence	Greater	Less
Doing things alone	Greater	Less
Willingness to help	Less	More
Team play/teamwork	Less	More
Physical distance from others	Greater distance	Closer distance

T6.1 Results of money priming experiments. (Vohs)[36]

These experiments suggest that while money can enhance personal motivation it may also diminish social bonds and reduce willingness to help others.

Commenting on these experiments Kahneman wrote, "Professor Vohs' experiments are profound – her findings suggest that living in a culture that surrounds us with reminders of money may shape our behaviour and our attitudes in ways that we do not know about and of which we may not be proud" (Kahneman)[37].

In the 1996 Everest expedition the tourist-climbers were all from an environment that was priming them to believe that money could buy anything – including a ticket to the top of Everest without training or experience.

Climber Nick Bullock in his blog post, "I want it and I want it now... I want it and I want it now... I want it and i wa..."[38] wrote, "We all need to strive for personal goals and have dreams and ambition, but what I don't understand is people who don't want to learn about climbing (and themselves in the process) through, the usual and accepted way. People attempting to climb 8000-meter mountains with very little actual climbing experience appear to think that handing over a load of cash will help when the crap hits the fan. No mountain guide or a Sherpa can guide [on] an 8000-meter mountain when the weather comes in it turns into every man and woman for themselves."[39]

Research by Kathleen Vohs should not be viewed as solely concerned with money but priming in general. Our goals are frequently shaped not by reasoning but by unconscious cues and social pressures.

Evolutionary perspectives

Why do cognitive biases and errors of judgment persist? Kahneman's System 1, fast, instinctive thinking, offered survival advantages in environments with sudden threats, where slow deliberation was risky.

Evolutionary biologist Robert Trivers proposed that humans deceive themselves to better deceive others. Self-deception, in this view, evolved to enhance social and survival advantages, embedding itself in our cognition (TRIVERS)[40].

Trivers cites warfare: a warrior who believes in his exaggerated strength becomes more fearless, convincing others of his superiority. This boosts morale and aggression, increasing chances of victory — even if the belief is false.

Yet Trivers warns that in modern high-stakes situations, unchecked self-deception can undermine rational thinking and lead to large-scale failures. What was adaptive in "primitive social structures" is now dangerous when it blocks individuals and institutions from confronting reality.

Mercier and Sperber argue that reasoning evolved not to seek truth, but to persuade others, strengthening social cohesion.[41] Once Kettering and Midgeley embraced the TEL belief, they became evangelical. Opposition only solidified their conviction, binding them closer through shared arguments.

Strategies for cognitive integrity

Can we overcome biases?

Kahneman was sceptical about overcoming cognitive biases. He wrote, "Changing one's mind about human nature is hard work, and changing one's mind for the worse about oneself is even harder" (KAHNEMAN)[42]. While awareness helps, real-time decision-making remains difficult. System 2 thinking is effortful and best reserved for major decisions. Yet, some strategies can improve judgment.

Checklists

Atul Gawande shows how checklists reduce errors and improve consistency in complex tasks.[43] Examples include:
- Pilots standardizing procedures to minimize error
- Hospitals lowering complications and mortality rates
- Construction teams coordinating steps across large projects

Teamwork

In 1996, despite paying Hall USD 70,000 each, four climbers turned back when risks became clear. Hutchison, Taske, and Kasischke formed a small team, discussed oxygen shortages, and turned around together. Their dialogue and trust likely saved their lives.

Teamwork also defined the 1953 expedition. Hunt emphasized collaboration and excluded himself from summit attempts, writing, "The ascent of Everest... demanded a very high degree of selfless co-operation... In this, and in the work of our Sherpas, lies the immediate secret of our success."[44]

Friendships, dissent, and trust

Kahneman's research shows biases persist despite evidence. But friendships help. An example is Daryl Davis, a Black musician, befriended KKK members, leading many to abandon their beliefs.[45]

Trust takes time. Hunt had the team trek together for weeks, building unity. He paired climbers to complement each other. Tenzing saved Hillary from a fall in the Khumbu Icefall, deepening trust. Bourdillon, known for "bulldog determination," was paired with the sensible Evans.[46] Finding our own "Evans" is one way to reduce decision-making risks.

Decisions made together

Hunt treated Sherpas as equals. Tenzing, with five prior Everest expeditions, was a leader. He was one of three summit pairs Hunt planned. Hunt welcomed input from all:
- Tenzing's terrain knowledge reshaped the route
- Ward's medical advice guided pacing and oxygen use

- Wylie's logistics ensured supplies
- Hillary's Icefall strategy improved safety
- Lowe refined the summit approach

Evaluate purposes and reflect

The 1996 expeditions were driven by commercial goals and prestige. In contrast, Hunt saw Everest as a challenge akin to science or art—an unknown to explore. He hoped the ascent would inspire others to climb their own "Everests."

Major life decisions are rarely clear-cut. We often choose based on how we wish the world to be. Readers may spot other biases in the Everest case, affecting judgment and risk. But perhaps the most illuminating insight from the two Everest stories is that we "see in ourselves one of the most internally conflicted animals ever to walk to Earth. It is capable of unbelievable destruction of both its environment and its own kind, yet at the same time it possesses wells of empathy and love deeper than ever seen before" (FRANS DE WAAL)[47].

Emotions as judgments

The 1996 and 1953 Everest expeditions can also be explored through the emotional dimensions of human endeavours. Philosopher Martha Nussbaum proposes that emotions are not mere feelings, but judgments about value, vulnerability, and ethical significance (NUSSBAUM)[48]. They are appraisals of what matters most.

Nussbaum identifies twelve core emotions: grief, fear, love, joy, hope, anger, gratitude, hatred, envy, jealousy, pity, and guilt, as representative lenses through which we interpret the world and are called to act.[49] Emotions are shaped by bodily needs, desires, imagination, memory, and narrative identity. Hunger, exhaustion, and vulnerability heighten emotional salience. Desire for achievement may evolve into envy or hope; memory triggers grief or gratitude; identity determines emotional significance.

Through this lens, climbers' decisions in both Everest expeditions become emotionally intelligible—even when they led to tragedy.

Emotions and judgments in 1996

Hall's decision to stay with Hansen beyond the turnaround time was animated by pity, hope, and fear. Hansen had returned for a second attempt, investing heavily. Hall may have feared failing him again and hoped to help him succeed. His final radio call to his wife was suffused with grief and love.

Fischer, charismatic and resilient, succumbed to altitude sickness. He may have felt envy toward Hall's success and guilt for hiding his condition. Tourist climbers, pushing past the turnaround time, were driven by hope, fear, and guilt—seeking transformation or redemption, fearing failure after significant emotional and financial investment.

Emotions and judgments in 1953

In contrast, the 1953 expedition showed how emotionally animated behaviour, grounded in shared purpose and rational discipline, can lead to triumph. Hunt's leadership was rooted in ethics and emotional intelligence. He fostered humility, respect, and interdependence, ensuring ambition never overrode judgment.

Hillary's account of the expedition reflects hope, gratitude, and love for the mountain and the team. Tenzing's emotional landscape, shaped by reverence for Everest and years of labour, reflected joy, pride, and pity for those who had failed before. Their partnership was marked by compassion, not rivalry, and their final ascent by restraint and caution.

Hunt understood that emotions must be tested for truth. In 1996, this did not happen, leading to tragedy. In 1953, emotional assumptions were tempered. Hunt anticipated emotional precursors like fatigue and ambition and built systems—rotating teams, turnaround protocols, psychological support—to counteract them. He understood each climber's narrative identity and framed the mission as a shared human achievement, channelling emotion into solidarity.

The lesson is clear: emotions are not the enemy of reason, but its companions. They spark and animate our endeavours but must be tested through science and reflection. The 1953 expedition shows that when this is done, humans can achieve the extraordinary—not by conquering nature, but by understanding ourselves.

This synthesis of emotion and empirically grounded reason defines humanism and forms a core thesis of this book.

N4 The Orange King

"The trouble with the world is that the stupid are cocksure and the intelligent are full of doubt" (BERTRAND RUSSELL)[50].

"Greater the artist, greater the doubt. Perfect confidence is granted to the less talented as a consolation prize."[51] (Robert Hughes)

Part 3

How the world IS

A Test of knowledge

How do we know
Science as a self-correcting feedback-based learning process

Tetraethyl lead in gasoline – The science
The Cold Fusion Fiasco

"While pursuing studies of the evolution of the Earth's mantle, my colleagues and I discovered, quite by accident, an apparent discrepancy in the form of a huge modern surge in the historical record of the flow of soluble lead from rivers through the oceans to sediments. This introduced the question of universal lead contamination of the Earth's atmosphere, biosphere, and seas. Other scientists and we discovered irrefutable proof that humans have poisoned the Earth's biosphere and themselves with excessive amounts of industrial lead."[1]

CLAIR PATTERSON

"It would require belief in miracle after miracle for one to accept their discovery [of cold fusion]."[2]

John Huizenga

7

Tetraethyl Lead in Gasoline – The Science

Lead poisoning

<u>Lead</u>

Lead (atomic number 82) was identified as a poison in 250 BCE by Nicander of Colophon, who linked it to colic and nerve palsies.[1] Yet ancient Rome used it widely—its symbol, Pb, comes from plumbum, reflecting its plumbing role. Lead poisoning became a disease of affluence, as the wealthy used it in utensils, wine urns, and water systems. Some historians even suggest it contributed to Rome's decline.[2]

The link between lead and illness remained contested for centuries. In 1767, Dr George Baker traced "Devonshire colic" in English cider drinkers to lead contamination, sparking backlash.[3]

By the late 19th century, industrial lead poisoning was well documented in smelting and battery work.[4] Most research focused on occupational hazards, with rare exceptions like Dr Lockhart Gibson's Australian studies (1903–1913), which linked lead paint to childhood poisoning. Still, lead-based paints remained common in U.S. homes and toys until the 1970s, despite a 1922 League of Nations ban.[5]

In the early 20th century, airborne lead was blamed on natural sources. Only in the 1930s did scientists suspect leaded petrol, though proof remained elusive for decades.

This case study examines the pivotal work of geochemist Clair Patterson in the 1960s, who showed that leaded gasoline was the dominant source of environmental contamination. It also highlights the contributions of Dr Herbert Needleman, Dr Philip Landrigan, and others in the 1970s, who revealed the cumulative effects of lead exposure, especially its role in childhood poisoning and developmental harm.

In contrast, studies by Robert Kehoe, an industry consultant, insisted leaded gasoline posed no significant public health risk, reinforcing decades of regulatory inertia.

Tetraethyl Lead

It was well known that in its pure form, tetraethyl lead (TEL) was a deadly poison[6]. The critical question was whether the use of TEL mixed with gasoline (leaded gasoline) was a risk to public health via lead oxides that emerged from automobile exhaust?

Midgley relied on occupational lead studies where it was believed that lead exposure below 80 μg/dL resulted in no measurable health effects.[7] Therefore, he concluded that leaded gasoline posed no threat to the public. Midgley did not consider that lead exposure could be cumulative over many years.

The Bureau of Mines experiment

Before 1925, the Bureau of Mines ran an experiment on leaded gasoline. After a short study, it deemed public risk from auto exhaust "seemingly remote," based on animals exposed for mere months. But the setup missed real-world conditions—animals lived in ventilated cages, with no lead dust buildup.[8]

Robert Kehoe

<u>The Kehoe Rule</u>

Despite strong recommendations from the Surgeon General's 1925 Committee to investigate the risks of leaded gasoline, no independent studies were pursued. This complacency was due to the influence of Robert Kehoe, a consultant to Ethyl Corporation and General Motors.

Kehoe wrote, "as it appears that there was no evidence of immediate danger to the public health, it was thought that these necessary extensive studies should not be repeated at present, at public expense, but that they should continue at the expense of the industry most concerned."[9] C3 S: Leaded gasoline and Ethyl Corporation explores Kehoe's background, including his reliance on funding from U.S. oil companies and GM for research related to leaded gasoline.

At the conference on May 20, 1925, Kehoe came up with the 'Kehoe Rule':

- Since every technology or product carried some inherent risk, the burden of proof did not lie with industry, but with those asserting that TEL was dangerous.
- If the dangers of TEL could not be demonstrated, then, given its significant economic benefit, Ethyl Corporation should continue production of leaded gasoline.

GM, Du Pont, and SONJ funded Kehoe, who dominated 'research' into leaded gasoline's effects on public health. Regulators and the government (the U.S. Public Health Service) and the American Medical Association supported both the Kehoe Rule and the findings that came out of the Kettering

Kehoe's model of normal levels

A key aspect of Kehoe's argument centred on his interpretation of "normal levels" of lead in the human body. Kehoe found measurable amounts of lead in the blood of Dayton workers who had no direct contact with TEL. He concluded that lead was a natural constituent of the human body, an assumption that shaped decades of regulatory policy and delayed recognition of widespread environmental contamination.[10]

When confronted by Yandell Henderson, of Yale's Applied Physiology Lab, and others that all the workers in the Dayton plant were probably exposed to TEL fumes, Kehoe studied lead levels in a remote village in Mexico. Kehoe's assumption was that the villagers were unexposed to industrial lead, making them an ideal control group. He took samples from the villagers' excreta, food and utensils.

Kehoe found high levels of lead in the villagers' excreta and concluded that this was the 'natural' lead level in human bodies. He therefore concluded that this provided evidence of no risk from lead in gasoline. What Kehoe failed to publicize or investigate was the presence of high levels of lead in the clay dishware, an overlooked source of exposure that likely skewed his conclusions about "normal" lead levels in the population. Traditional glazing techniques used lead while making the clay dishware and this was gradually dissolved and drawn out by contact with liquids during food preparation and then ingested by the villagers. The villagers were not an accurate control group.[11]

Kehoe also insisted that lead poisoning was a 'yes-no' phenomenon. "Affected people had to develop overt illness, a severe anaemia, or a neuropathy or encephalopathy before being recognized as being ill."[12] Kehoe reasoned that the body would excrete lead such that a balance back to normal levels was reached.

Data from the risks of occupational lead led Kehoe to use a level of 80 µg/dL as the 'normal' level of lead in humans. He had not seen workers with clinical signs of lead poisoning whose blood lead was lower than this threshold. Regarding any increase in "lead taken in from the atmosphere," Kehoe was adamant that "there is not the slightest evidence that there has been a change in this picture. Not the slightest."[13]

Clair Patterson

James Hutton and Deep Time

In the late 18th century, Scottish geologist James Hutton proposed that Earth's surface was shaped by slow processes—erosion, sedimentation, and uplift—unfolding over vast spans of time. His ideas, though largely overlooked in his lifetime, were later clarified by Charles Lyell (1830–33), laying the foundation for modern geology and enabling breakthroughs in dating Earth's age.

The age of the Earth

In 1953, thirty-six-year-old geologist Clair Patterson succeeded in measuring the age of the Earth, accepted as 4.54 billion years, with only slight revisions since. Patterson wrote, "The discovery electrified my soul... True scientific discovery...

forces the brain to thunder, 'We did it.'"[14] The "we" honoured the many scientists who paved the way.

Patterson's breakthrough came through using radiometric methods to measure lead isotopes in rock samples:
- Lead-204. A stable isotope present since Earth's formation, used as a baseline.
- Lead-206, 207, 208. Are products of uranium and thorium decay, occurring at known rates.
- Meteorites untouched by Earth's geological processes, served as pristine time capsules. By comparing lead ratios in meteorites and Earth rocks, Patterson calculated the time since the solar system's formation.

Initially, the task seemed simple. His professor, Harrison Brown, called it as easy as "duck soup."[15] But Patterson soon found his results skewed by environmental lead. Tim Lain described the challenge: "It was like trying to hear someone whisper from ten rows away at a rock concert."[16]

It took Patterson seven years. He created new lab protocols, purified air, water, reagents, and had researchers wear gowns and masks. Only then could he "hear the whisper" of ancient lead.

His perseverance revealed Earth's age and gave humanity a new perspective on its place in the cosmos.

Experiments on lead in the environment

Patterson was not initially drawn to the issue of lead pollution. In 1963, Patterson met Harriet Hardy, head of MIT's occupational medical service. Hardy had studied lead's effects on industrial workers and feared for children. Her plea was "You have to help me."

Patterson said, "There's a bewitching attraction to pure science... It remains an abstract, beautiful refuge disconnected from the dirty world. [However] my science got entangled with social problems. It threatened the beauty of my refuge. But there was no way out."[17,18]

From 1963 to 1981, Patterson conducted experiments and targeted observations to uncover environmental lead's source. His research revealed a staggering rise of up to 1,000 times in environmental lead levels from those in natural sources. Patterson showed that Kehoe's 'normal levels' of lead in humans were just the present 'typical levels' in humans who were all breathing in lead particles from the air. Patterson pointed out that the 'natural' level of lead in humans existed before the air was contaminated by industrial processes and the use of leaded gasoline.

Patterson argued that since the effects of lead were so pervasive, Kehoe would never find a control subject amongst contemporary humans. Patterson looked at:

- Patterson's analysis of 1,600-year-old Peruvian bones revealed that modern humans carry 3,000 times more lead than ancient Peruvians.
- Ice cores in Greenland showed that lead in the atmosphere was a recent occurrence. F7.1 shows the increase in lead concentrations over the last 2,800 years.

- Deep ocean tuna which contained significantly less lead than fish in shallower waters.

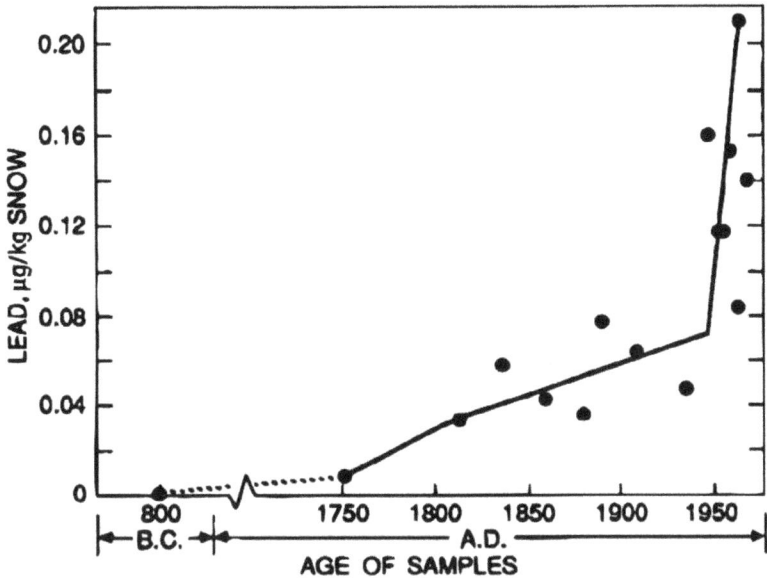

F7.1 Lead Concentrations in Polar Ice of Northern Greenland. Source: Environmental Protection Agency (EPA), *Air Quality Criteria for Lead* (1989)[19]

Patterson explained modern lead levels via a picture similar to F7.2. This shows lead content in humans prior to the iron age, in the 1960s, and under lead poisoning. Unlike Kehoe, who insisted that individuals were either lead poisoned or not poisoned at all, Patterson argued that there "is no abrupt change between a response and no response. Classical poisoning is just one extreme of a whole continuum of responses [of the] human organism to this toxic metal [lead]. There is no reason why this shouldn't be so."[20]

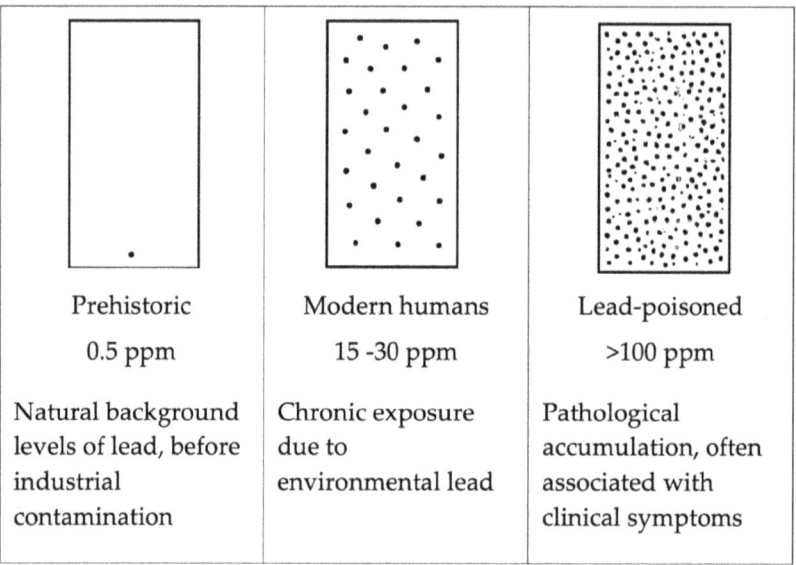

F7.2 Estimated lead accumulations across human populations. ppm (parts per million) refers to the concentration of lead in human bone tissue in the cortical bone. Adapted from Patterson (1985) and Needleman (1998)[21]

Patterson's research faced substantial opposition from oil companies and Kehoe's response to Patterson's 1965 article was patronizing.[22] Patterson was unambiguous regarding the causes of lead pollution. In his 1963 article, he wrote:

> "The concentrations of common lead in sea-waters off southern California [are] found to decrease abruptly from surface to deep waters. The high concentrations of lead in surface waters were ascribed to industrial contamination, mainly lead tetraethyl aerosols... [This] strongly supports the idea of contamination from leaded gasoline exhausts."[23]
>
> PATTERSON

Patterson's model of biological systems

Before beginning his experiments, Patterson was troubled by data showing unusual ratios of calcium to lead across objects and living organisms. Todd Hinkley explained that Patterson "got the idea that as a biological system evolved, it developed mechanisms to take the good metals [such as calcium] that it needs for nutrition to incorporate into its system and discriminate and reject similar related poisonous metals [such as lead]."[24]

Thus, as one goes from rocks to food sources and up the food chain to humans, the ratio of calcium to lead would change. There would be more calcium and less lead. However, when Patterson examined the ratio of lead to calcium in surface ocean water and in human tissues, he found that the ratios were virtually identical, suggesting environmental lead exposure was widespread and anthropogenic.

Patterson's model could be wrong or as Lydia Denworth explained, the data could be showing that "Like a bathtub drain overwhelmed by a running faucet, the natural rejection of lead couldn't keep up with the amount coming in. If no one turned off the tap, too much lead in our bodies was as certain as a flood on the bathroom floor."[25]

In the 1925 Surgeon General's conference, Hamilton had said, "Lead is a slow and cumulative poison that does not usually produce striking symptoms that are easily recognized."[26] Patterson's studies in the 1960s showed that lead in the modern environment (the 1960s) was much higher than in pre-industrial times and also in comparison to the periods just prior to the introduction of leaded gasoline.

The effects of lead in children

There are many epidemiological studies on the effects of leaded gasoline, particularly in children.[27] However, in the 1970s Dr. Needleman, Dr. Landrigan, and their colleagues first showed that children were most susceptible to lead poisoning. Needleman's breakthrough came from measuring lead levels in children's shed milk teeth, offering a non-invasive window into long-term exposure.

Needleman and his colleagues examined children's teeth and found a strong correlation between high levels of lead in children with learning disabilities, lower IQs, attention deficits, and erratic behaviour.[28]

Research has shown that lead poisons children at even minimal doses and that while this does not cause overt or immediate symptoms, the effects are severe and long-term. This is termed as an illness from 'silent doses.' The reason children are particularly susceptible to lead poisoning is their "rapid respiration and metabolism that are designed for sponge-like absorption of nutrients [thus retaining] a higher percentage of ingested lead than do adults."[29]

There were allegations by Claire B Ernhart[30] (who obtained a USD 375,000 grant from the International Lead Zinc Research Organization) that Needleman's research and statistical methods were biased.[31] Needleman's original article had some statistical errors that did not affect the conclusions, and Needleman issued a reanalysis.[32]

No natural level of lead in humans

F7.3 shows the sharp decline in the blood lead of Americans as leaded gasoline was phased out.

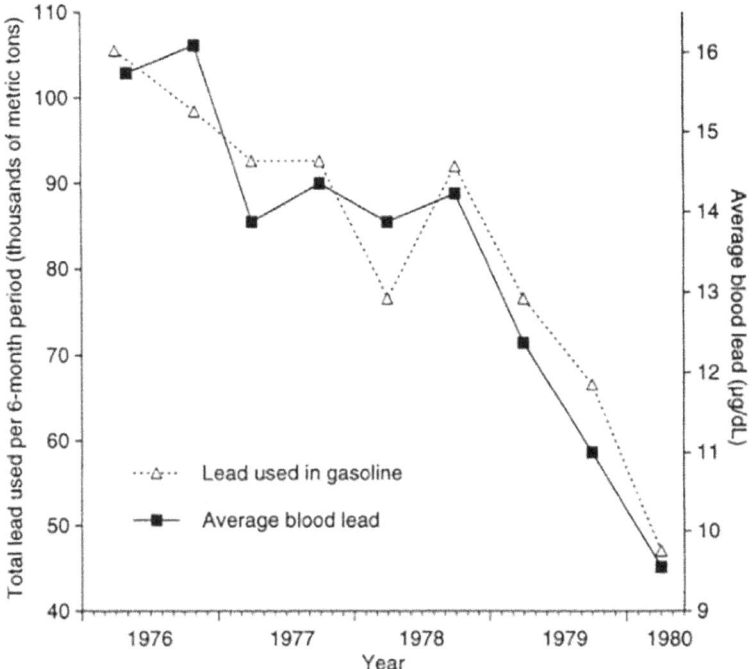

F7.3 Average blood lead in the U.S. and lead used in gasoline. Reprinted, with permission from John Wiley & Sons, Ltd. J.L Annest, "Trends in the blood lead levels of the U.S. population: The Second National Health and Nutrition Examination Survey (NHANES II) 1976-1980," in Lead Versus Health, M. Rutter and R.R. Jones, eds., New York: John Wiley & Sons. ©1983, John Wiley & Sons, Ltd.[33]

The effects of lead at the molecular level

Advances in molecular science have revealed how lead harms the body:
- Enzymes speed up chemical reactions. A meta-analysis showed the enzyme ALAD, with Zinc, makes heme—

vital for haemoglobin, which binds oxygen. Lead mimics Zinc, binds ALAD, blocks heme production, reducing haemoglobin and oxygen.[34]

- The blood-brain barrier protects the brain but can't distinguish Calcium ions from lead. Calcium enables neuron messaging; lead disrupts these paths and inhibits nerve cell cladding.[35]

The molecular mechanisms of lead as a potent poison were unknown until after the 1970s.

The use of probabilities and statistics

Research by Jessica Reyes[36] states that "changes in childhood lead exposure are responsible for a 56% drop in violent crime in the 1990s."[37] Reyes studied crime rates in U.S. states in a given year compared to childhood lead exposure in those states 22 years earlier. F7.4 displays the annual lead levels in the environment overlaid with violent crime data from 22 years later.

> "This paper shows a significant and robust relationship between lead exposure in childhood and violent crime rates later in life. The legalization of abortion ... [also] remains an important and significant factor. Thus, two major acts of government, the Clean Air Act and Roe v. Wade, neither intended to have any effect on crime, may have been the largest factors affecting violent crime trends at the turn of century. These results emphasize the importance of accounting for earlier life influences when explaining adult behaviour."[38]
>
> REYES

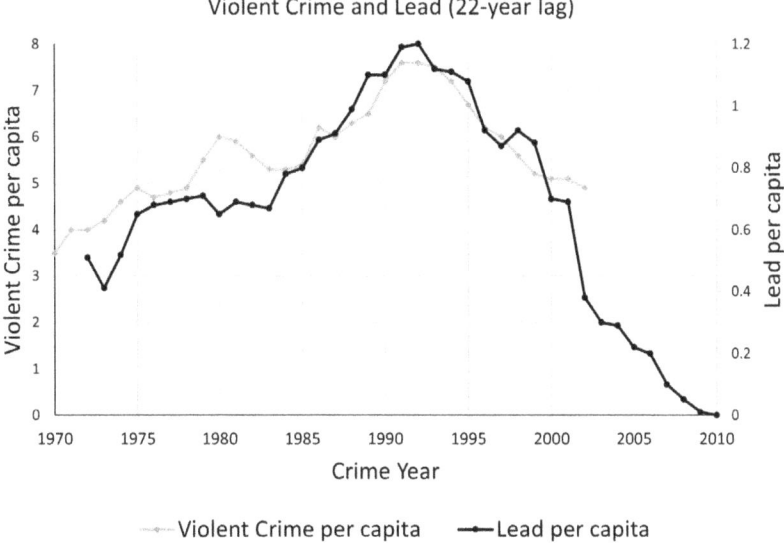

F7.4 Violent crime and lead in U.S.[39] states with high and low lead levels[40] Adapted from Jessica Reyes, *Environmental Policy as Social Policy? The Impact of Childhood Lead Exposure on Crime*, Working Paper 13097 (Cambridge, MA: National Bureau of Economic Research, 2007).

Note that in F7.4:

- Violent Crime Per Capita refers to the number of violent crimes committed per 100,000 people in a given year, allowing for comparisons across populations of different sizes.
- Lead Exposure Per Capita is based on grams of lead per gallon of gasoline, which Reyes uses as a proxy for environmental lead exposure. She then links this to childhood blood lead levels, estimating the average exposure per child in the population.

C30 S: Plausible reasoning briefly explains how Reyes reached her conclusions.

8

The Cold Fusion Fiasco

Scientific communication via press release

On March 23, 1989, two business newspapers, The Financial Times[1] and the Wall Street Journal,[2] announced the finding of room temperature fusion in a test tube. This was followed by a University of Utah press release[3] stating:

'Simple Experiment' results in sustained N-Fusion at room temperature for the first time. Breakthrough process has potential to provide inexhaustible source of energy...

Collaborators in the discovery are Dr. Martin Fleischmann, professor of electrochemistry of the University of Southampton, England, and Dr. B. Stanley Pons, professor of chemistry and Chairman of the Department of Chemistry at the University of Utah."

University of Utah press release, 1989[4]

Scientific research is usually presented via a research paper that is reviewed, vetted and changes made through a process of peer review. Pons and Fleischmann published their findings in the *Journal of Electroanalytical Chemistry and Interfacial Electrochemistry* a month after the press release, and without thorough peer review.[5] Their submission to the journal, *Nature*, was withdrawn after queries from the peer review process could not be addressed.[6]

Background science of nuclear fusion

Nuclear fusion is the coming together of the nuclei of two or more light elements with a release of energy. Nuclear fusion is understood to occur in the core of stars like our Sun, releasing an immense amount of energy.[7]

If we could reproduce and control fusion we would have, as the University of Utah press release stated, a truly "inexhaustible source of energy." In the hydrogen bomb, humans have reproduced fusion on Earth but in an uncontrolled manner. Fusion research has been ongoing for decades, focussed on the fusion of two hydrogen isotope nuclei; deuterium (which has one proton and one neutron) and tritium (having one proton and two neutrons).[8]

Fusion is notoriously difficult to control because positively charged atomic nuclei repel each other due to electrostatic forces. In the cores of stars, extreme temperatures around 16 million Kelvin (roughly the same in Celsius) and pressures more than 20 times the density of iron, create an environment where some hydrogen nuclei can overcome this repulsive barrier and fuse together.

When fusion occurs, the resulting nucleus has slightly less mass than the combined mass of the original nuclei. This 'missing' mass is converted into energy, as described by Einstein's equation $E = mc^2$. The energy released is immense and also produces a flux of neutrons, uncharged particles that escape the reaction zone and can interact with surrounding materials.

This neutron flux plays a crucial role in sustaining fusion reactions and has significant implications for fusion reactor designs and radiation shielding. The neutron flux, heat and pressure generated by initial reactions create conditions that enable further nuclei to collide and fuse, making stellar fusion a self-sustaining chain reaction, one that powers stars for billions of years.

The University of Utah announcement claimed to have achieved fusion without requiring such extreme conditions, making use of a simple catalyst (substances that facilitate chemical reactions) in a test tube.

Changing data

The original article in the "Journal of Electroanalytical Chemistry and Interfacial Electrochemistry" was followed by errata that was "25% of the length of the original paper,"[9] included a third author and contained different data. This pattern of providing different data and experimental outcomes continued.[10]

Secrecy

Pons and Fleishman did not provide details of the experiment. Instead, they conducted private conversations, such as a phone call with Edward Teller, proponent of the hydrogen bomb. Teller told the press he accepted the reality of cold fusion and was "extremely happy."[11]

Replicating cold fusion

Several experimental groups including at the Universities of Texas A&M and BARC (India) reported success using similar apparatus to Pons and Fleischmann.[12]

However, a set of experiments led by Nathan Lewis (Caltech) consisting of fifteen chemists and physicists showed an upper limit on neutron flux that 100,000 times smaller than what was reported by Pons and Fleischmann which did not exceed the background level. There was also minimal excess heat.[13,14]

R. D. Petrasso (MIT) provided evidence that Pons and Fleischmann measurements were unfounded and were "a factor of two smaller than their instrumental resolution would allow."[15]

The U.S. Department of Energy set up a Panel to investigate cold fusion claims. John R. Huizenga, Co-Chairman of the panel, wrote, "In none of our meetings with the various groups claiming positive results did we see an operating cell that was producing excess heat at a time of our visit!"[16]

Controls in experimentation

Harold Furth (Princeton University) suggested a crucial control experiment where results under the same conditions but using light water (H_2O) versus and heavy water (D_2O) (where D_2 is the hydrogen isotope Deuterium) should be compared.[17] Pons first implied such a control experiment had been conducted and then later said it was not necessary.

Models, theories and calculations

Models on fusion had been extensively developed and tested since the 1950s. however new theories were developed to fit Pons and Fleischman's claims, sometimes contradicted earlier models.[18]

Pons and Fleischmann did not check their calculations with those experienced in fusion research and included assumptions that had extreme values. For example, Pons assumed the attainment of extremely high pressures, more than 20 orders of magnitude (10^{20} - 1 followed by 20 zeros), far exceeding the values derived usual electrochemical equations.

Justification and suppression of dissent

Cold fusion supporters had explanations why experiments that did not validate Pons and Fleischmann occurred. These reasons included "it's hard for them to see effects because they are doing it wrong," "negative results can be obtained without skill and experience," and claims that cold fusion only worked intermittently.[19]

Cold fusion became a point of pride for the state of Utah. In July 1989, the State Fusion/Energy Advisory Council formally endorsed the claims made by Fleischmann and Pons. Following this, the National Cold Fusion Institute was established in Salt Lake City, and efforts were made in Washington to secure an additional USD 125 million in federal support to complement the USD 5 million already committed by the state.[20]

In a National Science Foundation and Electric Power Research Institute sponsored workshop in October 1989 few sceptics of cold fusion were invited. Huizenga described the atmosphere in the first annual conference on cold fusion, held in March 1990, as having a "religious-like fervour" where "most of the scientific talks were given by people who were believers." Dr. Fritz Will, Director of the National Cold Fusion Institute "lashed out" at those who dissented and "denounced critical questions."[21]

In 1990 when Michael Salamon from the University of Utah and nine collaborators found no evidence of fusion in Pons' own laboratory, they were threatened with lawsuits by Pons and Fleischmann's lawyers.[22]

Status of cold fusion in 2025

Douglas R. O. Morrison of CERN described the scientific community's reaction to cold fusion as unfolding in three distinct phases: "In Phase One the original report is quickly confirmed; in Phase Two there are about equal numbers of positive and negative results; and in Phase Three there is an avalanche of negative results."[23]

Pons and Fleischmann remained committed to their belief in cold fusion. In 1992, they relocated to France and established a laboratory funded by Toyota. However, after six years and expenditure of GBP 12 million, the lab was shut down in 1998.[24]

9

Science as a Test of Knowledge

An empirical basis and falsification

An empirical basis

Richard Feynman wrote, "The test of all knowledge is experiment. Experiment is the sole judge of scientific 'truth'" (FEYNMAN)[1]. He argued that hypotheses must be tested through observation and experimentation to understand the workings of the universe.

A powerful example comes from Wilbur and Orville Wright, who tested nearly 200 wing designs and conducted the first wind tunnel experiments to achieve heavier-than-air flight. In contrast, Samuel Langley of the Smithsonian spent ten years on desk work then his two attempts with a full-sized 'Aerodrome' ended in crashes.

Wilbur Wright compared learning to fly to riding a fractious horse: one must mount and learn through trial, not observe from a distance.[2,3]

Humans are swayed by delusions and fantasies. Empirical inquiry grounds us in reality and understanding.

This does not mean experiments yield easy or definitive answers. We must remain sensitive to inconsistencies between belief and evidence. Ambiguous data can lead to tentative beliefs. In such cases, scientists rely on falsification and methods like Bayesian reasoning, which allows graded belief revision.

Falsification

In 1925, the Bureau of Mines found the danger of leaded petrol "seemingly remote." Robert Kehoe claimed lead was a natural component in humans, based on observations near Mexico City. Clair Patterson reached a different conclusion, underscoring the importance of understanding how science works to assess the veracity of scientific claims.

Patterson kept open the possibility he was wrong, embodying Feynman's principle: "You must not fool yourself—and you are the easiest person to fool" (FEYNMAN)[4]. While dating the Earth, Patterson found elevated lead levels in samples. Rather than accept this, he spent years testing for contamination. He tried to falsify that high lead levels were natural.

Scientific theories like Newton's and Einstein's are testable and open to being wrong. Dogmas, by contrast, resist falsification. Many are framed to prevent questioning and inquiry is met with hostility.

Karl Popper formalised falsification, though it had been practised since Galileo. He wrote: "Every genuine test of a theory is an attempt to falsify it... confirming evidence should not count except when it is the result of a genuine test" (POPPER)[5].

Applying empirical inquiry and falsification

Dogmas: religious, Marxist, capitalist, often fail under empirical scrutiny. In the 1982 U.S. case *McLean v. Arkansas Board of Education*, the judge ruled against teaching creationism as science, citing its lack of empirical testability and falsifiability.[6]

Philosopher Larry Laudan agreed with the ruling but argued that creationism should be challenged by confuting its empirical claims. (Laudan in CURD, COVER, AND PINCOCK)[7].

Long before Laudan, Patterson practised this approach. He didn't dismiss Kehoe's claims outright but assembled diverse evidence, from ancient skeletons to ice cores and deep-sea tuna, all showing a recent surge in environmental lead.

Pons and Fleischmann claimed to discover cold fusion but altered data without trying to falsify their claim. Nathan Lewis at Caltech and R. D. Petrasso at MIT conducted experiments that falsified Pons and Fleischmann's findings.[8,9,10]

Thinking backwards: inversion in empirical inquiry

Inversion, popularised by Charles Munger, is a powerful tool while approaching falsification. Munger said: "All I want to know is where I'm going to die, so I'll never go there."

Applied to lead poisoning, inversion asks: 'How can we maximise lead exposure for the largest number of people?' This reframing could have led researchers to test whether car exhaust might be a widespread delivery mechanism. Inversion sharpens thinking, exposes hidden assumptions, and guides more targeted, falsifiable inquiry.

Bayesian plausible reasoning

Application of Bayes' Theorem

Falsification, however, does not tell us how strongly we should believe a theory that survives experimental tests. Bayesian reasoning fills this gap, offering a framework for updating confidence in a hypothesis based on new evidence. It assigns degrees of plausibility to theories.

Empirical tests are rarely clear-cut and depend on auxiliary assumptions—instrument reliability, background conditions—sometimes failed predictions may reflect flaws in setup rather than the theory. This challenge, known as the Duhem-Quine problem, shows that theories are rarely tested in isolation.

Bayes' Theorem resolves this by combining prior beliefs with new evidence to produce a revised belief, or posterior probability. If a theory predicts a rare outcome and it occurs, plausibility increases; if not, plausibility drops, but the theory is not accepted outright. Bayesian reasoning accommodates uncertainty, allowing rational belief even with incomplete or ambiguous data (See C30).

Patterson's use of plausible reasoning

Clair Patterson's investigation into environmental lead pollution exemplifies Bayesian reasoning. He did not rely on a single measurement but assembled diverse evidence reinforcing the hypothesis that elevated lead levels were modern and unnatural.

Patterson's confidence grew through converging evidence, not a single decisive experiment. In contrast, Robert Kehoe's defence of leaded petrol rested on present-day blood lead levels, interpreted as 'normal' without testing alternatives or historical baselines. His approach lacked the iterative revision that Bayesian reasoning demands.

Jessica Reyes's use of plausible reasoning

Jessica Reyes's research on environmental lead and crime, showing a 22-year lag between childhood exposure and violence, illustrates Bayesian reasoning. Building on medical and psychological studies, she hypothesised that reduced lead exposure would lower crime rates decades later.

Reyes tracked cohort-level crime changes across U.S. states with varying lead removal timelines. By controlling for income, policing, and demographics, she strengthened the plausibility of a causal link. Her modelling updated belief in the hypothesis as independent evidence converged. Like Patterson, Reyes showed that scientific knowledge advances through iterative belief revision, not single experiments.

Bayesian inference complements falsification

Bayesian reasoning aligns with philosophical insights from Lakatos and Kuhn. Lakatos viewed science as advancing through research programmes with a central 'hard core' and flexible 'protective belt.' Bayesian models can assign plausibility to entire frameworks. Kuhn's paradigm shifts, where dominant worldviews are replaced after accumulating anomalies, can be seen as Bayesian shifts in belief. Confidence shifts gradually, then rapidly, as evidence mounts.

Bayesian inference quantifies how much failed predictions reduce confidence and how much successful ones increase it. It supports replicability: repeated confirmations strengthen a theory's plausibility. Critics argue Bayesian reasoning relies too much on subjective priors, but these can be constrained and updated transparently. In complex, noisy conditions, Bayesian methods often outperform binary falsification.

Rather than replacing falsification, Bayesian reasoning complements it, offering a quantitative, evidence-sensitive framework for belief revision that reflects how science actually works: critically and probabilistically.

Feedback loops and living with uncertainties

A core feature of science is self-correcting feedback loops. Scientific beliefs are provisional, revised when faced with contradictory evidence (F9.1).

This feedback loop, self-correcting learning is illustrated by the Wright brothers who tested nearly 200 wing designs. When a wing failed to produce the expected lift, they discarded it and tried another. This iterative process of testing, learning, and refining exemplifies the backbone of scientific learning.

Cosmology has advanced through a feedback loop—observation, hypothesis, and revision—that continuously refines our understanding of the universe. Ancient models like Ptolemy's geocentric system dominated until Copernicus proposed heliocentrism, and Galileo's telescope revealed that celestial bodies were not fixed 'heavenly bodies' and observed that the Milky Way was composed of countless faint stars. Later, in the 18th century, astronomers catalogued fuzzy objects called nebulae, which were later shown to be entire galaxies beyond the Milky Way. Today, we observe billions of galaxies across cosmic distances, and modern models incorporate the Big Bang and an expanding universe to explain this large-scale structure. Each discovery corrected prior assumptions, demonstrating how empirical data reshapes cosmological models through a feedback-based learning process.

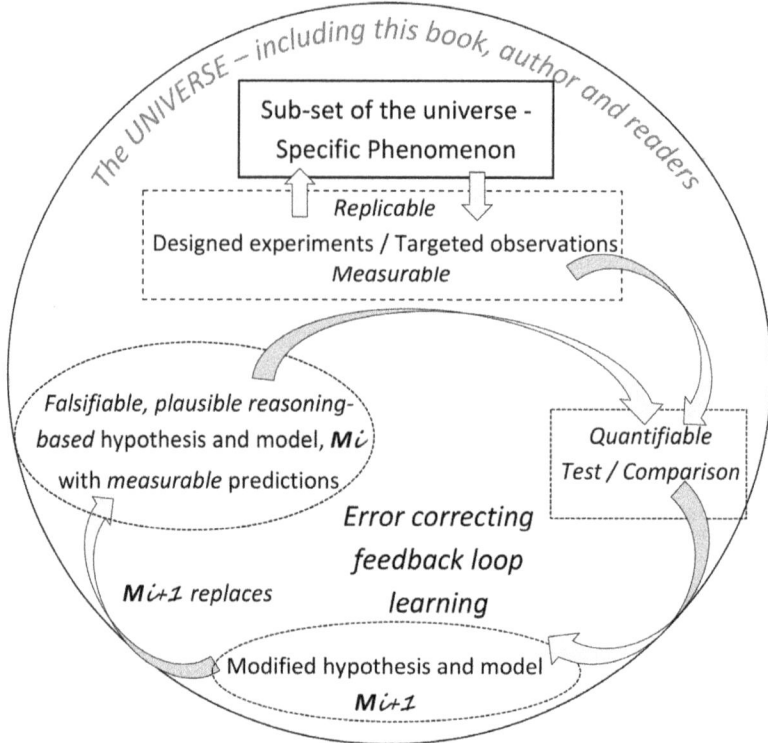

F9.1 Conceptual framework for the scientific learning process. Adapted from Box, Hunter, and Hunter (1978, Figure 1.3, p. 4),[11] with modifications.

In this way, scientific findings help mitigate uncertainty. What gives science its moral power is not certainty, but humility: the willingness to accept uncertainty and to revise beliefs through this self-correcting feedback loop grounded in evidence. This dynamic process, of testing, learning, and adapting, propels science forward on its continuous journey in search of truth.

Instruments

<u>Why instruments matter in science</u>

Empirical science goes far beyond our five senses. While observation is vital, instruments, measurements, and methodologies allow us to detect patterns and phenomena our senses cannot grasp. Instruments and statistical models refine and extend our observational reach, reducing bias and improving precision.

Morris Kline warns against relying solely on sensory perception: "Major phenomena of our physical world are not perceived at all by the senses... our senses are not only limited but also deceiving."[12]

To counter illusions and erroneous intuitions, Kline asks: "What is our recourse?"[13] Instruments are one answer. They quantify and measure phenomena, making observations precise, comparable, and replicable.

<u>Instrument function, resolution, and quantification</u>

Understanding how instruments work, and their limitations, is essential. Without this, they can mislead or offer partial views.

Ancient Chinese astronomers likely saw the Crab Nebula in 1054. In 1731, John Bevis gave it a precise position using an optical telescope. Its name comes from a drawing resembling a crab. High-resolution photos later showed it expanding, leading to its classification as a supernova. But only in 1969, with radio telescopes, were flashes detected, now understood as a rotating neutron star.

Philip and Phylis Morrison wrote: "It has been [flashing] for centuries... no one ever looked in a way that could catch so fast a flash... Like our eyes and ears, scientific instruments, too, can be fooled." (Philip And Phylis Morrison)[14]. Every instrument has limitations—the flip side of its strengths.

Instrumental limits and cold fusion

Instrument resolution and quantification are critical to scientific integrity. In the cold fusion controversy, R. D. Petrasso found that Pons and Fleischmann's reported measurements were "a factor of two smaller than their instrumental resolution would allow."[15]

To understand what this means imagine a ruler marked in centimetres and millimetres (F9.2).

F9.2 Ruler showing centimetres and millimetres

You can measure a line as 9.6 cm, but claiming it is 9.642 cm goes beyond the ruler's resolution—it's guesswork. Petrasso argued that Pons's reported signal was not evidence of a 2.224 MeV gamma ray, but an instrumental artefact, an error from misinterpreting equipment limits.

The design of experiments and the use of controls

The Bureau of Mines conducted an experiment using animals housed in well-ventilated cages, where lead dust was prevented from accumulating.[16] As a result, any lead emitted from nearby car engines during the short duration of the experiment was removed, making it impossible to measure its impact. This illustrates how experimental design can unintentionally obscure the effects it seeks to observe.

Thoughtful use of controls and comparisons is important in experimental science. Controls allow researchers to isolate variables and assess causality. These comparisons can be made across similar experiments or across samples separated by time. Clair Patterson, for example, compared lead levels in humans from the 1960s with those found in ancient Peruvians and in ice cores that preserved atmospheric records from thousands of years ago. This temporal comparison revealed the dramatic rise in environmental lead due to industrial activity. In contrast, Robert Kehoe relied solely on contemporary lead levels, ignoring historical baselines and failing to use controls, thereby weakening the foundation of his conclusions.

A similar issue arose in cold fusion claims. Pons and Fleischmann used a test tube filled with heavy water (D_2O) and attributed the observed heat increase to a fusion reaction involving palladium electrodes. To test this hypothesis, a control experiment using regular water (H_2O) could have been conducted. If similar results were observed with H_2O, it would suggest that the heat was not due to nuclear fusion involving heavy water, but rather some other variable in the setup. Without such controls, the interpretation of results remains speculative and prone to error.

An amusing story of how 'controls' can be misused come from E. Bright Wilson where he recounts:

"A story is told of a test of a seasickness remedy in which samples of the drug were given to a sea captain to test during a voyage.... When the ship returned, the captain was highly enthusiastic about the results of the experiment. 'Practically every one of the controls (those not given the drug) was ill, and not one of the subjects (those who had been given the drug) had any trouble. Really wonderful stuff.' However, a sceptic present asked how he had chosen the controls and subjects. 'Oh, I gave the stuff to my seamen and used the passengers as controls.'"

(E Bright Wilson, 1952)[17]

Targeted observations

Kehoe measured blood lead levels in the 1950s and concluded that levels up to 80 µg/dL were 'normal.' But he overlooked a factor: historical baselines. By ignoring pre-industrial lead levels, Kehoe succumbed to the survivorship bias—analysing surviving cases, excluding what was missing.

A useful example comes from the Challenger disaster. After the shuttle broke apart in 1986, engineers examined O-ring failures at low temperatures—but only from problematic flights. This limited dataset (F9.3) showed no clear pattern. When data from all launches was included, a correlation emerged between cold temperatures and O-ring failures (F9.4).

F9.3 reflects Kehoe's narrow analysis and F9.4 mirrors Patterson's historically informed method. With the full dataset, the conclusion becomes clear.

Science as a Test of Knowledge

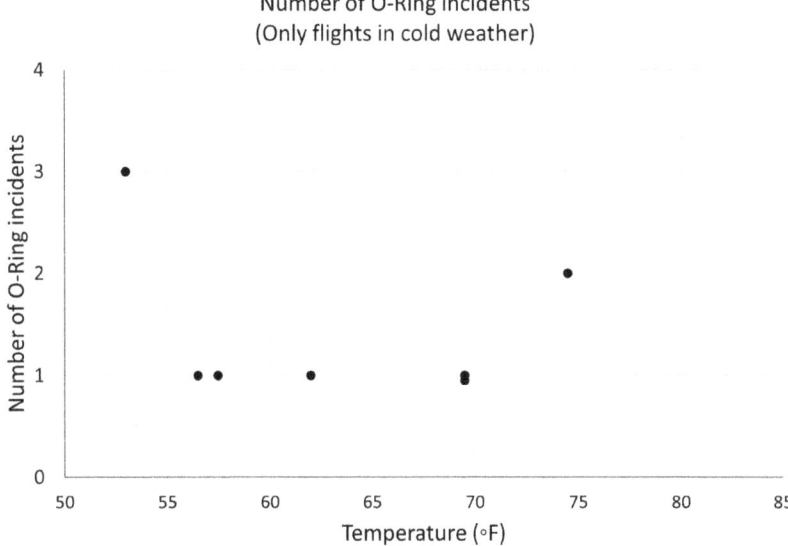

F9.3 Data on O-Rings from previous flights in cold weather. Adapted with revisions from Dawes, R, 2019[18]

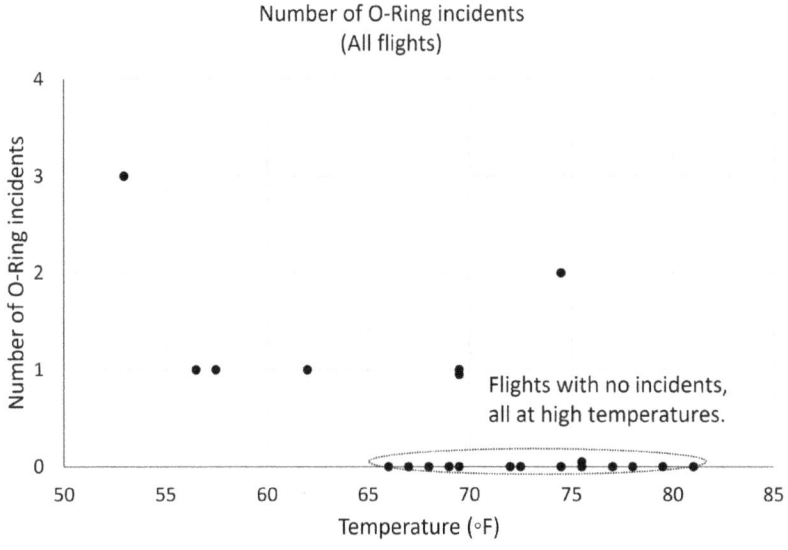

F9.4 Data on O-Ring incidents from all flights. Adapted with revisions from Dawes, R, 2019[19]

115

Replication, universal findings, and no authorities

Science does not allow findings to remain private, regardless of the source. Personal claims, whether from a purported prophet, a novice, or a Nobel laureate, must be replicable. Replication transforms isolated observations into universal knowledge.

This principle was tested in the cold fusion episode. Replication attempts quickly failed. Nathan and Petrasso found no evidence of excess heat or fusion byproducts. Their results showed the original findings were not robust.

Despite its importance, replication is often undervalued. Journals and funders favour novelty over confirmation. After the 1989 cold fusion claim, failed replications like those by Nathan and Petrasso were initially overlooked, though they were crucial in preserving scientific integrity.

Even successful replications rarely receive recognition. This imbalance discourages researchers from pursuing replication, though it is vital to science. Replication is not technical; it is a moral stance against intellectual privilege and epistemic isolation.

Importantly, replication must not mean blindly repeating flawed experiments. E. Bright Wilson warned: "A hundred repetitions of a biased experiment is merely one hundred times more misleading... it is usually more fruitful to comb the design and procedure for bias before continuing."[20]

Theories, models, and stories

<u>The roles of theories, models, and stories</u>

Theories, models, and stories serve distinct roles:
- Theories offer broad explanatory frameworks that generate predictions and guide inquiry.
- Models use calculations and abstractions to simulate reality, simplifying complexity to test outcomes.
- Stories provide narrative explanations, connecting ideas and observations in an intuitive, coherent manner.

While theories and models aim for precision, stories provide context and meaning. Kehoe explained blood lead levels with a narrative that they were 'normal,' lacking a consistent model or theoretical grounding. Patterson integrated theory, model, and story. His theory drew on radiometric dating and geochemistry. His model linked lead concentrations in ancient remains, ice cores, and ocean depths to trace historical exposure. His story: elevated lead levels in the 1960s were not natural, but a consequence of industrial pollution.

F9.5 shows Kehoe's model resembled a pale with a steady water level, 'normal' blood lead. F9.6 shows Patterson's abstraction was a running tap overflowing the pale, lead continuously entering the body from the environment.

F9.5 Kehoe and Patterson's abstractions

Stories as explanatory narratives

Lewis Carroll Epstein calls scientific stories "myths," narratives that must "save the phenomenon." That is, they must explain what's found in nature and predict what will be found. Stories help us feel at home in the world.[21] Patterson's model saved the phenomenon; Kehoe's contradicted historical data.

Models as abstractions

Timothy Gowers writes that useful models of phenomena are (i) adaptable, allowing them to be applied to new circumstances (ii) prioritize understanding over exact replication and (iii) abstract and simplify a complex system.[22] The goal of a model is not to create a perfect replica of phenomena but to develop a framework for analysis.

Three key features of models

Garrett Hardin identifies three essential aspects of accurate models:[23]

Literacy. Models must be explained clearly and grounded in data. An example of what can go wrong when clarity is lost is Alan Sokal's hoax. Sokal submitted a deliberately nonsensical article, titled "Transgressing the Boundaries: Toward a Transformative Hermeneutics of Quantum Gravity," to *Social Text*, a prominent social science journal.[24] The article was well received until Sokal pointed out his hoax. This reveals the danger of ideological beliefs and meaningless complex writing.

Numeracy. Theories should be quantitative with clear assumptions using statistical and probabilistic methods.

Hardin defines numeracy as a disposition: "The numerate temperament… looks for approximate dimensions, ratios, proportions, and rates of change."[25] With effective education, this mindset is widely attainable.

Statistical reasoning is important for uncovering causal links in complex systems. Jessica Reyes's work on lead exposure and crime shows how empirical modelling turns data into actionable insights. Without statistical logic, such connections would seem implausible.

Not all science requires advanced mathematics: E. Brian Davies cites evolution and plate tectonics as "least mathematical" yet enduring theories, offering consistent explanations of vast facts.[26] But even in evolutionary theory mathematical models play a crucial explanatory role.

Ecolacy. This involves asking what happens next and how a theory affects broader systems. In the cold fusion case, Pons and Fleischmann failed to predict a surge in neutron flux, a key consequence of nuclear fusion. They only addressed this after criticism.

Patterson exemplified ecolacy. Discovering high lead levels in meteorites, he asked: Why? What does it mean? What's causing this? What must be done? These questions led him to identify leaded petrol as the source.

Models as deductive systems

Scientific models function as deductive systems, built from axioms, principles, and logical rules. If internally consistent and complete, they guide reasoning and prediction. Though assumptions are rarely stated explicitly, models embed background knowledge and causal mechanisms that must cohere logically.

This distinction is clear in Patterson vs. Kehoe. Patterson hypothesised that elevated lead levels were due to pollution and tested this across multiple domains. Lines of evidence reinforced his model's consistency. Kehoe's model lacked transparency and ignored environmental sources and biological accumulation of lead. He treated a single observation as a conclusion, bypassing the rigour science demands.

The reductionism – emergence spectrum

In the 1970s, Dr. Herbert Needleman and Dr. Philip Landrigan examined children's teeth to assess lead's impact on cognitive development and behaviour. By the 1990s and 2000s, research had shifted to the molecular scale, exploring how lead affects cellular processes and biochemical pathways.

These examples show that science operates across a spectrum, from social behaviours to molecular interactions. Each level of analysis contributes meaningfully; it is not a matter of one being more important than another. This leads to two broad conceptual approaches:

- *Reductionism*, or the 'Breaking Things Down' approach, seeks to explain phenomena by analysing their smallest components. In the context of lead exposure, this includes studies at the molecular and cellular level.
- *Emergence*, or the 'New Properties Arise' approach, emphasizes that complex systems can exhibit behaviours that are not predictable from their individual parts. This is exemplified by findings that elevated lead levels in children correlate with learning disabilities and increased aggression, outcomes that cannot be fully understood by molecular analysis alone.

The terminology, 'reduction' and 'emergence,' can be misleading. It may suggest that studying complex systems is less rigorous, or that emergent phenomena spontaneously 'come into existence,' which is not always the case. Both approaches are essential, and together they offer a more complete understanding of how scientific knowledge is constructed across scales[27]

The science of human organs, for example, is as important as the study of their constituent molecules and atoms. There is no rigid boundary between 'fundamental' levels and the range of emergent properties that arise from them. F9.6 illustrates how a person's health can be shaped both by gene-level disruptions cascading upward to influence environmental interactions, and by environmental factors cascading downward to affect gene expression.

Science is based on the assumption that all phenomena including "human action comprise events of physical causation."[28] But this does not imply that insights lacking reductionist explanations are of lesser value. Understanding complex systems often requires engaging with multiple levels simultaneously, and findings at higher levels of organization can be just as scientifically valid and impactful.

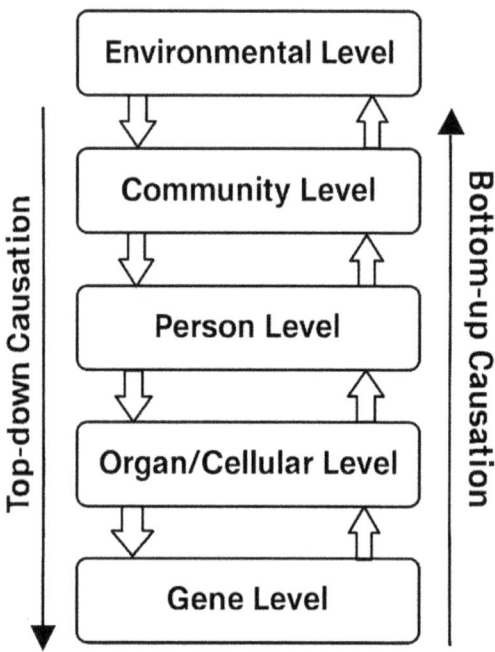

F9.6 Top Down and Bottom Up effects. Adapted and revised from Fig 4 Sturmberg *et al.*, 2019[29]

Questions science addresses and those it cannot

Science does well at answering questions of fact – the IS questions. It can tell us that environmental lead pollution exists, and that leaded gasoline was a major contributor.

But science cannot answer OUGHT questions: moral or normative judgments like 'should we continue using leaded gasoline?' That requires ethical reasoning, value judgments, and societal deliberation. The boundary between IS and OUGHT is often blurred in practice: scientists may advocate policy based on their findings as if the policy is a scientific finding, and moral philosophers may invoke scientific evidence to support ethical claims. Yet the distinction remains crucial. A more detailed discussion of the conflation by moral philosophers on this issue is provided in the C31.

Non-specialists - the challenge of scientific understanding

Feynman reminds us to remain sceptical of authority and rely on experimentation. There are two aspects to this caution. First, it is not an invitation to become an "anti-expert expert" who disregards scientific findings and replaces them with fantasies and conspiracies of one's own making.[30] Second, mastering any domain—science, arts, or service—takes years. How should non-specialists approach scientific claims?

Within a field, peer review allows experts to assess research validity. For non-specialists, peer-reviewed work offers a baseline of credibility. Consider Robert Kehoe, who claimed unmatched expertise on lead in petrol, boasting, "It so happens that I have more experience in this field than anyone else alive."[31] His work lacked independent review and replication, a red flag. In contrast, Clair Patterson's findings on Earth's age have been repeatedly validated. Pons and Fleischmann's cold fusion claims received mixed replication results, warranting caution and further inquiry which showed them to be wrong.

Funding matters. In "Merchants of Doubt,"[32] Naomi Oreskes and Erik Conway show how tobacco and oil companies funded misinformation campaigns on climate change and public health. Check who funded the research.

When evaluating science as a non-specialist, ask: Was the claim falsifiable? Was it replicated? Was the full dataset shared? Were instruments explained? Did the communication include clear explanation, numerate reasoning, and ecolate awareness of broader consequences?

Challenges persist:
- John Ioannidis warns that many studies reflect bias more than truth.[33]
- Machine learning can fabricate citations and obscure reasoning.
- Research is hard to access and interpret.
- Experts may disagree and enter arcane discussions.

There are no easy answers. But we must stay sceptical and embrace science's self-correcting mindset. Otherwise, we risk surrendering judgment to those driven by self-interest.

In 1995, a year before he passed away, Carl Sagan, wrote:

"Science is more than a body of knowledge; it is a way of thinking…when the people have lost the ability to set their own agendas or knowledgeably question those in authority…

we slide, almost without noticing, back into superstition and darkness."[34]

SAGAN

N5 Galileo Fatwaed

"The suppression of uncomfortable ideas may be common in religion or in politics, but it is not the path to knowledge" (SAGAN)[35].

Part 4

How the world IS

Evolution and Chance

What do we know - Origins

An understanding of evolution and considering the roles of progress and chance

A brief history of life

"I have called this principle, by which each slight variation, if useful, is preserved, by the term of Natural Selection."[1]

CHARLES DARWIN

"Man at last knows that he is alone in the unfeeling immensity of the universe, out of which he emerged only by chance. Neither his destiny nor his duty have been written down. The kingdom above or the darkness below: it is for him to choose."[2]

JACQUES MONOD

10

A Brief History of Life

The nature of life

<u>What is life?</u>

In 1944, physicist Erwin Schrödinger posed the question, "What is Life?"[1] He described it as an ordered system with a genetic foundation, introducing genes as carriers of heredity. Schrödinger speculated that molecules behave differently in living organisms than in inanimate matter, and that consciousness might grant life properties beyond physical laws.

Jaime Gómez-Márquez offers a modern view: life is not a static state but a dynamic process within highly organized organic structures. He identifies four defining features:[2]

- Preprogrammed nature. Genetic instructions in DNA (deoxyribonucleic acid) or RNA (ribonucleic acid) guide cellular function, development, and reproduction.
- Interactivity. Organisms respond to stimuli and maintain internal balance (homeostasis).
- Adaptation. Life evolves through natural selection, leading to extinction and emergence of species.
- Constant change. Mutations, recombination, and environmental pressures drive diversification.

Biologists define cells as the smallest units of independent life. They reproduce, carry genetic material, interact with their environment, and evolve.

Gómez-Márquez argues that viruses should also be considered alive. Though biologically distinct, they share key traits: a protein or lipid coat and either DNA or RNA (never both). While reliant on host cells to reproduce, viruses influence evolution and exhibit organized complexity, supporting the idea that life is a process.[3]

The origins of life on Earth - last universal common ancestor

Clair Patterson first measured Earth's age, refined to 4.54 billion ±50 million years.[4] Life may have begun soon after:[5]

- Fossil Evidence. Stromatolites, layered microbial structures, found in 3.7-billion-year-old Greenland rocks.
- Biogenic Carbon Signatures. Carbon isotopes linked to life dating back 3.7–4.1 billion years.
- Microfossils in Hydrothermal Deposits. Life-like forms in 4.28-billion-year-old deep-sea vents.

Fossils form when organic material is replaced by minerals. Dating methods include:

- Radiometric Dating (see C7 S: Clair Patterson)
- Electron Spin Resonance. Measures trapped electrons from natural radiation.
- Thermoluminescence. Detects light released from minerals when heated, revealing last exposure to heat or sunlight.
- Palaeomagnetism. Uses magnetic minerals to track Earth's geomagnetic reversals and date rock formation.
- Stratigraphy. Studies rock layers to determine relative age.

Each method has uncertainties and cross-checking improves accuracy.

All known life shares genetic and biochemical traits, pointing to a single origin: the Last Universal Common Ancestor (LUCA), around 3.7–4.1 billion years ago.[6,7]

How did life begin? Experiments offer clues:
- Miller-Urey (1953).[8] Simulated early Earth with water, methane, ammonia, and hydrogen. Electrical sparks produced amino acids, life's building blocks.
- Bada *et al.* (2008).[9] Reanalysis of Miller-Urey suggested early Earth was even more conducive to life.
- Meng-Zare (2025).[10] Found that microelectric discharges between water droplets, common in ocean spray, waterfalls, and rain, can form organic molecules, offering a frequent and accessible pathway for life's emergence.

Is life unique to Earth?
- The universe contains at least 100 billion galaxies, possibly up to 2 trillion (2025 estimates).[11]
- Thousands of exoplanets lie in habitable zones, some showing atmospheric gases like oxygen and methane—potential biosignatures.
- Amino acids and organic compounds found on meteorites and in interstellar clouds suggest life's building blocks are widespread.
- Extremophiles—organisms thriving in harshest environments—show that life can exist in extreme conditions.

Classification, mass, and number of organisms

Cells are either prokaryotes (without a nucleus) or eukaryotes (with a membrane-bound nucleus). This distinction forms the three domains of life:

- Bacteria. Single-celled prokaryotes.
- Archaea. Single-celled but with molecular similarities to eukaryotes.
- Eukarya. Complex, nucleus-containing cells—plants, animals, fungi, and protists.

DNA is present in all domains (except some viruses) and built from four nucleotides: adenine (A), cytosine (C), guanine (G), and thymine (T). Despite unique arrangements, the double helix structure and core functions—genetic storage and replication—are universal. RNA helps transfer genetic information and build proteins.

Viruses, though classified separately, share DNA/RNA structures with cellular life. They carry either single or double-stranded DNA or RNA and replicate by infecting host cells. Viruses are evolutionary influencers and possibly contributed to the rise of eukaryotic cells.[12]

The three domains are further divided based on shared traits and evolutionary links. The lowest taxonomic rank is species, groups that can interbreed and produce fertile offspring. For asexual organisms, species are defined using genetic and ecological markers. T10.1 shows the taxonomy for humans.

Category	Name	Type
Domain	*Eukarya*	Organisms with complex cells and a nucleus
Kingdom	*Animalia*	Multicellular organisms that cannot produce their own food
Phylum	*Chordata*	Animals with a spinal cord
Class	*Mammalia*	Warm-blooded vertebrates with fur and mammary glands
Order	*Primates*	monkeys, apes, and humans
Family	*Hominidae*	The great apes – humans, chimpanzees, bonobos, gorillas, orangutans
Subfamily	*Homininae*	Humans, chimpanzees, bonobos, and gorillas
Tribe	*Hominini*	Humans, chimpanzees, bonobos
Genus	*Homo*	Humans (and the extinct relatives like the neanderthals)
Species	*Sapiens*	Humans

T10.1 Taxonomy for humans

Abundance and biomass

Louca et al. (2019)[13] estimated 2.2–4.3 million species of bacteria and archaea. Earth hosts $\approx 5 \times 10^{30}$ (5 nonillion) microbes. In contrast, there are $\approx 10^{18}$ (quintillion) animals, including 8.2×10^9 (8.2 billion) humans (2025).

The number of bacteria is staggering—many live inside multicellular organisms. The human gut alone contains 10–100 trillion bacteria, essential for digestion, immunity, and mental health.

Bar-On, Phillips, and Milo estimate Earth's biomass at ≈550 GtC (Gigatons of carbon). 1 GtC = 10^9 tons = 10^{12} kg.[14] (T10.2). Their methodology included:

- Reviewing hundreds of studies
- Linking biomass to environmental factors
- Global extrapolation
- Refining estimates with geometric means

Category	Biomass (Gt C)
Plants	≈ 450
Bacteria	≈ 70
Archaea	≈ 7
Fungi	≈ 12
Protists	
(type of eukarya)	≈ 4
Animals	≈ 2
Of which humans	≈0.06
Marine Biota	≈ 6
Total	≈ 550

T10.2 Biomass of life on Earth (Bar-On, Phillips, and Milo, 2018)[15]

The Cambrian explosion

Earth's early atmosphere had little free oxygen. Cyanobacteria, through photosynthesis, released oxygen into oceans and air.[16] Iron-rich rocks formed as oxygen reacted with dissolved iron, providing evidence of this shift.

Oxygen, highly reactive, was a biochemical breakthrough. For anaerobic life, however, it was toxic. Around 2.4–2.3 billion years ago, the fossil record shows a sharp decline, up to 80%, in anaerobic microbes. Some survived in oxygen-deprived niches, like deep sediments and hydrothermal vents.

Yet oxygen offered an advantage: aerobic respiration yields up to 18 times more energy from glucose than anaerobic methods. This leap in efficiency enabled the rise of complex multicellular life.

The oldest eukaryotic fossils, red algae, date to 1.6 billion years ago. Alongside bacteria and archaea, the fossil record begins to show multicellular microbes and soft-bodied organisms. As oxygen levels rose, more complex life emerged, including early animals. (See F10.1 for Earth's geological ages and life timelines.)

The Cambrian Period (≈540–480 million years ago) marks a time when oxygen reached near-modern levels. Around 541 million years ago, the Cambrian explosion occurred, a pivotal phase when fossil evidence of many modern phyla first appears.[17]

A key fossil site is the Burgess Shale in the Canadian Rockies (GOULD)[18]. Here, soft-bodied organisms were exceptionally preserved, buried quickly by mudslides and sealed under calcium carbonate. This allowed detailed study of ancient anatomy, including gills, appendages, and digestive systems. The Burgess Shale reveals immense diversity: over 200 species, many unlike anything alive today. More than 80% went extinct without descendants.

Throughout the Cambrian and beyond, bacteria and archaea remained abundant and diverse in the fossil record.[19]

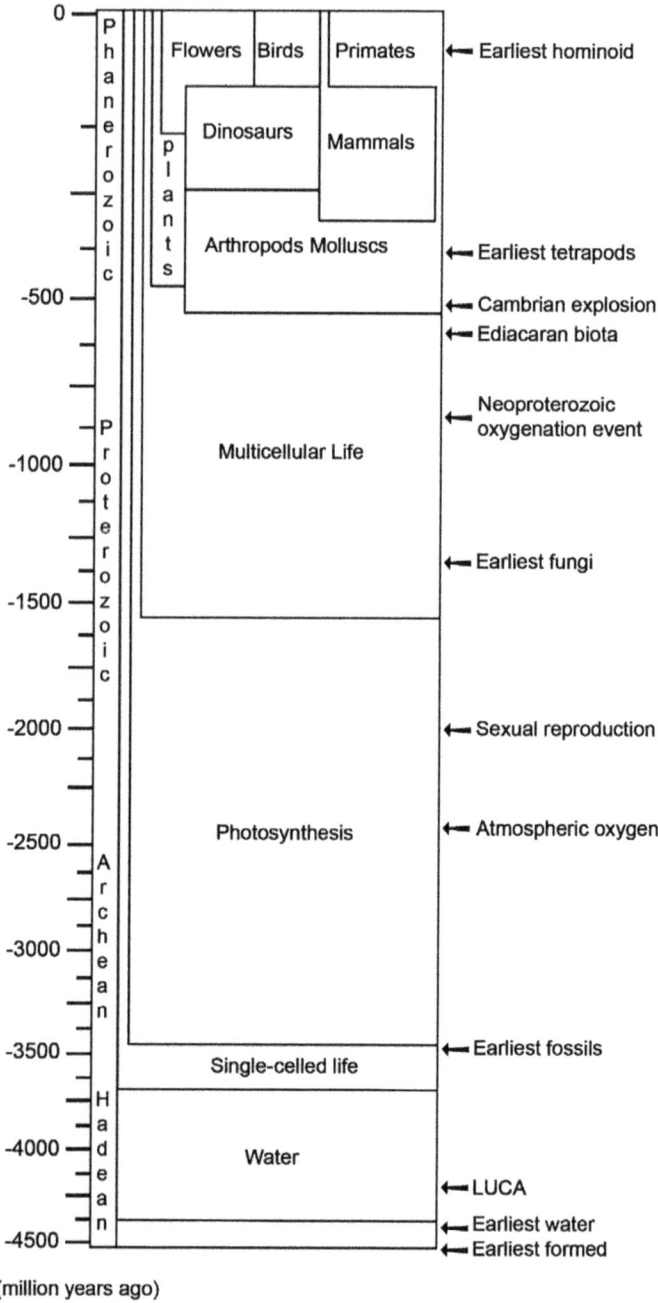

F10.1 Geological ages of Earth (GOLDSROBI, CC0, via Wikimedia Commons)[20]

Evolution by natural selection

The Earth has been in constant flux. Considering changes over tens and hundreds of millions of years, the fossil record shows that species become naturally extinct and new species occur. Charles Darwin's theory of natural selection explains this by describing four processes (i) adaptation (ii) secular change (iii) speciation and (iv) extinction (HARDIN)[21].

The features or traits in a population of a species vary widely. For any particular trait, say the height of a human, the frequency with which this is seen in a population follows a bell-shaped curve, also called a Gaussian or Normal curve. (See C30 S: The normal distribution.) This is illustrated in F10.2.

<u>Adaptation</u> occurs when a species develops traits that enhance survival and reproduction in its environment. The top curve in F10.2 shows an adaptation for a particular trait in a particular environment and its spread around the best fit trait (the mean).

<u>Variation</u> (the spread) around the mean occurs via mutations of genes over generations. These mutations result in features of the trait spreading out (the bottom curve in 10.2). The environment is best suited to those individuals close to the mean and this 'pushes' the trait's bell-curve to have a smaller spread (second curve in F10.2)

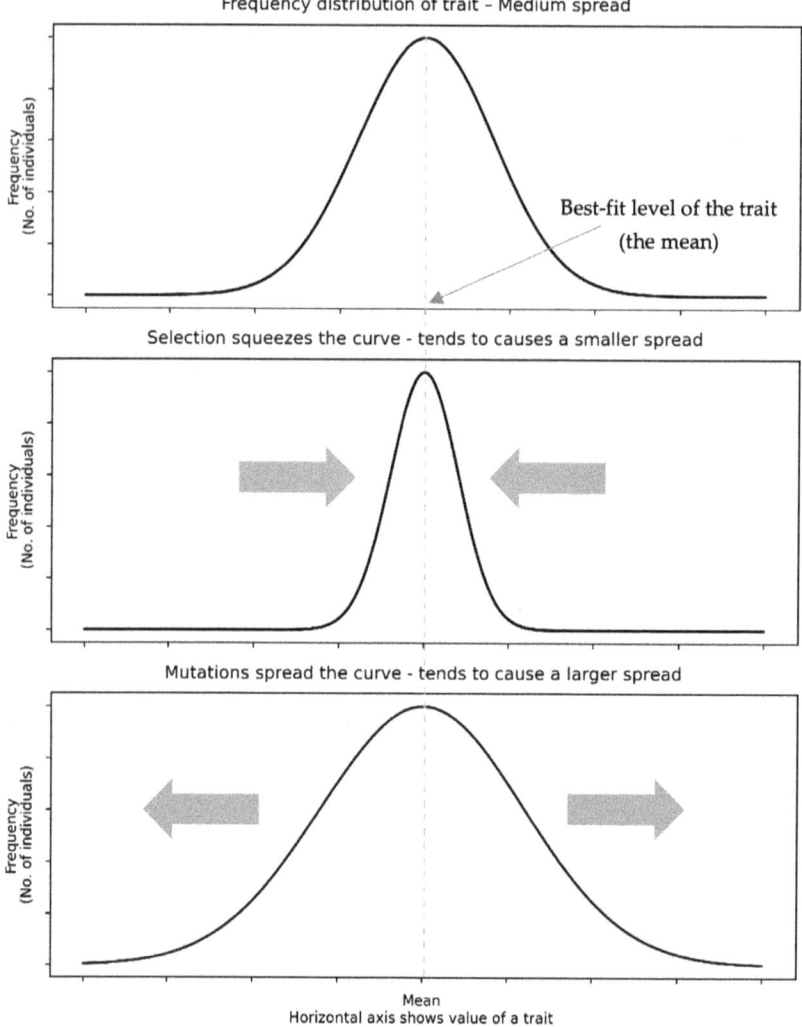

F10.2 Bell-curves of a trait in a population showing variation and the effects of mutations and selection. Adapted with revisions from Garrett Hardin, *Nature and Man's Fate*, Fig. 3-6[22].

When the environment changes, the individuals in the species with variations that are better adapted to the new environment leave more offspring. There is a shift in the population around this new "best-fit" set of traits. This is shown in F10.3.

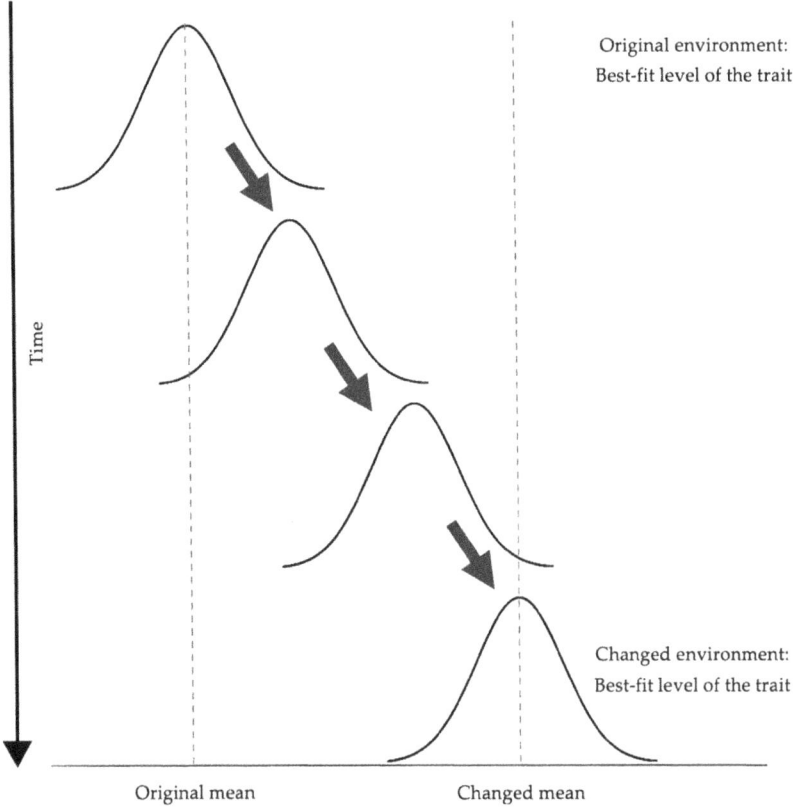

F10.3 Evolution in terms of change and adaptation. Adapted with revisions from Garrett Hardin, *Nature and Man's Fate*, Fig. 3-7[23]

Speciation occurs when, as shown in F10.3, individuals better adapted to the new environment leave more offspring. The environment can 'tolerate' only a certain amount of variation in a species' traits. Those that make individuals maladapted do not survive in the long term. There is a shift in the population around this new 'best-fit' set of traits. When the shift is sufficiently different and the populations diverge genetically from the original, new species form.

Extinction occurs when species that fail to adapt to environmental changes in their particular habitat die out.

Multitude of traits and local optimizations All species have a multitude of traits. Mutations, selection and adaptation act on the combination of these traits. Instead of a 2-D bell curve, a population of a species can be thought of as a 3-D bell composed of these many traits, as shown in F10.4.

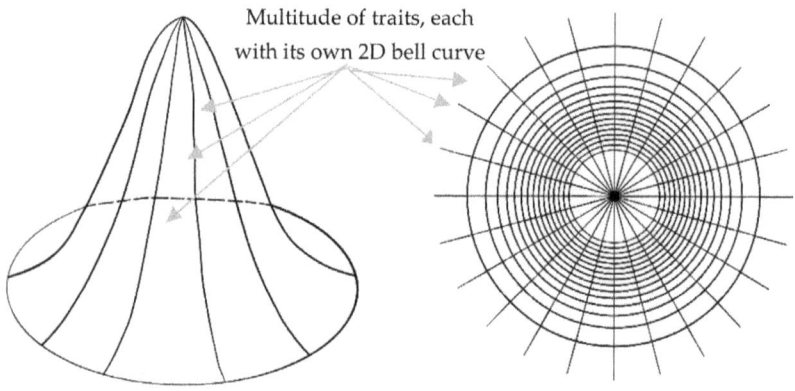

F10.4 3-D Bell-curve multitude of traits in a population. The figure on the left is a side view of the 3-D bell and the figure on the right is a top view. These show an array of traits in a population of a species.

Evolution by natural selection is driven by an optimization considering the combination of all traits - in a many dimensional space, made up of multiple traits; eyes, locomotion, defence mechanisms etc.

It is difficult to picture multidimensional space optimization. With one trait there was an identified 'best fit' (mean). Considering just two traits (X and Y) there are several 'local maxima'. These are local best-fits - better than other nearby combinations of the two traits. When mutations occur and there is a spreading out in the features of traits, it is possible that the population ends up in a local maximum not at the most optimized position.

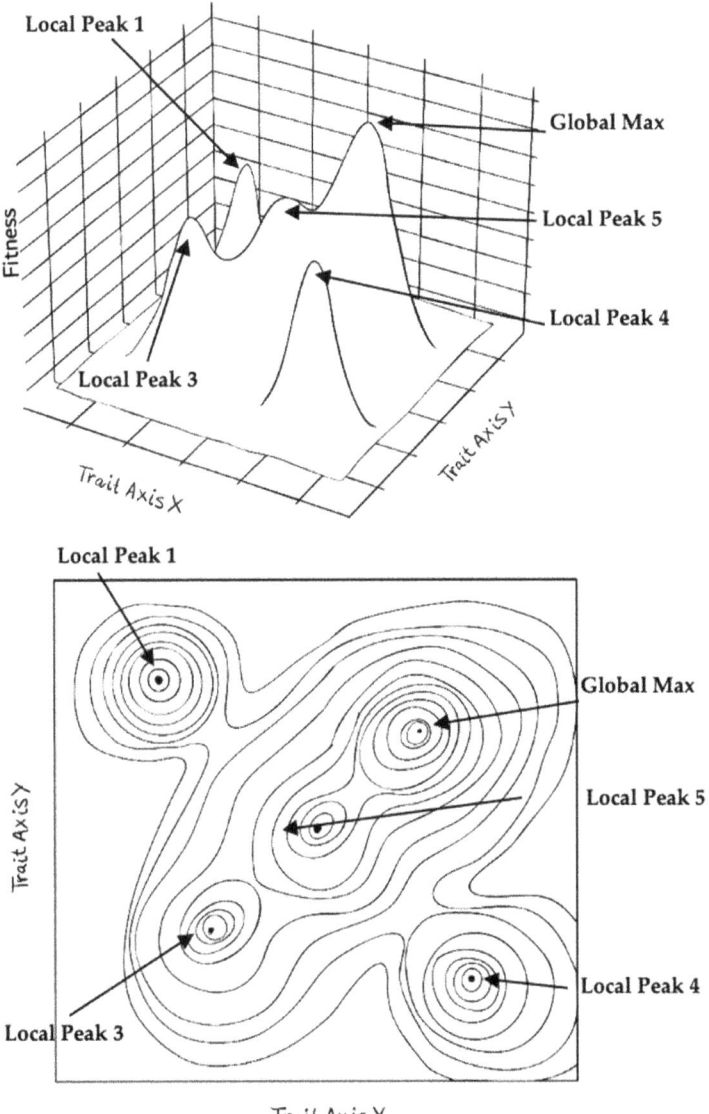

F10.5 Adaptation of two traits showing several local optimizations. Side view above and top view below.

This is shown in F10.5. If mutations in the two traits arise such that they are closer to 'local peak 4', then natural selection will drive variation such that this local optimization will occur. The optimization at the 'Global Max' is too far away. It is possible that if selection pressures are not intense, then the population can spread out more widely. It may then overlap with another peak. Reaching close to, say, local peak 5, it will, by natural selection, spread to the higher local peak 5. If selection pressures become high, resulting in a narrow spread from local peaks 4 and 5, then there will no longer be an overlap, and two separate species will arise where there was one before (HARDIN[24]).

<u>Running to stand still – The Red Queen hypothesis</u>

The 'Red Queen' hypothesis, proposed by Leigh Van Valen, explains why species must constantly evolve to survive.[25] Van Valen suggests that species are locked in an endless evolutionary arms race, where one organism's success forces others to adapt or perish. He named it after the Red Queen in "Through the Looking-Glass:" "It takes all the running you can do, to keep in the same place." In other words, species must keep evolving just to maintain their fitness in a changing world.

This concept helps explain coevolution, like hosts and parasites evolving in response to each other, and supports the advantage of sexual reproduction, which creates genetic diversity and flexibility. Extinction is driven by ecological pressures, including competition. This hypothesis also clarifies why speciation and extinction often rise together: as species diversify into new niches, they gain opportunities but also face new threats. Interestingly, this model has been used to describe political power dynamics. (See C24 S: The Social Contract and boundaries of power.)

However, the Red Queen hypothesis may overemphasize competition and predation, while downplaying climate, geology, and genetic drift. It also overlooks cooperation, like the mutual benefits between humans and their gut microbiota, where both adapt to enhance health. Evidence for the hypothesis varies: some species show little sign of evolutionary arms races, while others, like cheetahs and gazelles, fit the model perfectly, evolving greater speed to survive. It's useful to see adaptation and extinction as shaped by a mix of environmental forces.

Evolution is net-like, not just tree-like

Evolution isn't a tree with neat branches. It's more like a tangled net, as shown in F11.6, a phylogenetic tree diagram illustrating how species emerge and sometimes interbreed.[26] For example, Homo sapiens (humans) and Homo neanderthalensis (Neanderthals) interbred. Evolution is complex, with species diverging, reconnecting, and influencing each other in unexpected ways.

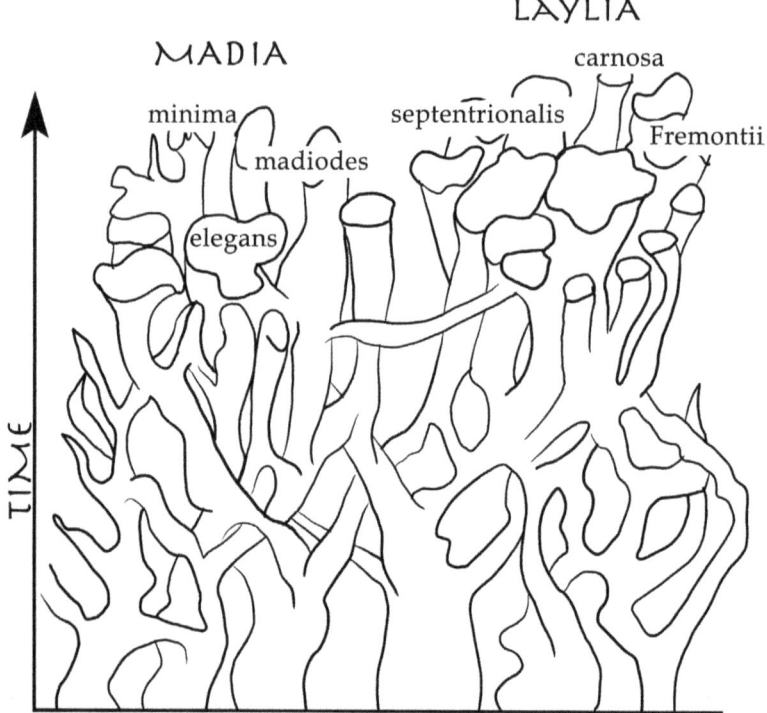

F10.6 Evolution is net-like not branch like. Adapted from Clausen, Stages in the Evolution of Plant Species[27]

Mendel's experiments and findings

While Darwin published "On the Origin of Species" in 1859, Gregor Mendel was conducting pea plant experiments (1856–1863) that would later provide the genetic foundation for Darwin's theory. His 1866 paper, "Experiments on Plant Hybridization," offered an account of heredity, though it went unnoticed until 1900.

Mendel studied how traits like seed colour and plant height passed from one generation to the next, showing that traits are controlled by genes, DNA segments with different versions called alleles. He proposed three laws: dominance (some traits mask others), segregation (each parent passes one allele), and independent assortment (traits are inherited separately).[28,29] Later discoveries (1953–1965) revealed the structure and role of DNA and RNA, completing the picture Darwin had started.

Decades later, Mendel's data came under scrutiny when Ronald Fisher's 1936 statistical analysis suggested the results were "too perfect." However later studies showed unconscious bias or repeated trials could plausibly explain the data. This is a case of scientific scrutiny and highlights the value of peer review. It also shows how statistical methods can illuminate historical science. Mendel's laws were validated by molecular biology and genetics, confirming their accuracy and enduring significance.

The modern synthesis

Modern biology has expanded Darwin's theory using genetics, molecular biology, and population dynamics. Key additions include:

- Genetic Drift. Random shifts in allele frequencies (e.g., blue vs. brown eyes), especially in small populations.
- Gene Flow. Movement of genes between populations, like human migration and intermarriage, reducing genetic differences.
- Horizontal Gene Transfer. Microbes exchange genes directly. For example, antibiotic resistance spreads rapidly when bacteria pass resistance genes to others via contact.[30]
- Punctuated Equilibrium. Evolution can occur in bursts, not just gradually. Both patterns are supported by evidence.[31]

Epigenetics plays a role. Environmental factors such as pollution or stress can alter gene expression without changing DNA, affecting immunity and disease susceptibility.[32,33]

Species evolution and extinction are shaped by both heredity and environment. Genetic variation within populations from spontaneous mutations during DNA replication or caused by radiation and chemicals, drives evolution. New species emerge as better-adapted groups evolve or become reproductively isolated.

These processes produce a background extinction rate estimated by Jablonski at 0.1–1.0 extinctions per million species-years (E/MSY).[34] At 1 E/MSY, one species goes extinct per year per million species. In 2015, De Vos et al. refined this using fossil and genetic data, lowering the estimate to 0.023–0.135 E/MSY.[35]

Mass extinction events

The fossil record reveals sudden, widespread losses of families and genera, called mass extinctions. These events show extinction rates far above the background level. Over the past 500 million years, 12 major events occurred, with five standing out (see T10.3).[36] Only one, the End-Cretaceous, is linked to an asteroid. The rest were driven by environmental and geological upheaval, showing how climate change can devastate life.

After the End-Permian extinction, dinosaurs emerged and dominated for 165 million years as top predators and herbivores, though bacteria and archaea remained the most numerous. Dinosaurs evolved into giants like Brachiosaurus and Allosaurus, peaking long before their extinction. Supersaurus survived about 8 million years.[37]

An asteroid impact 66 million years ago triggered wildfires, a blast wave, and an "impact winter" that blocked sunlight. Most large land animals and plants died out.[38]

Some species survived - flowering plants with dormant seeds, small mammals, and theropod dinosaurs (with reptilian and avian traits) - by adapting and exploiting vacant niches. Genetic studies and fossils show birds are descendants of theropods.[39,40] Around 55 million years ago, the first primates appear in the fossil record.[41]

Million years ago	Geologic period name	Possible cause
444	End-Ordovician	Climate change due to glaciation and sea level drop
360	Late Devonian	Ocean anoxia (lack of oxygen) and volcanic activity
252	End-Permian	Climate shifts and ocean acidification due to volcanic eruptions (Siberian Traps)
201	End-Triassic	Global warming due to volcanic activity from the Central Atlantic Magmatic Province
66	End-Cretaceous	Asteroid impact (Chicxulub crater) combined with volcanic activity (Deccan Traps)

T10.3 Big five extinctions over the last 500 million years (Rampino *et al.* 2019)[42]

Homo sapiens

Hominins – 7 million to 1 million years ago

Chimpanzees, bonobos, and hominins share an ancestor from 5.5 to 7 million years ago. Between 4.4 and 7 million years ago, the fossil record shows four hominin species with traits linking them to the human lineage, though evidence is sparse and incomplete.[43] From 4 to 2 million years ago, the record strengthens, revealing three major genera:[44]

- Australopiths. Fossils like Lucy (*Australopithecus afarensis*) and the Taung Child (*Australopithecus africanus*) show early bipedalism.
- Robust australopiths (*Paranthropus*). Found in East and South Africa, their skulls and jaws reveal adaptations for heavy chewing.
- Humans (*Homo*). The *Homo* lineage shows larger brains, upright walking, and tool use.

Around 2 million years ago, two species appear:[45]

- *Homo habilis* ("Handy Man"). Lived 2.4–1.4 million years ago. Had smaller teeth and face, long arms, an ape-like look, and brain size of 500–600 cc. Stone tools suggest a mixed diet of plants and meat.
- *Homo rudolfensis*. Lived 1.9–1.8 million years ago. Larger brain (~775 cc), longer face, and bigger molars, possibly more plant-based diet.

It's unclear whether either species led to *Homo erectus* or *Homo antecessor*.

Homo erectus appears 2 million years ago in East Africa and becomes the first hominin to migrate widely, reaching Southern and Northwestern Africa, Asia, and Europe. With brain sizes from 550 to 1,250 cc and human-like body proportions, they were likely omnivorous, used hand axes, controlled fire, and lived in cooperative groups.

They were among the longest-surviving hominins. African populations may have gone extinct 500,000–600,000 years ago, while Asian groups lasted until 300,000–400,000 years ago, and in Java until 100,000 years ago.

Though *Homo erectus* is considered a direct ancestor of modern humans, it's possible that only African populations contributed to our lineage.

Homo antecessor lived in Europe between 1.2 million and 800,000 years ago. Though like *Homo erectus* in brain size and build, genetic studies suggest it was a sister species to the common ancestor of modern humans.

A Near Extinction Event ~930,000 Years Ago

A 2023 study by Hu *et al.*[46] suggests a population bottleneck around 930,000 years ago, affecting *Homo erectus* or a related species. Genetic data and simulations indicate that the number of breeding individuals dropped to around 1,280, a near extinction event. This aligns with:
- A gap in the fossil record, possibly due to small, scattered populations.
- Climate records showing environmental stress between 900,000–950,000 years ago.

The bottleneck likely lasted ~117,000 years, with a population low of ~1,280 individuals.[47]

From 1 million to 40,000 years ago

Homo sapiens is the only surviving member of Tribe Hominini and Genus *Homo*, yet the fossil record from 1 million to 40,000 years ago reveals several related species across Africa and beyond:[48]

Genetic studies show interbreeding among species— *Homo sapiens* with *Homo Neanderthalensis* (Neanderthals) and *Homo Denisova* (Denisovans)[49] —suggesting evolution was a complex web of migrations, interactions, and extinctions. Some species, like Denisovans, may have been absorbed into *Homo sapiens* populations. Others likely vanished due to isolation or environmental stress. Neanderthals may have succumbed to climate shifts and competition.

As Hardin noted, species can only coexist if they occupy different ecological niches. If they compete in every respect, only one survives.[50] *Homo sapiens* may have had advantages: better climate adaptability, larger cooperative groups, and genetic mixing that boosted immunity. Extinctions may also have been triggered by disease or natural disasters.[51]

Vestigial organs

The earliest accepted *Homo sapiens* fossils were found at Jebel Irhoud, Morocco, dating to ~300,000 years ago. These remains, five individuals: skulls, jawbones, and skeletal fragments, show a mix of archaic and modern traits.[52]

By 30,000–20,000 years ago, *Homo sapiens* were the only surviving *Homo* species.

Humans retain vestigial organs, structures that have lost their original function (see T10.4).

Structure	Original function	Current role / status
Appendix	Digestive aid	Regulate gut bacteria
Coccyx (Tailbone)	Balance and mobility via ancestral tails	muscle attachment point
Wisdom Teeth	Chewing tough food	problematic
Ear Muscles	Moving ears to detect sound direction	No function
Third Eyelid	Eye protection / moisture	No function
Body Hair	Insulation	Cosmetic
Palmar Grasp	Clinging to mother	Fades with growth

T10.4 Examples of human vestigial organs / features

The Human Brain

Though the "triune brain" model is oversimplified, the human brain does reflect evolutionary layers:

- Reptilian structures. Brainstem and basal ganglia control survival functions like breathing and instinct
- Mammalian structures. The limbic system (amygdala, hippocampus, hypothalamus) governs emotion, memory, and bonding
- Primate structures. The prefrontal cortex enables advanced thinking, decision-making, and language.[53]

The human brain is complex with ~86 billion neurons and ~100 trillion synapses.[54] A complete map of its structure and function remains incomplete, though a landmark study of the mouse visual cortex has linked neural architecture to function, offering a model for human brain research.[55]

The historical record

Modern humans began migrating out of Africa 60,000 to 70,000 years ago. Cave art, burial rituals, and complex tools became widespread in the last 50,000 years, agriculture 12,000 years ago, and written language 5,000 years ago. If Earth's history were compressed into a 24-hour clock, humans would appear in the last few seconds before midnight.

Written records come from the Sumerians and Babylonians around 3000 BCE. These are cuneiform inscriptions on clay tablets dealing with administrative, economic, and religious items.

Thinkers who argue that history follows a purpose driven trajectory of intellectual, or moral advancements include Marx[56] (class struggles), Pinker[57] (reason driven progress across the board), and Fukuyama[58] (liberal democracy).

Those who reject directed progress include Karl Popper[59] (who criticized historicism) and Jared Diamond. Diamond's thesis reframes the questions of history in terms of geography and ecology.[60] He argues that the success of Eurasian civilizations was largely due to access to domesticable plants and animals, a favourable east-west axis that allowed ideas and crops to spread, and development of agriculture that led to food surpluses, specialization, and complex societies. These factors enabled the rise of technologies, immunity to deadly diseases, and infrastructure that allowed some societies to dominate others, not through directed progress, but through geographic luck.

11

Evolution as Chance and Necessity

Evolution without purpose

Stephen Jay Gould points out staggering numbers, 5×10^{30} bacteria and archaea versus 10^{18} animals, to argue evolution isn't a march toward complexity, but a statistical spread dominated by simplicity (GOULD)[1,2,3]. Since organisms can't be simpler than a single cell (or virus), mutations mostly result in life forms staying simple. Complexity arises occasionally, but far less frequently.

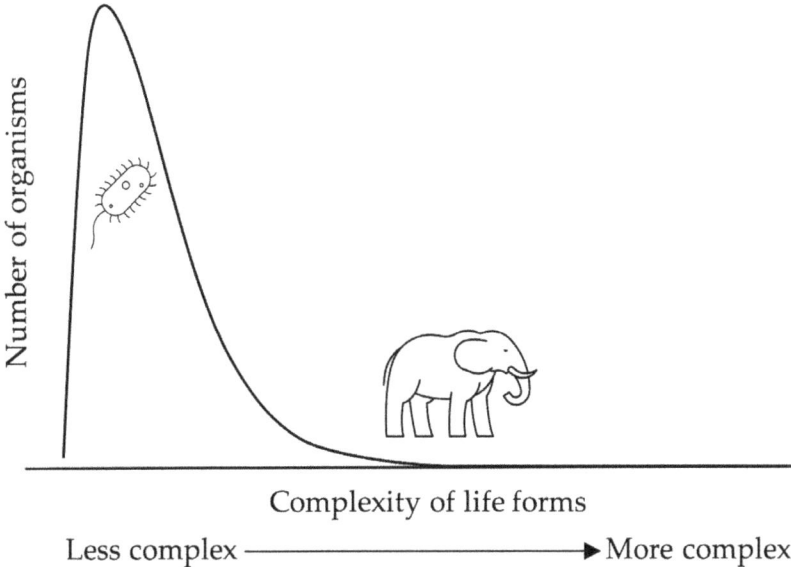

F11.1 Evolutionary Complexity. Adapted and revised from Stephen Jay Gould, *Full House: The Spread of Excellence from Plato to Darwin*[4]

This pattern resembles a 'drunkard's walk,' illustrated in F11.1, a right-skewed distribution where simple organisms vastly outnumber complex ones. Although each evolutionary step is random, complexity can only increase or stay the same—never decrease—creating a directional bias toward simplicity at one end. Though bacteria and archaea have evolved over billions of years, they remain the most dominant forms of life.

Bacteria thrive because they're supremely adapted to their niches. Their simplicity is not a flaw but a winning strategy. The illusion that evolution favours intelligence or complexity comes from focusing on rare outliers at the far end of the distribution.

Veit *et al.*[5] challenge Gould, claiming complexity can be measured and that evolution trends toward it. Critics argue their definition is subjective and ignores overwhelming dominance of simple life.

In some cases, natural selection may favour complexity, like in host-pathogen arms races, but these are exceptions.

Gould's model looks at the big picture, where simplicity consistently outperforms complexity. His argument passes Hardin's three filters against folly: it's well-reasoned, backed by data, and offers insight into both our past and our future.[6]

Evolution without direction

The question whether evolution is directed differs from whether it represents progress toward complexity. Creationism, for instance, claims humans were designed by god, with evolution serving a purposeful path culminating in us.

Simon Conway Morris argues for convergent evolution, where unrelated species evolve similar traits due to shared

environmental pressures:[7]
- Complex eyes evolved independently in vertebrates, cephalopods, and arthropods.
- Flight arose separately in birds, bats, and insects.
- Advanced cognition developed in cephalopods, primates, and birds like crows and parrots.

Richard Dawkins counters that natural selection alone explains life's complexity.[8] He uses the eye as an example of cumulative selection:
- Light-sensitive cells offered basic survival advantages.
- Indentations formed shallow cups, improving light direction detection.
- Deeper cups became pinhole eyes, like in the Nautilus.
- Transparent coverings evolved into primitive lenses.
- Lenses refined to focus light, enhancing vision.
- Over time, mammalian eyes emerged with corneas, irises, and retinas.

Dawkins acknowledges convergent evolution but sees it as natural selection responding to similar challenges, not evidence of directionality.

Despite fossil and genetic evidence, creationists argue for human uniqueness, often citing the brain. Yet neuroscientist Suzana Herculano-Houzel refutes this, showing the human brain is a scaled-up primate brain with high neuron density.[9]

Human brains weigh 1.3–1.4 kg, but other animal brains are larger:
- Dolphins (1.6 kg) show social intelligence and vocal learning.

- Elephants (4.5–5.4 kg) display empathy and memory, though most neurons are in the cerebellum.
- Sperm whales have the largest brains (up to 7.8 kg), mainly for motor control and echolocation.

The human brain:
- Has ~86 billion neurons, mostly in the cerebral cortex.
- Consumes 20% of body energy, proportional to neuron count.
- Has a large cortex, but not disproportionately among primates.

Other animals also have large numbers of neurons:
- Dolphins (~42 billion neurons)
- Gorillas (~33 billion)
- Orangutans (~32.6 billion)
- Chimpanzees (~28 billion)

Herculano-Houzel's work challenges the myth of human neurological exceptionalism, showing our brains follow primate scaling laws.

Insights into adaptations

Stephen Jay Gould cautions against overreliance on adaptationist explanations for traits like the human brain. While humans possess extraordinary abilities, these capacities aren't necessarily the result of direct selection for intelligence. Other animals also show aspects of these traits even though not with the same abstraction or complexity.

Gould introduces two key ideas: spandrels and exaptations. A spandrel is a trait that arises as a byproduct of other adaptations. Abstract reasoning or music may have emerged from a brain evolved for survival tasks like tool use or social coordination.

Exaptation refers to a trait evolved for one function but later repurposed. Gould suggests the brain may have developed for basic tasks, but its structure was later co-opted for functions like language and mathematics.

These abilities weren't built by natural selection for their own sake, they emerged from the brain's architecture. Gould even extends this to human aggression and destructiveness, not as selected traits, but as unintended consequences of complexity. As he puts it, "Selection builds brains for a reason, and then by virtue of its structural complexity it can do many other things... that fall outside the realm of evolutionary reasons for its construction" (GOULD)[10].

Evolution shaped by contingency

Evolution may not be progressive or directed, but is it determined? Two lenses explore this: the fossil record and models of trait optimization through mutation.

The asteroid impact 66 million years ago wiped out large dinosaurs, opening ecological niches for mammals. Without that chance event, primates and humans might never have evolved.

Stephen Jay Gould's study of the Burgess Shale fossils revealed greater anatomical diversity in the Cambrian than today. Many body plans vanished not due to poor adaptation, but random events (GOULD)[11]. Examples include:
- *Hallucigenia*, once thought to walk on its spines, now seen as having defensive spikes.

- *Opabinia*, with five eyes and a proboscis, had a radically different body plan.
- *Pikaia*, an early chordate, survived not because it was superior, but due to luck—its survival paved the way for vertebrates.

A 2020 study by Nanglu *et al.*[12] supports Gould's view. They found Burgess Shale ecosystems were patchy, with survival depending on location:
- Oxygen or sediment changes favoured species in certain areas.
- Localized disasters wiped out geographically restricted species.
- Predators and food availability varied by site.
- Storms or ocean shifts could eliminate isolated populations.

Gould's "contingency argument" suggests evolution's outcomes hinge on chance. Rewind the tape of life, and the result could be entirely different.

Garrett Hardin adds that evolution optimizes traits for survival, not perfection (HARDIN)[13]. Examples:
- The panda's thumb, evolved from a wrist bone, is "good enough" for grasping bamboo.
- The primate brain is powerful but energy-intensive.
- Mutations yield incremental improvements, like the eye evolving over 600 million years.
- Optimization occurs in a multidimensional trait space. F10.5 showed how mutations near Local Peak 4 led to local optimization, while the global maximum remained unreachable, purely due to chance. (See C10 S: Evolution

by natural selection.)

Each of the peaks shown in F10.5 denotes a local optimization based on two traits. As seen, there are many such peaks at different values of the model two traits. It is a matter of chance where the two initial mutations arise. In this example they occurred near Local Peak 4, one amongst many optimization points. The chanciness of this process is further complicated by there being a variety of traits that are all optimized in combination.

The last common ancestor of mammals and cephalopods lived over 600 million years ago, likely with no brain or eyes. Yet both lineages evolved these features independently.

Octopuses (a genus in the class cephalopoda) have ~500 million neurons, mostly in their arms, enabling problem-solving, camouflage, and communication. Some live only a year and reproduce once. Philosopher Peter Godfrey-Smith writes, "Cephalopods are an island of mental complexity... evolution built minds twice over. This is probably the closest we will come to meeting an intelligent alien."[14]

It could have been that only cephalopods survived, or another lineage evolved brains and eyes differently.

Carl Sagan argued that the emergence of intelligent life is fundamentally influenced by random biochemical events, suggesting that if Earth's evolutionary process were restarted, the resulting intelligent beings might be unlike humans.[15]

Chance shapes our lives too. In the 1996 Everest disaster, sudden bad weather, an unpredictable event, amplified altitude effects, hindered rescue, and led to tragedy. Chance also operates within us. Millions of neurons fire unpredictably, shaping decisions. In the 1996 Everest disaster, guide Rob Hall may have made a rare decision not to turn back, possibly influenced by random neural activity.

In history too, chance shows its strong hand. On 20 July 1944, a group of German officers attempted to assassinate Hitler. Colonel Claus von Stauffenberg placed a bomb in a meeting room, but a series of chance events led to failure.[16] Though four people died and Hitler was injured, he survived, largely due to luck.

The nature of chance

Defining chance

L. Rastrigin identifies three dimensions of chance:[17]
- *Quantum uncertainty.* The Heisenberg principle shows that certain pairs of properties, like position and momentum, cannot be precisely measured at the same time. This introduces inherent probabilistic dynamics, particularly in systems such as atomic clocks or quantum computing.
- *Chaos theory.* Edward Lorenz demonstrated that even simple weather models can behave unpredictably due to extreme sensitivity to initial conditions, the "butterfly effect."[18] Nonlinear systems can produce emergent patterns. Catastrophe theory adds that small changes can trigger abrupt shifts between stable states.
- *Measurement limits.* The universe's vastness and our limited instruments and intelligence mean we can only speak in probabilities.[19] We measure fragments, not the whole, and randomness arises from this partial view.

Despite their unpredictability, chaotic systems follow deterministic rules (unchanging physical laws). But chanciness means long-term forecasting is nearly impossible. This applies to weather, biology and evolution.

Chance and free will

Science and modern humanism embrace probabilistic reasoning, acknowledging randomness without mysticism or dogma.

Robert Sapolsky argues that although the universe is governed by chance, human behaviour is shaped by prior causes—leaving no room for free will.[20]

Daniel Dennett proposes that determinism and free will can coexist (DENNETT)[21]. Humans, as intentional agents, can anticipate outcomes and adjust behaviour, which he sees as sufficient for moral responsibility. He distinguishes between determinism (events shaped by causes) and inevitability (events that must happen).

I lean toward Dennett: we possess a 'meaningful form of free will' that enables moral decision-making. Yet, as Ivar Ekeland observes, though we can plan and adapt, randomness ensures that no model can prescribe a single optimal way to live or organize society.[22]

Chance and Necessity

Evolution's implications for human identity are profound. In a purposeless universe, we are shaped by physical laws and random events, one species among millions.

Jacques Monod writes in "Chance and Necessity:" "Chance alone is at the source of every innovation... but the edifice itself is governed by rigorous necessity."[23]

This combination, randomness and necessity, forces us to rethink meaning and morality. Liberated, as Sam Harris writes, from "ludicrous and divisive doctrines [of] religious [and other] beliefs that are morally and intellectually grotesque," we are free to rationally explore a fascinating universe and define our own purposes (HARRIS)[24].

From this perspective, values like justice and fairness become deliberate moral choices, essential companions to understanding. This sets the stage for the foundational moral choices of modern humanism.

N6 Octopus - Cogito, Arm Sum

"If we wish to understand alien intelligence, we should look no further than the octopus."[25] (Peter Godfrey-Smith)

Part 5

Humanism's choice -
How the world OUGHT to be

Humanist Principles and Processes

Foundational building blocks of modern humanism

The eradication of smallpox and the COVID-19 crisis

"Men and women with open minds worked to combat smallpox, the common enemy. They shared ideas with researchers from enemy nations and insisted that medical progress in service to humanity had no price and no borders."[1]

Steven M. Opal and J.M. Opal

"COVID-19 has demonstrated in stark terms that a pandemic is a complex phenomenon... and revealed deeply rooted structural weaknesses and insufficiencies in our health, socioeconomic, environmental and political systems."[2]

Tedros Adhanom Ghebreyesus

12

The Eradication of Smallpox and the COVID-19 Crisis

Smallpox and COVID-19 – postmortem questions

This case study provides a foundation for reviewing the principles and standards used, or overlooked, by public health and governments during the smallpox and COVID-19 responses. It prompts reflection on core elements of modern humanism's ethical choices and values.

Topics covered include:
- Historical context of the diseases and their viruses
- Strategies for outbreak control
- Political and economic pressures on public health
- Interactions between scientists and governments
- Crisis communication approaches

Infectious diseases and major pandemics

Diseases arise from various causes: genetic mutations, environmental toxins, lifestyle choices, autoimmune reactions, and infectious agents. Among these, infectious diseases, caused by invading organisms, have had the most devastating global impact.

In our bodies, bacterial and archaeal cells, together with phages (viruses that replicate in bacteria) outnumber human cells.[1] Most of these microbes aid digestion, nutrient absorption, and immune regulation, while also defending against pathogens. Yet some microbes evolve into deadly pathogens. T12.1 summarizes history's most destructive pandemics.

The Eradication of Smallpox and the COVID-19 Crisis

Pandemic	Years (Regions affected)	Pathogen	Deaths (mn)
Plague of Justinian	541-549 (Europe, North Africa, Western Asia)	Yersinia pestis	30–50
Black Death	1334-1353 (Europe, North Africa, Asia)	Yersinia pestis	75-200
New World Smallpox	1520-Early 1600s (Americas)	Variola Major/Minor	25-55
The Third Plague	1855-1959 (Worldwide)	Yersinia pestis	~ 12
Spanish Flu	1918-1920 (Worldwide)	H1N1 Influenza	50-100
AIDS	1981-present (Worldwide)	HIV	27-48
COVID-19	2020-present (Worldwide)	SARS-CoV-2	19-36

T12.1 The most harmful pandemics/ epidemics in human history (Source: Statista)[2] All these were caused by viruses except the plague.

- *Plague of Justinian* (541 CE).[3] Spread via grain shipments carrying rats and fleas. Caused by Yersinia pestis, it killed up to 10,000 daily in Constantinople. The outbreak weakened the Byzantine Empire, derailing Emperor Justinian's ambitions.
- *Black Death* (1347–1351 CE).[4] Possibly entered Europe via Crimean trade ships. Mongols may have used plague-infected corpses as early biological weapons. Up to 50 million died, half of Europe's population.

- *Spanish Flu* (1918–1920).[5] Caused by H1N1 influenza A virus, likely originating in birds and pigs. It infected ~500 million people and killed 25–50 million. The pandemic had three waves and was worsened by wartime censorship. Spain's open reporting led to the misnomer.
- *HIV/AIDS* (1980s–present).[6] Caused by HIV, a retrovirus that attacks immune cells. Originated in chimpanzees, it entered humans via exposure to infected blood. HIV spreads more easily through anal sex due to the vulnerability of rectal tissue. The virus has killed millions and remains incurable due to latent reservoirs. Antiretroviral therapy (ART) suppresses the virus, but eradication will require breakthroughs in vaccines and gene therapy.

Modelling the spread of pandemics

Epidemiological models are vital tools for forecasting outbreaks, guiding interventions, and allocating resources.

A key formula for herd immunity is $f = 1-(1/R_0)$. Here f = fraction of the population that needs to be immune and R_0 = the basic reproduction number (the average number of people one infected person transmits the disease to in a fully susceptible population). Fraser *et al.* showed that control depends not just on R_0, but also on how much transmission occurs before symptoms or via asymptomatic carriers, making containment harder even with moderate R_0.[7]

Modelling faces major challenges:
- Data quality. Incomplete testing and delayed reporting skew predictions.

- Human behaviour. Responses to mandates vary, creating feedback loops that models struggle to capture.
- Immunity. Waning immunity and variants complicate forecasts.
- Population diversity. Age, health, mobility, and social networks affect transmission.
- Pathogen evolution. Viruses change, altering spread dynamics.

Effective modelling requires integrating math, biology, behaviour, and policy, not to predict perfectly, but to guide decisions amid uncertainty.

Smallpox

<u>Before smallpox vaccines</u>

Palaeomicrobiologists (who study disease-causing microorganisms in archaeological sites) provide evidence smallpox dates back to Egyptian mummies from 1580–1100 BCE, including Pharaoh Ramses V.[8] Epidemics followed in Africa, Europe, and Asia, becoming endemic by the 11th century.

In the 18th century, smallpox was a constant threat. Gareth Williams notes it affected all classes, with one in three Europeans infected and one in five dying.[9] Globally, it claimed one in twelve lives, with 15 million deaths over 25 years. Children under ten were especially vulnerable due to lack of immunity.[10]

Smallpox shaped history. In 1519, Spanish invaders brought the virus to Mexico. The 1520 outbreak killed nearly half of Tenochtitlan's population, easing the conquest of the Aztec and later Inca empires.[11,12]

Societies blamed supernatural forces. Smallpox gods, miasma theory, and rituals like wearing red or offering sacrifices persisted into the 20th century. Treatments ranged from herbs to bloodletting, but only isolation helped until variolation emerged in 1550s China.[13]

Variolation involved inserting or inhaling smallpox material to induce mild infection and immunity. Though effective, it risked spreading the disease if not carefully managed. Isolation during recovery was crucial.

Variola and pathways of transmission

Variola major and minor are viral siblings—closely related strains with genomic differences that shape their replication and virulence. Variola major caused severe smallpox with ~30% fatality; Variola minor was milder (~1% fatalities). The virus, 270–400 nm in size, likely evolved 3,000–4,000 years ago in Egypt or the Near East.[14,15,16,17]

The virus enters via the respiratory tract, incubates for 10–14 days, then spreads through lymph nodes and blood. Symptoms include fever, and pustules.[18] Severe cases led to haemorrhagic smallpox (90% fatality), with survivors often scarred or blinded.

Contagion begins with mouth sores and lasts until scabs fall off.[19] By 1967, transmission was understood to occur via droplets and contaminated materials, rarely airborne. By the 18th century, it was clear smallpox had no animal reservoir.[20]

The smallpox vaccine

In 1796, Edward Jenner tested the belief that milkmaids infected with cowpox were immune to smallpox. He inoculated cowpox pus into eight-year-old James Phipps. Six weeks later, Jenner variolated Phipps, who showed immunity. Though not the first to explore vaccination, Jenner's method was safer and widely promoted, transforming public health.[21]

Despite opposition rooted in superstition and confusion about how vaccination worked, the practice spread rapidly worldwide.[22] Its simplicity allowed use without medical supervision. Jenner's approach faced three challenges:
- Cowpox was scarce, mostly found in Europe and transportation was difficult.
- Immunity waned over time.[23]
- Poor hygiene during vaccination risked bacterial infections.

As vaccination expanded, so did anti-vaccination resistance. Gareth Williams writes, "If smallpox had been a sentient predator... it might have been puzzled by the spectacle of its victims fighting amongst themselves."[24]

Some countries made vaccination voluntary; others mandated it. Epidemics became less frequent and severe, but vaccination systems struggled due to limited cowpox supply.[25]

Labs began storing what they believed was cowpox, but mutations and cross-species transmission likely introduced other poxviruses. In 1939, Allan Watt Downie identified the vaccinia virus, not cowpox, as the vaccine's active agent. Its origins remain unclear, possibly linked to horsepox.[26]

What is astonishing is that for a long time no one knew that the virus used in the smallpox vaccine was no longer cowpox but the vaccinia virus.

In the 1950s, Dr. Leslie Collier developed a freeze-dried vaccinia vaccine, stable for a month at 37°C and easily distributed. It could be produced without animals.

By 1967, 77 labs worldwide were producing vaccines, though quality varied, only 10% of vaccines outside the superpowers were estimated to meet efficacy standards.[27]

Vaccination techniques improved. Jet-injector guns were effective but fragile. In 1967 the "ultimate vaccination solution" was developed by Dr. Benjamin Rubin of Wyeth Laboratories.[28] This was a bifurcated needle that was simple to use, easily sterilized, and with a high success rate.[29] Soon thereafter a plastic needle dispenser was also designed by Dr. Ehsan Shafa.[30] While Rubin and Wyeth Laboratories patented the bifurcated needle design they waived royalties for its use in smallpox vaccination.[31]

Approaches to smallpox eradication

Two main strategies emerged:
- *Mass Vaccination.* Initially, WHO aimed to vaccinate 80% of each country's population to achieve herd immunity. This target was based on field experience.[32] Later, 90% coverage was found feasible with community support.
- *Surveillance and Containment.* Introduced in 1967 by Dr. William Foege in Nigeria, this strategy focused on ring vaccination around outbreaks. It proved more effective than mass campaigns but required timely, accurate data, often lacking due to poor infrastructure and political suppression.[33]

D. A. Henderson

Dr. Donald Ainslie Henderson led WHO's smallpox eradication effort from 1967 to 1977 (HENDERSON)[34]. He had previously directed virus surveillance at the CDC and helped launch USAID-backed programs in Africa. His vision and field expertise laid the groundwork for the WHO's eradication effort, which became one of the greatest health triumphs in history. After the success of smallpox eradication, Henderson continued to shape global health policy, serving as Dean of the Johns Hopkins School of Public Health and later advising on bioterrorism preparedness and pandemic response.

WHO resolution and politics

In 1959, the USSR proposed a global smallpox eradication plan, partly in response to U.S. support for malaria campaigns. Despite Cold War tensions, the U.S. backed the proposal, driven by domestic outbreaks and advocacy from Dr. Fred Soper, who believed eradication was feasible due to:

- Existing technology
- Human-only transmission
- Potential for strict management

Marcelino Candau, WHO Director-General from 1953 – 1973, opposed the plan, citing impracticality and political limitations. As a result, early funding was minimal.

In 1965, the U.S. launched a smallpox program in West Africa, hoping to redeem its failed malaria campaign. Henderson's CDC team proposed global eradication, which gained White House approval.[35] In 1966, WHO voted: 60 nations supported the plan, 20 opposed, and 12 abstained. Candau insisted Henderson lead the program, expecting failure.

Organizational challenges

Henderson's Geneva team had fewer than 10 staff. National governments were expected to fund 70% of operations. WHO provided guidance and vaccines, but implementation relied on regional offices and national programs. This created fragmented efforts across countries, each with its own approach.

Henderson's humanist approach

Henderson's success lay in his humanist approach:
- Field-first philosophy. He believed in direct engagement with affected communities and hands-on problem-solving based on science. Initially focused on mass vaccination, he embraced surveillance and containment when field data from Nigeria showed its superior effectiveness.
- Local capacity-building, education and empathy. Henderson empowered volunteers to explain vaccination at the grassroots level, fostering trust and understanding in communities unfamiliar with interventions.
- Global equity. His work was driven by the conviction that every human life—regardless of geography or wealth—deserved protection from diseases.
- Was inclusive, building diverse coalitions, respecting contexts, and prioritizing collaboration over hierarchy.

Henderson understood that eradicating smallpox required a combination of scientific rigor with a shared purpose and moral clarity. He, and the world, were also lucky. Both the U.S. and USSR cooperated, none of the world leaders suggested faith-based or alternative remedies, and there was a not-for-profit approach by vaccine manufacturers.

Status from 1967 to 1980

By 1967, smallpox was rare in Europe, North America, and parts of South America due to mass vaccination and prevalence of the milder Variola minor strain. Elsewhere, it remained endemic but due to Henderson and his team's efforts the last natural case occurred in Somalia in 1977. The last fatal case, from lab exposure, was in the UK in 1978. WHO declared smallpox eradicated in 1980.

COVID-19

Coronaviruses and associated epidemics

Coronaviruses were first identified in 1931 in chickens.[36] The first human strain, 229E, was discovered in 1965 and caused respiratory illness.[37] (While some cold symptoms stem from coronaviruses, most are due to rhinoviruses.) Coronaviruses share traits:

- Single-stranded RNA genome
- Lipid envelope surrounding genetic material
- Spike proteins forming a crown-like surface
- Many originate in animals and can jump to humans (zoonosis)

Before COVID-19, two major coronavirus epidemics occurred:

- *SARS (2002–2004).* Originated in China, spread to 29 countries, infected ~8,000 people, with ~10% fatality. Transmission occurred via droplets and contaminated surfaces.[38,39] China delayed sharing outbreak details, but containment through travel restrictions and quarantines eventually succeeded.

- *MERS (2012–present).* First identified in Saudi Arabia, spread mainly via camels. Human-to-human transmission was limited. Fatality rate ~35%. The largest outbreak outside the Middle East was in South Korea (2015), with 186 cases and 38 deaths.[40]

Origins and features of SARS-CoV-2

The first COVID-19 case outside China was reported in December 2019, but the true origin remains debated.[41] Theories include natural spillover (e.g., from bats) and accidental lab leak from the Wuhan Institute of Virology (WIV). Gain-of-function research—genetically enhancing viruses— conducted in multiple labs, including WIV was also suspected.

China's secrecy, delayed alerts, and suppression of researchers complicate origin investigations. WIV announced a new coronavirus on 30 December 2019, but the full genome wasn't released until 10/11 January 2020.

Key traits of SARS-CoV-2:
- Enveloped RNA virus with spike protein binding to ACE2 receptors in human cells
- Suppresses interferon response, weakening early immune defence
- Reaches peak viral load before symptoms appear
- Stable in aerosols, enabling airborne transmission
- Mutated into variants like Alpha, Beta, Delta, and Omicron, some more transmissible

Transmission pathways and WHO response

Initial reports from WIV circulated on 30 December 2019.[42] On 14 January 2020, WHO stated there was no clear evidence of human-to-human transmission.[43] Days later, Wuhan hosted a mass banquet.[44] Lockdowns in China began on 23 January, but international flights continued until 28 January.[45,46]

WHO delayed declaring an emergency until 30 January and labelled COVID-19 a pandemic on 11 March.[47,48] Early guidance emphasized droplet and surface transmission, downplaying airborne spread. This led to a focus on handwashing and surface cleaning, while ventilation and masks were underemphasized.

In March 2020, aerosol scientists proposed that the virus might linger in the air, suspended in fine particles. Despite this, the dominant framework for a long time was that respiratory viruses were primarily transmitted through droplets, not airborne pathways.[49] It was only starting in October 2020 and then more clearly in December 2021 that WHO changed its statements to say that the virus could be spread as an aerosol.

Spread of COVID-19

The disease spread rapidly:
- *December 2019.* China (Wuhan) reported the initial cases.
- *January 2020.* Thailand (13 Jan), Japan (16 Jan), South Korea (20 Jan), United States (21 Jan), Vietnam and Singapore (23 Jan), France (24 Jan), Australia and Canada (25 Jan), Germany (27 Jan), UAE (29 Jan).
- *February 2020.* Russia, Spain, Sweden, Belgium, and several others.
- *March 2020.* The virus reached nearly every continent, including most African and Latin American countries. By the end of March, over 170 countries had reported infections.

Strategies to deal with the disease

Countries adopted different strategies in combating the disease.

- *Zero-COVID strategy.* China and New Zealand enforced strict border closures, mass testing, and lockdowns. Effective early on, but hard to sustain long-term.
- *Mitigation strategy.* Germany and India used distancing, masks, phased lockdowns, and vaccines to manage healthcare capacity. India faced challenges due to its population and uneven compliance.
- *Herd immunity / minimal intervention strategy.* Sweden and Brazil relied on voluntary measures and focused on vulnerable groups. Sweden saw high elderly mortality, especially in care facilities, and ranked 30th out of 47 in Europe for per capita deaths.[50]
- *Technocratic containment strategies.* South Korea and Singapore used digital tracing, targeted lockdowns, and real-time data. Efficient but raised privacy concerns.
- *Fragmented governance strategies.* The U.S. and Australia used decentralized responses, with state-level decisions leading to varied outcomes and locally tailored strategies.

Excess deaths due to COVID-19

Governments often underreported COVID-19 deaths by counting only those with a confirmed test result, excluding suspected cases, deaths outside hospitals, or those attributed to other causes like pneumonia or heart failure. A more reliable estimate comes from excess mortality — the number of deaths above what would be expected under 'normal' conditions, based on data from years prior to the pandemic (T12.2).

The data in T12.2 was based on the "World Mortality Dataset" developed by Ariel Karlinsky and Dmitry Kobak. They modelled expected deaths using historical data from 2015–2019 and used statistical techniques to arrive at a range of deaths with upper and lower bounds. The wide range reflects uncertainty in the data from national statistical offices.

(Figures in mn)	World	India	China	U.S.	Brazil	Nigeria
Upper bound	36.29	11.82	7.44	1.61	0.93	0.74
Central estimate	27.31	6.09	2.64	1.46	0.93	0.53
Lower bound	19.25	2.81	0.48	1.31	0.89	0.003
Reported	7.05	0.53	0.12	1.19	0.70	0.003

T12.2 Reported and excess deaths due to COVID-19 Source: OurWorldInData[51]

Many of the deaths were due to an overwhelmed healthcare infrastructure, disproportionately affecting poorer people. Older adults faced higher COVID-19 risks due to common comorbidities and age-related inflammation, which may have triggered immune responses.[52] Children, possibly exposed to other coronaviruses, might have had partial immunity. Additionally, older individuals have more ACE2 receptors—the virus's entry point—making them more vulnerable than children.

COVID-19 vaccine development

Traditional vaccine development takes 10–15 years, progressing through exploratory research, animal testing, phased human trials, regulatory approvals, and large-scale production.[53] COVID-19 vaccines were developed and deployed at speed.

Three vaccine types were used:
- *Legacy* (inactivated/live-attenuated) - were less effective over time and against variants.[54]
- *Conventional* used adenovirus vectors to deliver spike protein genes, triggering immunity.[55]
- *mRNA* instructed cells to produce spike proteins, prompting immune response. They were highly effective and adaptable to new variants.[56]

The efficacy shown for vaccines in T12.3 is for the original variant of the SARS-CoV-2 virus. Efficacy for later variants were lower, and the vaccines had to be continually updated.

Vaccine	Vaccine Type	Deployment	Efficacy
Sputnik V	Legacy	5 Dec '20	No peer review
Pfizer-BioNTech	mRNA	8 Dec '20	95%
Moderna	mRNA	21 Dec '20	93% to 94%
Sinopharm	Legacy	Dec '20	78% to 79%
Oxford–AstraZeneca	Conventional	4 Jan '21	76% to 81%
Sinovac	Legacy	Jan '21	51% to 84%
Covaxin	Legacy	16 Jan '21	78%
J&J (Janssen)	Conventional	2 March '21	66% to 74%
Novavax	Legacy	July '22	90%

T12.3 Vaccine type, roll out and efficacy[57]

COVID-19 vaccine deployment

All the vaccines underwent accelerated trials and emergency authorizations. In the U.S., Operation Warp Speed invested USD 18 billion to fund production and logistics before trials ended, coordinating across government agencies and firms.

Globally, countries like the UK, India, and EU members fast-tracked vaccine production. The Jenner Institute partnered with AstraZeneca and the Serum Institute of India, which began manufacturing doses mid-2020, before Phase 3 results.

The rapid delivery of vaccines was built on decades of research and the contributions of countless scientists across fields ranging from fundamental biology to technology. While the rollout was often pitched as a groundbreaking new discovery, every vaccine, including mRNA-based ones, was rooted in work by thousands of unsung researchers over many years.

Two excellent resources for understanding how thousands of scientists contributed can be found in the article by Elie Dolgin, 'The tangled history of mRNA vaccines,'[58] and in the book, Vaxxers, by Sarah Gilbert and Catherine Green.[59]

Most of the vaccines were provided at low costs, except for the Pfizer and Moderna vaccines which were sold at high prices. Pfizer-BioNTech's vaccine generated revenues of over USD 36 billion in 2021 and USD 11 billion in 2023 and Moderna's vaccines generated USD 18.4 billion in 2021 and USD 19.3 billion in 2022.[60] The Oxford-AstraZeneca vaccine was initially provided on a not-for-profit basis but later transitioned to a profit model generating USD 4 billion for AstraZeneca in 2021.[61,62]

Political pressures

During the COVID-19 pandemic, Anthony Fauci was Director of the U.S. National Institute of Allergy and Infectious Diseases (NIAID) and Chief Medical Advisor to the President, who led the U.S. response to the COVID-19 pandemic. Fauci and other scientists faced political, economic, and social pressures.

The U.S. President, Donald Trump publicly clashed with Anthony Fauci and other members of his COVID task force, giving a confused message to the public.[63] This included Trump:
- Minimizing the threat of COVID-19, saying in January 2020, "We have it totally under control. It's going to be just fine."[64]
- Comparing COVID-19 to the flu and repeatedly claiming that COVID-19 was "no worse than the flu."[65]
- Claiming that the virus would disappear, saying in February 2020, "It's going to disappear. One day, it's like a miracle—it will disappear."[66]
- Questioning mask effectiveness and discouraged mask-wearing.[67]
- Halting WHO funding in April 2020 and blaming WHO for the pandemic.[68]
- Promoting treatments such as hydroxychloroquine[69] (used for malaria) and staying silent when some senior politicians from his party endorsed ivermectin (used for livestock).

Governments in other countries also restricted or distorted information. Examples include:
- China. The Government silenced whistleblower doctors like Li Wenliang and censored online discussions. Millions

of posts containing keywords related to COVID-19 were removed from online platforms.
- Brazil. President Jair Bolsonaro downplayed the virus, promoted treatments like hydroxychloroquine, and launched an anti-lockdown campaign called "Brazil Can't Stop."
- Egypt. Journalists were arrested and news outlets blocked for reporting on prison conditions and questioning official COVID-19 statistics.
- North Korea. Officials claimed zero COVID-19 cases, despite scepticism and reports of internal outbreaks.

Economic pressures

Many businesses opposed mask mandates and lockdowns, arguing that they harmed the economy. Some industries lobbied against restrictions, pressuring officials to ease health measures.

Some workers refused vaccination due to personal beliefs or misinformation, leading to workplace conflicts over vaccine requirements.

Social and anti-vax pressures

Anthony Fauci was targeted by conspiracy theories, including claims that he was profiting from vaccines or that COVID-19 was a hoax.

Vaccine scepticism was also partly driven by a strong anti-government authority sentiment particularly in the U.S., where people resisted government intervention in health decisions.

13

The Building Blocks of Modern Humanism

Characterizing the building blocks

Parts 2 to 5 underscored science's role in shaping our understanding of reality, while offering insights into the forces that drive human behaviour and evolutionary change. This chapter marks a transition from how the world is to how it ought to be. It examines the moral decisions humanists face: the life questions, the purposes, the worldviews, and the social, political, and economic structures.

Using the case study "The Eradication of Smallpox and the COVID-19 Crisis," humanist values are distilled into core principles and processes, the practical moral choices that form the foundation for broader ethical frameworks.

These foundational building blocks of how humanism believes the world 'ought' to be helps clarify competing choices while inviting reflection, revision, and experimentation toward a more humane and resilient society.

Importantly, humanism doesn't aim to define a perfect world. Its principles lead to varied, practical arrangements. This flexibility is a strength: it allows science to counter delusion, offers a vision of flourishing that resonates emotionally, and supports diverse paths toward individual and planetary well-being.

T13.1 provides a summary of these foundational building blocks.

	The basics of knowledge and belief
1	The epistemic commitment: choosing science over myth
2	Living with uncertainty and embracing transparent experimentation
3	Emotions as moral catalysts
4	Staying informed: engaging with scientific knowledge
	Ethical processes and decision-making
5	Rejecting authority, embracing dissent and collaboration
6	Evaluating and applying trade-offs
7	Including stakeholders and generational time scales
	Philosophical boundaries and moral humility
8	No utopias, No taboos
9	The moral relevance of chance
10	Humility and openness to changing minds
	Human-centred values and systems
11	Kindness, compassion, and human dignity
12	The ethics of fair processes
13	Elemental roots of human identity
14	The responsibility to make moral decisions for planetary flourishing

T13.1 Humanism's building blocks

The basics of knowledge and belief

<u>1. The epistemic commitment — choosing science over myth</u>

For centuries, diseases were seen as divine punishment. Even during COVID-19, religious leaders and politicians opposed vaccines, invoking spiritual objections that led to exemptions, lawsuits, and rallies. Yet science, which identifies viruses, traces their evolution, and explains how vaccines work, was rarely defended. Anthony Fauci, for instance, avoided challenging religious beliefs directly, focusing instead on institutional stances.

Before science, knowledge was observation and myth. Mysteries were filled with gods. That tension persists: many still cherry-pick when to apply science. Humanism rejects this. It embraces the epistemic imperative, a moral commitment to justified belief, testing, falsification, and revision. As Julian Huxley wrote, science must apply to all aspects of life.[1]

Carl Sagan warned against invoking gods for knowledge gaps. Bertrand Russell's regressive method — tracing beliefs to their underlying assumptions and testing them — remains vital. For humanists, applying science is a matter of integrity: careful observation, courageous truth-seeking, and rejecting comforting myths.

This isn't just theory. In *McLean v. Arkansas* (1982), creationism was ruled unscientific. Larry Laudan argued empirical evidence is the strongest refutation.[2] Claims of a young Earth and flood are falsified by fossils and genomes. Similarly, religious objections to vaccines such as purity and divine will, are biologically and empirically untenable.

William Clifford's 1877 principle still holds: "It is wrong... to believe anything upon insufficient evidence."[3] Humanism chooses science not as method, but as a moral stance, the most reliable way to understand reality and uphold truth.

Humanism makes a foundational moral and epistemic choice: to treat science as the most reliable method for understanding reality. This commitment rejects myth and dogma in favour of evidence-based inquiry.

2. Living with uncertainty and embracing transparent experimentation

Humanism and science don't promise certainty, only evidence-based, revisable answers. COVID-19 vaccines, though safe, caused rare adverse effects. We can't yet predict who will react negatively, but we understand the trade-offs. That's humanist reasoning.

Variolation worked long before its mechanism was known. Smallpox eradication, too, succeeded despite uncertainty: multiple strains, patchy data, and uneven vaccine efficacy. Adaptation and coordination made it possible.

Humanism embraces principled experimentation: transparent, consensual, and respectful of life. It's a moral method based on testing, learning, and adapting toward a more humane society.

Sweden's COVID-19 strategy valued trust and autonomy but lacked democratic endorsement and clarity, especially for the elderly. A humanist approach would have made uncertainties explicit, encouraged smaller-scale trials, and ensured informed consent. This is where science informs, ethics evaluates, and humanism guides.

Humanism means living with uncertainty where evidence is ambiguous, and embracing experimentation — but always with transparency and the agreement of those affected.

3. Emotions as forms of judgement

Martha Nussbaum writes, "The story of an emotion is the story of judgments about important things... our neediness and incompleteness before those elements we do not fully control"

(Nussbaum)[4]. Emotions are not irrational; they are moral judgments.

Smallpox eradication was driven not just by science, but by empathy and the desire to relieve suffering. It required imagining a world without disease, then pairing that vision with data, surveillance, and strategy. This fusion of aspiration and discipline is powered by emotion.

During COVID-19, vaccines, lockdowns, and messaging lacked emotional nuance. Artists, educators, and activists could have helped shape narratives and resilience. Humanist ethics insists emotions be communicated clearly, adjusted transparently, and engaged thoughtfully.

The pandemic raised far-reaching questions: how to protect the vulnerable, grieve collectively, and envision recovery. Mask mandates curbed transmission but triggered fear and defiance. Humanism doesn't dismiss these emotions, it sees them as meaningful, but insists they be accountable to truth. Through engaged dialogue and persuasion, it aligns emotion with evidence.

New Zealand's early lockdowns paired restrictions with empathetic leadership. Jacinda Ardern's emphasis on kindness and shared responsibility helped citizens process hardship, turning compliance into moral commitment.

Emotions are also connected with stories that help us make sense of suffering, imagine recovery, and connect personal experience to collective meaning. Local volunteers during the smallpox eradication campaign provided humanist stories to explain vaccinations - not as myth-making, but as a way to align emotion with truth and foster moral imagination.

When emotions are acknowledged and guided by evidence, public health becomes not just effective, it becomes humane.

Humanism's moral choices are sparked and animated by human emotions and imagination, but both tempered and accountable to empirical tests and the findings of science.

4. Staying informed — engaging with scientific knowledge

Resistance to the smallpox eradication program by WHO leaders wasn't purely scientific, it reflected cognitive biases. Sunk cost bias made shifting focus from malaria politically difficult. Loss aversion raised fears of reputational damage if eradication failed.

Such biases are shaped by evolutionary pressures, memetics, neuroscience, and behavioural economics—sciences that explain what we do and why.

We often frame viruses as adversaries, but they aren't strategic agents. They're evolutionary phenomena shaped by selection pressures and host environments. Understanding this helps us respond effectively. Evolutionary theory also sheds light on human behaviour, from cooperation to resistance.

Humanism insists we stay informed. It's not enough to value science abstractly, we must engage with its findings to guide our beliefs and decisions.

Humanism requires ongoing engagement with the findings of science — from evolution to behavioural science — to ensure beliefs and actions remain aligned with the best available understanding.

Ethical processes and decision-making

5. Rejecting authority, embracing dissent and collaboration

In both smallpox and COVID-19, flawed assumptions about transmission persisted. Early models downplayed airborne spread. Though William and Mildred Wells proposed a 100-micron threshold in 1955, CDC's Alexander Langmuir promoted a 5-micron standard — shaped by Cold War fears.[5,6] Even after recognizing his error, the narrower threshold stuck, delaying recognition of aerosol transmission in both pandemics.

Donald Milton's 2012 review showed smallpox could spread across hospital floors and downwind from treatment centres, signs of airborne transmission.[7] In COVID-19, the same misunderstanding led to misplaced emphasis on distancing over ventilation and masks.[8] Crises demand re-evaluating assumptions, although that is hard when authority goes unchallenged.

Political ideologies can be even more dangerous. China's suppression of COVID-19 data and punishment of dissenters like Zhang Zhan stifled inquiry.[9] History offers darker lessons: Stalin's embrace of Lysenko's pseudoscience and Mao's adoption of the same led to catastrophic famines, killing tens of millions.

Physicist Robert Laughlin warns that inquiry today is constrained by secrecy, patents, and economic control.[10] In fields like biotech, curiosity itself can be punished. He fears a "Dark Age" where access, not data, is the barrier.

Humanism rejects dogma: religious, political, commercial, or scientific. It values dissent, transparency, and collaboration.

Debate expands knowledge and reduces suffering. The Human Genome Project, a global effort marked by ethical debate and open data, exemplifies this spirit.[11]

Humanism is the rejection of authorities and dogmatic ideologies — religious, commercial, political, or military — and a journey of collaboration, vigorous debate, and the encouragement of dissent.

6. Evaluating and applying trade-offs

Ignoring trade-offs in public health is a moral and scientific failure. In 1973, India vaccinated 100 million people, yet smallpox persisted. WHO shifted to 'search and containment,' targeting infections and contacts. This trade-off, broad coverage vs. resource efficiency, led to eradication by 1977.

COVID-19 posed different dilemmas. UK lockdowns cut transmission by 75% but triggered a 9.7% GDP drop and a 27% rise in anxiety disorders. These weren't just epidemiological trade-offs, they were economic, psychological, and generational.

Vaccine deployment required balance. AstraZeneca's rare clotting risk was outweighed by its benefits in high-transmission zones. As Marc Lipsitch said, "Every intervention has a cost. The question is… how to act wisely."[12]

Trade-offs also shape how we apply the precautionary principle and assess dual-use technologies. Greater precaution during COVID-19 might have accelerated protective measures. Yet full precautions, like universal N95 use or negative-pressure rooms, carry their own burdens. These must be weighed against evidence.

Critics argue precaution stifles innovation, but it shows it can balance safety and progress. The Wright brothers tested cautiously; Otto Lilienthal died from a stalled glider. Boldness paired with caution is responsible innovation.

Climate change presents hidden trade-offs: expanding disease habitats, zoonotic spillover, and thawing permafrost. These risks by themselves may justify drastic fossil fuel reductions.

Dual-use technologies demand scrutiny. Gain-of-function research may aid preparedness but risks accidental release. Nuclear energy offers low-carbon benefits but carries catastrophic potential and waste challenges. Should we accept these risks to decarbonize?

Humanism begins not with ideology, but with evaluating trade-offs. Moral choices may diverge, but the starting point must be a shared commitment to evidence-based inquiry.

Humanism recognizes that trade-offs are inherent in all decisions and are essential pre-requisites to moral decision making — including evaluating dual-use technologies and applying the precautionary principle.

7. Wide circle of stakeholders and generational time scales

Pandemic responses often begin with institutional actors, WHO, CDC, governments, pharma companies, NGOs. Yet the most critical stakeholders are the most vulnerable: young children, the elderly, and communities in developing regions.

Smallpox threatened children most, who were prioritized vaccination. The elderly, often immune from prior exposure, played key roles in outreach and mobilization, bridging generations.

COVID-19 reversed this. The elderly faced the highest mortality risk and were vaccinated early. Children, less biologically at risk, endured long-term consequences: disrupted education, economic instability, and mental health challenges that continue for decades.

Reframing stakeholders reveal an essential truth: moral decisions must weigh not just immediate risks, but long-term generational impact. Public health must be inclusive: biologically, socially, and temporally.

Humanism demands that decisions reflect a wide circle of stakeholders, including the vulnerable, and consider consequences across generational time scales.

Philosophical boundaries and moral humility

8. No utopias, no taboos

Taboos delay progress. Silence on sexuality and drug use hindered HIV/AIDS responses. Menstrual taboos keep girls from school. Mental illness remains stigmatized, leading to suffering.

COVID-19 exposed similar risks. The idea of keeping schools open while shielding the vulnerable was rarely explored. Countries like Uruguay and Taiwan tried alternatives, but elsewhere, dissent became risky. The success of mRNA vaccines, once dismissed, shows the power of open inquiry.

The Great Barrington Declaration proposed "focused protection." It was criticized as unfeasible, but its dismissal without engagement created the appearance of taboo.[13] Scientists should have publicly addressed its flaws. Avoiding debate damages trust and stifles innovation.

However, Garret Hardin reminds us: in emergencies, not all ideas deserve equal attention. Extraordinary claims require evidence. Rejecting taboos means all ideas must be testable, falsifiable, and accountable, not endlessly entertained.

Humanism also rejects utopias. Science doesn't promise perfection; intervention carries trade-offs. Vaccine backlash shows the danger of demanding flawlessness. AstraZeneca's rare clotting risk and mRNA-related myocarditis were real but minimal compared to lives saved.

Smallpox eradication succeeded through humility and adaptation. COVID-19 needed flexibility, but dissent was often silenced. History warns us: utopian visions, from the Taiping Heavenly Kingdom to Jonestown, Mao's Great Leap, and Soviet Lysenkoism, have led to catastrophe.

Humanism investigates without taboos and acknowledges no utopias.

9. The moral relevance of chance

Humanism acknowledges the role of chance, whether inherent in the universe or emerging from complexity and ignorance, as a force shaping our lives.

Zoonotic spillovers are profoundly chancy events—rare, unpredictable, and often triggered by trivial interactions—yet their consequences can be catastrophic, as seen with coronaviruses and HIV.

This awareness demands moral humility. A wrong turn sparked World War I. Typhoons derailed Mongol invasions and reshaped Japan. A cosmic collision ended the dinosaurs, clearing the path for human evolution. Even well-intentioned actions unfold in unpredictable contexts.

Humanism urges caution, foresight, and readiness to adapt when outcomes diverge from intentions. It recognizes that uncertainty is not failure, it is part of being human.

Humanism takes into account an awareness of how chance, whether an inherent part of the universe or due to complexity and our ignorance, plays a significant role in our lives.

10. Humility and openness to changing minds

Humanism values the courage to revise beliefs in light of better evidence. Before 1967, mass vaccination was the dominant smallpox strategy. Henderson, learning from West Africa, shifted to targeted containment, a moral and intellectual choice to prioritize results over pride.

In contrast, General MacArthur's conduct during the Korean War shows the cost of refusing to change course. After Inchon, he pushed deep into North Korea, ignoring warnings about Chinese intervention. Convinced of his infallibility, he dismissed intelligence. When China counterattacked, tens of thousands died, and the war descended into stalemate. His hubris caused immense suffering.

Henderson's legacy is one of humility, willingness to listen, learn, and adapt. MacArthur's failure is a caution against dogmatism and the illusion of certainty.

Humanism sees life as a journey where we need to maintain humility and openness to changing our minds based on science and reason.

Human-centred values and systems

11. Kindness, compassion, and dignity

Smallpox eradication succeeded not just through science, but through compassion. Health workers in India and West Africa went door to door, listening to fears, respecting customs, and building trust. Henderson emphasized dignity: "We had to win the confidence of the people... not just deliver vaccines" (HENDERSON)[14].

Kindness and compassion also shape policy, in prioritizing the vulnerable, ensuring access, and recognizing the psychological toll of isolation and grief. These are moral imperatives.

Christina Feldman describes compassion as a deliberate choice to meet suffering with presence rather than avoidance. It is not sentimental or passive, but a courageous response rooted in clarity and care.[15] Compassion, she argues, is a capacity we all share—and one that deepens when we attend to pain with intention.

In public health, service, humanitarian work, knowledge-sharing, this view reinforces the need to face all aspects of life directly, without denial or detachment. Compassion becomes a form of moral engagement: not just feeling for others, but acting with integrity and resolve to reduce harm.

Humanism considers kindness and compassion as essential aspects of inclusive knowledge-sharing and interactions based on human dignity.

12. The ethics of fair processes

Fairness is moral and emotional. People feel outrage when rules are broken and acceptance when fairness is upheld, even if outcomes aren't in their favour. Studies suggest fairness evolved to support trust, reciprocity, and group survival.

COVID-19 exposed failures of fairness. In the UK, Boris Johnson's government enforced strict lockdowns, yet officials attended parties — "Partygate" mocked public sacrifice and shattered trust. In the U.S., Trump promoted unsupported treatments, yet received elite care when infected. The contrast was stark.

Sometimes evaluating fairness is complicated. Scientists must balance presenting evidence with advocating decisions. If roles blur, bias creeps in; if too rigid, science goes silent in moral crises.

Smallpox eradication offers a better model. Henderson's team adapted strategies based on field data, not ideology. They listened to local health workers, adjusted tactics, and made context-sensitive decisions. Fairness meant integrity, responsiveness, and trust.

Humanism is practiced with fair processes — decisions made and implemented with integrity, where dissent is heard, and those impacted can trust the systems that govern or affect their lives.

13. Elemental roots of human identity

Humanism recognizes our connection to the cosmos, not as metaphor, but as truth. We are made of atoms born in three cosmic events: hydrogen from the Big Bang, carbon and oxygen in stars, and heavier elements like iron from supernovae. As Morgan Freeman put it, "We are the children of stars."[16]

This shared origin carries moral weight. All life is kin, forged from the same cosmic processes. Our bodies are tuned to planetary rhythms: circadian cycles, seasonal biology, even lunar tides. These are adaptations to celestial mechanics.

Even the microbes within us, many symbiotic, are ancient agents that shaped our biology. They remind us that life is interdependent and shaped by forces beyond our control.

Understanding our cosmic heritage expands moral imagination. It dissolves illusions of separateness and intensifies our commitment to empathy, sustainability, and awe.

Humanism is aware of our cosmic heritage and that humans are expressions of the universe, capable of reflection, creativity, and care. This perspective encourages reverence not just for all life, but for the conditions that make life possible.

14. The responsibility to make moral decisions for planetary flourishing

Science tells us what is, not what ought to be. In moments of uncertainty, we must act on tentative findings, guided not by dogma but by empathy, reason, and reflection.

Sweden's COVID-19 strategy prioritized minimal disruption, but the elderly suffered disproportionately. Science can't resolve moral dilemmas like whose interests to prioritize. That responsibility belongs to us.

Humanism rejects promises of afterlife rewards. It focuses on alleviating suffering here and now and building conditions where life flourishes. That flourishing includes all life forms and the ecosystems that sustain them.

We no longer see disease as divine punishment. We fight it, as with smallpox, because we believe suffering can be reduced. Pandemics reveal our interdependence with all life. Zoonotic diseases like Ebola, SARS, and avian flu emerge from ecological disruption: habitat loss, industrial farming, global movement. COVID-19 likely did too.

These are ecological crises. Human health is inseparable from planetary health. Stewardship of life on Earth is practical and a moral imperative.

Humanism places the responsibility for moral purpose — here and now — for individual and planetary flourishing.

N7 Trade-offs and the precautionary principle

"We are not only responsible for what we do, but also for what we fail to do."[17] Molière

Part 6

Consequences of how the world IS and OUGHT to be

Needs, Purposes, and Activities

A humanist worldview

Tim Berners-Lee and Mark Zuckerberg

"At pivotal moments, generations before us have stepped up to work together for a better future. Now too, as the Web reshapes our world, we have a responsibility to make sure it is recognized as a human right and built for the public good.

The Web is for everyone, and collectively, we hold the power to change it. It won't be easy. But if we dream a little and work a lot, we can get the Web we want."[1]

TIM BERNERS-LEE

"i have over 4000 emails, pictures, addresses. people just submitted it. i don't know why. they 'trust me.' dumb fucks"

Mark Zuckerberg's message to a friend when operating the precursor to Facebook at Harvard University[2]

"Domination"

Mark Zuckerberg – Pumping his fists and yelling while ending meetings during the time Facebook obtained more than 5.5 million users.[3]

14

Tim Berners-Lee and Mark Zuckerberg

Questions on needs, purposes, and activities for flourishing

This case contrasts Tim Berners-Lee and Mark Zuckerberg—two figures who reshaped our digital world with radically different motives. Berners-Lee built the Web to foster openness and cooperation. Zuckerberg, guided by venture capital, thrives on competition and monetization and built Facebook into a data-driven platform. Key questions the case brings are:

<u>Purpose</u>

Q1. What purpose drives your life and actions?

Q2. Are our life purposes shaped by early influences, immediate environments, or evolutionary heritage?

Q3. Can we develop new purposes beyond inherited traits?

<u>Human needs and capabilities</u>

Q4. Which needs and capabilities do Berners-Lee and Zuckerberg most target?

Q5. Do their platforms promote meaningful creativity or reinforce meaningless work?

<u>Cooperation or competition</u>

Q6. In our own lives, is cooperation viable—or is aggressive competition the norm? Can we flourish through collaboration?

Q7. What unites and divides life purposes rooted in humanism, art, science, and engineering?

Q8. How do differing purposes shape visions for the future?

Tim Berners-Lee

Upbringing and influences

Born in 1955 to mathematicians who helped build the first commercial stored-program computer, Berners-Lee grew up immersed in conversations about machines and minds. He tinkered with electronics; at Oxford, he built a computer from a discarded TV. He sought the intersection of hardware and software.[1]

Three key thinkers shaped his vision for the Web:[2]
- Vannevar Bush, who foresaw personal computing[3]
- Ted Nelson, who coined "hypertext" and "hypermedia"[4]
- Doug Engelbart, a pioneer in human-computer interaction[5]

Each bridged technical innovation with social impact—an ethos Berners-Lee would carry forward.

CERN and the culture of collaboration

Berners-Lee invented the Web at CERN, Europe's particle physics lab founded in 1954 to promote postwar scientific cooperation. Shaped by the horrors of WWII and totalitarianism, CERN's culture emphasized open dialogue, shared knowledge, and international collaboration. Its founders, de Broglie, Bohr, Bloch, embedded these values into its DNA.

The Internet's early challenges

By the late 1980s, millions of computers were connected via the Internet using TCP/IP protocols. But two problems remained:
- Computers couldn't interpret each other's data formats;
- Users couldn't easily search or retrieve information.

Berners-Lee saw the need for a universal system to bridge these gaps.

From ENQUIRE to the Web

In 1980, while consulting at CERN, Berners-Lee built a tool ENQUIRE to track projects based on how humans naturally link ideas. Though unused by others, it planted the seed for the Web.

Returning to CERN in 1984, he proposed a "networked hypertext system" to manage CERN's knowledge. In 1990, with Robert Cailliau, he built the architecture and software for what became the World Wide Web—released to the public in 1991.

The purposes behind the invention of the Web

Berners-Lee didn't just solve a local problem—he built for the world. His Web was open, scalable, and designed for universal compatibility. MIT's Michael Dertouzos noted that while many saw hypertext and networks, only Berners-Lee fused them into a transformative system.

His invention became a global information commons, free, interoperable, and built for cooperation. The Web now allows collaboration, shared understanding, and human connection. It is a ubiquitous information marketplace where individuals and organizations buy, sell, and exchange information and services.

Technological progress often outpaces ethical reflection. To address this, Berners-Lee founded the World Wide Web Consortium (W3C), ensuring the Web could evolve within diverse policy frameworks while promoting inclusion and accountability.

His vision rested on four pillars:
- Cooperation and collaboration
- Free and open access
- Decentralized architecture
- Technology aligned with human flourishing

Berners-Lee fought to keep the Web royalty-free. CERN agreed in 1993, making it a public resource. He rejected personal profit, believing the Web should empower users, not enrich gatekeepers. As James Gleick put it, the Web is "The Patent That Never Was."[6]

Berners-Lee urged Web creators to establish norms of accountability and resist opportunities that threatened its independence.

A crossroads in the early Web

In the early 1990s, the Web's future was uncertain. Competing models emerged:
- Marc Andreessen's Mosaic tried to rebrand the Web and later dominated via Netscape.
- Gopher, a menu-based system, introduced licensing fees.
- AOL built a gated community.
- Microsoft's Internet Explorer integrated the Web into its operating system.

Without Berners-Lee's intervention, the Internet could have splintered into proprietary silos. Instead, he created open standards and a royalty-free Web as a universal commons.

Founding the W3C

In 1994, Berners-Lee partnered with MIT's Michael Dertouzos to establish the World Wide Web Consortium (W3C)—a standards body modelled after the Internet Engineering Task Force (IETF). Unlike the volunteer-run IETF, W3C has members and full-time staff, ensuring continuity and global coordination.

Key collaborators - Cailliau and Dertouzos
- Robert Cailliau shared Berners-Lee's vision and helped advocate for a royalty-free Web within CERN.
- Michael Dertouzos, MIT professor and LCS director, provided the institutional support to launch W3C. He believed technology must serve humanity, not dominate it, and envisioned an "Information Marketplace" where people freely exchange ideas and work. (See C25)

Impact, challenges, and future directions for the Web

The Web has become a transformative force. Daniel Weitzner calls it "the most transformative invention of our time," advancing free expression, knowledge, and democracy.[7]

But its success has also amplified inequality and surveillance. Berners-Lee founded the Web Foundation to advocate for policies to protect privacy and access. The Foundation reports:
- 65% of people live under online censorship;
- Fewer than 30% of non-European countries have strong data protection laws.

Berners-Lee said:

"I was devastated…The centralization of the Web has ended up producing – with no deliberate action of the people who designed the platform – a large scale emergent phenomenon which is anti-human."[8]

"Facebook, Google, Amazon now monopolize almost everything that happens online along with a handful of powerful government agencies they are able to monitor, manipulate and spy in once unimaginable ways."[9]

A quiet legacy

Despite his monumental impact, Berners-Lee remains largely unknown. In a 2017 documentary, most people couldn't name the Web's inventor. Jeff Jaffe, W3C's CEO, remarked, "There are thousands of people who are probably more famous than Tim Berners-Lee and yet not one of them has had the impact of Tim."[10] Berners-Lee says, "I am very happy that I am not [well known]."[11]

Mark Zuckerberg

Upbringing and influences

Born in 1984 Zuckerberg attended the elite Phillips Exeter. He began coding, taught by his father and a private tutor.

In his Harvard University application, Zuckerberg stated that he could read and write Ancient Greek, French, Hebrew, and Latin. However, a college friend said, "I never saw Mark reading a book or expressing any interest in books. He definitely didn't have a broader interest in philosophy, political thought, or economics."[12]

Zuckerberg's formative influences were Peter Thiel and Marc Andreessen, both champions of aggressive market dominance. Thiel invested early in Facebook and served on its board for 16 years. Andreessen, co-creator of Mosaic and co-founder of Netscape, joined the board in 2008.

Thiel's philosophy, outlined in Zero to One, celebrates monopolies as engines of innovation. Zuckerberg embraced this ethos, ending team meetings with the chant: "Domination!"[13]

Facebook's origins

At Exeter, Zuckerberg's peer Kristopher Tillery launched an online student directory called "The Facebook." It quickly gained traction but caused disruptions including browser crashes linked to code planted by Zuckerberg.[14]

At Harvard, Zuckerberg built CourseMatch to track course selections and Facemash, a site ranking student photos. Facemash was shut down for violating data policies.

In 2004, Zuckerberg and Eduardo Saverin launched thefacebook.com. When Saverin hesitated on investment, Zuckerberg wrote, "I'm just going to cut him out."[15] A later settlement made Saverin a billionaire, but excluded him from operations.

Zuckerberg dismissed Harvard's progress on a universal student directory, saying, "I can do it better... in a week."[16] He was accused by three students of stealing their idea and hacking into their email accounts. A lawsuit led to a settlement of 1.2 million Facebook shares.[17]

Aaron Greenspan, another student with a similar site, clashed with Zuckerberg over merging their platforms, but no agreement was reached.

Silicon Valley and expansion

Facebook's adoption at Harvard where over half the student body joined within a month caught the attention of Sean Parker, founder of Napster and Plaxo. Parker became Facebook's president and introduced Zuckerberg to Thiel and other investors.[18] Thiel invested, and Andreessen became a key advisor, helping Zuckerberg focus on scale and strategy.[19]

Zuckerberg dropped out of Harvard and pursued growth, expanding Facebook to major U.S. universities and eventually to anyone over 13. By age 23, he was the world's youngest self-made billionaire.

Facebook - from directory to dominance

Though Facebook was not the first social media site, Zuckerberg grasped its commercial potential. By 2005, influenced by Silicon Valley mentors, Facebook's strategy crystallized around three pillars:

- A platform for social engagement and entertainment
- Aggressive data collection and monetization
- Relentless user growth

He described Facebook as a place to "waste time," aspiring to make it "the new MTV."[20],[21] Features were designed to be addictive, and by 2020, one-third of Americans got their news from Facebook.[22]

During the 2016 U.S. election, Facebook embedded staff in Trump's campaign to help target ads, raising questions about political neutrality and platform responsibility.

Data as power

Facebook's model hinges on harvesting user data—likes, browsing habits, and preferences—often via cookies that track users across the web. In an April 2018 testimony in front of a U.S. Senate committee, Zuckerberg admitted that Facebook collects data from users even when they're not on Facebook. He said, "If you're logged into Facebook and visit a website with the Like button, your browser sends us information about your visit. If you're logged out or don't have a Facebook account and visit a website with the Like button or another social plugin, your browser sends us a more limited set of info."[23]

Facebook's data infrastructure, the Hive, stores hundreds of petabytes and categorizes users into over 50,000 segments. It is estimated that in 2021 Facebook generated 4 petabytes of data per day.[24] One petabyte can be compared with 500 billion pages of printed text. Algorithms determine what users see, enabling advertisers and political groups to target messages with precision. Data is sold via auction to the highest bidder.

This model is shared by other tech giants, but Facebook's role in the Cambridge Analytica scandal, where data from 87 million users was used for political profiling, highlighted the risks.[25] Though Facebook knew, the breach only came to light in 2018.

Growth at all costs

Zuckerberg believes in "network effects"—the idea that a platform's value grows with new user. Facebook engineers are rewarded for user engagement and retention.

In 2006, Zuckerberg rejected Yahoo's USD 1 billion buyout offer, despite Facebook being unprofitable.[26] His confidence led to internal dissent but also attracted top talent from Google and Microsoft.

Facebook expanded aggressively, acquiring 91 companies by 2021 including Instagram, WhatsApp, and Oculus. Though it promised to keep these platforms independent, it later integrated messaging across them.

By the end of 2021, Facebook had:
- 2.91 billion monthly active users

- USD 117 billion in annual revenue
- Ownership of the four most downloaded apps

Mission, power, and influence

Facebook's mission - "to give people the power to share and make the world more open and connected"—has been interpreted by some insiders with stark clarity. Andrew "Boz" Bosworth, a senior executive, wrote in a 2016 memo, "We connect people. That can be bad... Maybe someone dies in a terrorist attack coordinated on our tools. And still, we connect people."[27]

Augustus Caeser

Zuckerberg has admiration for Augustus Caesar, whom he admits "had to do certain things"[28] in order to secure the peace, but that "Basically, through a really harsh approach, he established two hundred years of world peace."[29]

In 31 BCE Augustus became the dictator of Rome and ended the democracy of early Rome. Edward Gibbon describes Augustus as having "A cool head, an unfeeling heart, and a cowardly disposition, prompted him at the age of nineteen to assume the mask of hypocrisy, which he never afterward laid aside," and that "Every barrier of the Roman constitution had been levelled by the vast ambition of the dictator."[30]

Zuckerberg holds majority voting control at Meta and oversees operations with intense scrutiny. Facebook's internal security unit had access to all employee communications and activity logs.[31]

Apologies and accountability

Zuckerberg's history of public apologies, from FaceMash at Harvard to data scandals, follows a familiar pattern. As Zeynep Tufekci argues, these apologies often serve as ritual responses rather than catalysts for structural change. The company's core incentives remain intact.[32]

Lobbying and political influence

Facebook invests heavily in lobbying. During Zuckerberg's 2018 congressional testimony, the company had donated to over half the lawmakers questioning him highlighting its entanglement with political power.[33]

Philanthropy and public image

In 2020, Zuckerberg and his wife pledged their wealth to the Chan Zuckerberg Initiative (CZI). When challenged about Facebook's practices and CZI's values, Zuckerberg defended the alignment, suggesting both stemmed from shared principles.[34]

Sheryl Sandberg

From 2008 to 2022, Sheryl Sandberg served as Facebook's COO, helping scale the company from USD300 million to USD 117 billion in annual revenue. A former Google executive and Washington insider, Sandberg brought networks and operational expertise.

Critics like Shoshana Zuboff say that Sandberg played "the role of Typhoid Mary, bringing surveillance capitalism from Google to Facebook when she signed on as Mark Zuckerberg's number two."[35] Her tenure cemented Facebook's data-driven advertising model, which accounts for 97% of its revenue.[36]

User wellbeing

From 2017 to 2021, Americans spent an average of 33 minutes daily on Facebook. A longitudinal study published in the "American Journal of Epidemiology" found that increased Facebook use correlated with declines in physical health, mental wellbeing, life satisfaction, and increases in obesity.[37]

In 2017 Facebook acknowledged that "passive consumption"[38] can make people "feel worse," although it argued that "more engagement could improve wellbeing." A former Facebook executive wrote, "The short-term, dopamine-driven feedback loops that we have created are destroying how society works. No civil discourse, no cooperation, misinformation, mistrust."[39]

Privacy and competition

In 2020, the FTC and nearly every U.S. state sued Facebook for anti-competitive practices and privacy violations.[40]

Co-founder Chris Hughes called for Facebook's breakup, arguing that Zuckerberg's concentrated control prevents reform.[41]

Looking ahead

Zuckerberg continues to expand Facebook's reach, backed by billions in reserves. His attempt to enter cryptocurrency failed amid regulatory pushback. In 2021, he rebranded the company as Meta, signalling a pivot toward the metaverse, a virtual space for work and play. As of mid-2025, this venture has yet to succeed.

Facebook is refocusing on AI and younger users. Executives describe this demographic as the company's "North Star," shifting away from older audiences.[42]

Cultural legacy

The 2010 film "The Social Network" captured the polarizing image of Zuckerberg. Jesse Eisenberg, the actor playing Zuckerberg, said that "I was asked by older people again and again how I could play a character who is capable of being so mean, as if I were almost condemned by this role. But young people never had that reaction. They kept saying, 'This guy was a genius. Look what he has created.'"[43]

15

Needs, Purposes and a Humanist Worldview

Evaluating the actions of Berners-Lee and tech billionaires

Bullshit jobs vs creative projects

Anthropologist David Graeber's study "Bullshit Jobs" reveals that vast numbers of people spend their lives performing tasks they believe are meaningless—a condition he called "profound psychological violence" (GRAEBER)[1].

Alongside this, surveys show people spend 3 to 6 hours daily on digital infotainment across countries like Nigeria, India, China, and the U.S., a life shaped by distraction. (See T22.1)

In contrast, psychologist Mihaly Csikszentmihalyi studied "creative" individuals in the arts, music, literature, the sciences, business, and politics.[2] These people enter new situations, break boundaries, and share traits like:

- Deep knowledge paired with childlike curiosity
- Experimental spirit with focused persistence
- Imaginative thinking grounded in execution
- Humility alongside pride in contributions
- A blend of masculine and feminine traits

When immersed in their work, they experience 'creative flow,' an optimal state where creativity thrives.[3]

Needs, capabilities, environments, and worldviews

Philosopher Simon Blackburn writes, "We craft our morality according to what we know about the world."[4] Worldviews—

our frameworks of belief and value—shape how we perceive truth, purpose, and morality. Choosing between Graeber's 'bullshit jobs' and Csikszentmihalyi's creative projects depends on the worldview we inhabit.

Worldviews influence the environments we build, the needs we prioritize, and the capabilities we nurture. But the relationship is reciprocal. As explored in C6 S: Priming, our surroundings also shape the worldviews we adopt. Needs and capabilities reflect and reshape these worldviews. This interdependence is shown in F15.1

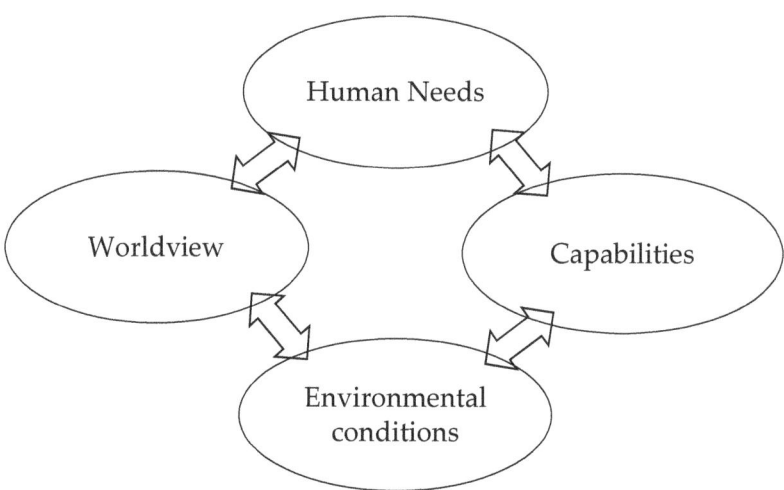

F15.1 Interplay between Needs, capabilities, the environment and worldviews

The worldview of tech billionaires

Mark Zuckerberg once said, "Facebook is more like a government than a traditional company."[5] This worldview turns tech platforms into quasi-sovereign entities where algorithms and unelected boards decide what billions can say, see, and share.

Tech leaders often treat social issues as engineering problems, solvable through code and data. Shoshana Zuboff contends Facebook's purpose is 'behavioural extraction,' transforming human experience into data to predict and manipulate future behaviour for profit.[6]

Research shows Facebook's News Feed prioritizes emotionally charged content to boost engagement, fuelling polarization, misinformation, and mental health decline. Internal studies leaked in 2021 revealed Instagram worsened body image issues for one in three teenage girls, yet reforms were delayed despite clear harm.[7]

Zuckerberg has positioned himself as a global actor, meeting heads of state and engaging in diplomacy exchanges with leaders including Donald Trump, Emmanuel Macron, Narendra Modi, and Xi Jinping.[8]

Berners-Lee's worldview

Berners-Lee envisioned the Web as open, designed to empower individuals by supporting their agency, dignity, and practical reasoning. His vision centres on applications serving individuals rather than extracting value. This fosters autonomy, free expression, and engagement without coercion. This supports psychological and social needs: connection, competence, and self-expression. The Web, as he imagined it, is a space for creativity, participation, and community.

Rather than nudging users toward outcomes, Berners-Lee designs systems that expand choice and protect cognitive liberty. His critique of surveillance capitalism reveals a concern for preserving human decision-making, freedom, diversity of thought, and the cultivation of capabilities that allow individuals to flourish on their own terms.

A model for comparing Zuckerberg and Berners-Lee

Artificial intelligence (AI) is often modelled on how developers think the human brain works. We can, however, use models of how artificial intelligence is set up to better understand our beliefs, decision making and actions. T15.1 uses Murray Shanahan's model for realising artificial general intelligence[9] to interpret human behaviour and its consequences.

Shanahan's model for AIs	Interpreted for humans
What is the AI's reward function?	What are our purposes? Do they balance ends with their means?
How are AI learning techniques implemented	How are we living while pursuing our purposes - through cooperation, widespread consultation or via closely held power and in authoritarian societies?
What optimization method maximizes the AI's reward function and learning techniques	What societal, economic, political structures help us live flourishing lives? Do we use iterative feedback, inclusive discussion, debate, and dissent or one-track beliefs in the rightness of one's actions?

T15.1 Shanahan's AI model used to interpret human behaviour.

This chapter explores how worldviews shape the needs and capabilities we cultivate or constrain. It then reverses the lens, showing how the deprivation of needs and capabilities influence the evolution of worldviews, impacting broader social and ethical consequences.

Worldviews

The following is a selective overview of worldviews and the purposes, needs, and capabilities they most influence. Societies often embody multiple worldviews, with considerable overlap between them.

Instrumental-commercial worldview

Clair Patterson argued that widespread lead poisoning reflected a cultural mindset: valuing knowledge for utility over truth.[10] In post-WWII America, science was subordinated to productivity and profit. Even the age of the Earth became a statistic, stripped of wonder. This worldview encourages exploitation and short-term gain, stifling curiosity and ethical reflection.

Market fundamentalist

Market fundamentalism sees free markets as morally superior and efficient. Milton Friedman championed it; critics like Stiglitz called it "quasi-theological." Keynes warned against assuming private interests serve public good. Piketty's data shows this worldview fuels inequality and weakens democracy, wealth concentration today mirrors pre-WWI levels.

Religious-theistic

Truth is divinely revealed. The universe has purpose—salvation or cosmic justice. This worldview has inspired both compassion and cruelty. Western Christians condemn Taliban restrictions, yet Christianity's history includes persecution. Despite secular shifts, the number of officially religious nations has remained stable. Even secular states often favour dominant faiths, as seen in the 2025 revival of the White House Faith Office.

Mythic-nationalist

Nationalism centres on cultural superiority and destiny. Citizens are expected to sacrifice for the nation. Putin's remark to a grieving mother, "Some people die in road accidents, others from alcohol – when they die, it's unclear how. But your son lived, do you understand? He fulfilled his purpose,"[11] captures this ethos. In the U.S., flag worship and military interventions reflect exceptionalism.

Lineage-ancestral duty

In some Hindu and Confucian traditions, sons are expected to perform funeral rites and continue the family name. This has led to gender imbalances, 22 million more men than women in India, 30 million in China.[12,13] These traditions often suppress autonomy, enforcing paths in marriage, career, and life.

Marxist-Leninist

Communism aims for a classless society through centralized control and forced transformation. Stalin and Mao believed history justified their brutality. Their worldview included totalitarianism, collectivization, personality cults, and threats. The cost was staggering, Matthew White estimates 20 million deaths under Stalin, 40 million under Mao, through purges, famines, and repression.[14]

The origins and spread of worldviews

Worldviews arise from an interplay of evolutionary, cultural, cognitive, and environmental forces. Recurring belief patterns may reflect inherited traits, while cultural evolution explains their diversity and change. Cognitive and emotional factors shape how beliefs are internalized and defended. Two influences—evolutionary dispositions and priming—help explain why certain worldviews persist or spread.

From an evolutionary perspective, worldviews may serve group or individual survival, act as cultural parasites, or be byproducts of other adaptations.[15] Some may have been adaptive in ancestral environments but are less relevant today. Worldviews often blend adaptive and nonadaptive elements, shaped by biology, culture, and psychology.

Priming helps explain how worldviews spread. Kathleen Vohs' research shows that subtle money cues shape thoughts and behaviour.[16] The importance of Vohs' work is that priming is pervasive and not just about money. Exposure to stimuli unconsciously shapes responses from birth. Vohs writes, "Democracy is valuable because it doesn't think of itself as finished or perfect… and neither does science" (VOHS).[17] Vohs urges psychologists to explore how priming influences not just economics, but the dogmas we live by.

Sources of priming - exemplars and the environments

Zuckerberg's exemplars: Augustus, Thiel, Andreessen, share a drive for power and control. He emerged from elite schools and Silicon Valley circles that view technology as a universal fix, with users as monetizable passive recipients. This worldview promotes societal change led by tech elites, often bypassing democratic checks.

Berners-Lee's exemplars: Bush, Nelson, Engelbart, championed inclusive, human-centred tech. Bush saw science as a public good. Nelson opposed hierarchy, empowering individuals. Engelbart prioritized collaboration and uplift, treating technology as a moral tool.

Berners-Lee's formative environments, home, university, CERN, were steeped in inquiry, shared knowledge, and community-building. His worldview reflects a commitment to enhancing life for all.

Needs and capabilities

Maslow's model of needs

In his 1943 model of human needs, Abraham Maslow aimed to establish a theory of human motivation by identifying a set of basic needs that drive behaviour.[18] His model arranged human needs in a hierarchy where more basic needs dominate behaviour until they are satisfied, after which higher needs emerge.

Maslow later extended his model and this is shown in T15.2.[19] Maslow emphasized that: motivation is dynamic and contextual, needs are never fully satisfied as they recur and shift, and the hierarchy is not rigid. Maslow distinguished between the first four "deficiency needs" (D-needs), which arise from lack, and the last four "being needs" (B-needs), which emerge from a desire to grow.

Need category	Description	Behavioural implication
Physiological	Basic survival: food, water, shelter, sleep	Survival prioritization: A hungry person will prioritize food over all other goals.
Safety	Security, stability, protection from harm	Individuals seek predictable environments, routines, and protection from harm.
Love/Belonging	Affection, relationships, group inclusion	People strive for friendships, family bonds, and acceptance in social groups.
Esteem	Self-respect, recognition, achievement	Drives ambition, status-seeking, and desire for competence.
Self-Actualization	Realizing personal potential, creativity, growth	Individuals pursue meaning, purpose, and peak experiences.
Cognitive	Desire for knowledge, understanding, curiosity	Pursuing education, skill development, and intellectual challenges.
Aesthetic	Appreciation of beauty, balance, and form	Interest in music, visual arts, architecture, and natural landscapes.

Need category	Description	Behavioural implication
Transcendence	Connecting to something beyond the self—altruism, peak experiences	Pursuit of values beyond the individual self, such as unity, service, or "mystical experience".

T15.2 Maslow's hierarchy of needs

It is worth noting that recent research has expanded on Maslow's foundational ideas by grounding human motivation in evolutionary psychology. Researchers from HSE University and the London School of Hygiene and Tropical Medicine recently proposed an evolutionary framework for human motivation.[20] They identified fifteen core motives grouped into five foundational categories: environmental, physiological, reproductive, psychological, and social. These categories reflect adaptive strategies that helped early humans survive and thrive in complex environments. This model complements Maslow's hierarchy by emphasizing how motives like status, play, and comfort shift across the lifespan.

A capabilities framework

Philosopher Martha Nussbaum developed a list of ten capabilities essential to human dignity and political justice (T15.3) (Nussbaum)[21]. Her framework builds on Amartya Sen's work in welfare economics, which emphasized assessing development through freedoms and people's capabilities to lead lives they value, rather than economic output alone (Sen)[22].

Sudhir Anand later translated Sen's ideas into policy tools that formed the foundation for the UN's Human Development Index (HDI).[23]

Nussbaum's framework treats each capability as a non-hierarchical, necessary condition for full human functioning, regardless of culture or economy. Philosophically grounded and globally informed, it offers a powerful lens for evaluating whether individuals have the freedom to live well-considered lives.

It is important to note that her capabilities are not limited to human flourishing. In "Frontiers of Justice," Nussbaum goes beyond calls for kindness or welfare protections for animals, insisting instead that they are rightful subjects of justice. This means not only preventing cruelty but actively enabling animals to flourish—through habitat preservation, the reform of industrial farming, and legal recognition of their rights.

Traditional theories of justice, particularly Rawls's social contract model, are inadequate in this regard. By grounding justice in mutual advantage and reciprocity, they exclude those who cannot participate in such agreements—animals, individuals with severe disabilities, and people outside national borders. Nussbaum challenges this exclusion, asserting that dignity and moral worth do not depend on bargaining power—all sentient beings flourish according to their specific capacities.

Modern Humanism resonates with Nussbaum's vision of planetary justice but draws inputs from science, not just philosophy. It emphasizes that evolution by natural selection is contingent and undirected, shaped by chance and necessity. From this perspective, humans are not entitled to special prerogatives. Ecology further reinforces this view, revealing the profound interdependence of all life on Earth (C10 and C20 S: Conservation, S: Planetary flourishing).

Capability	Explanation
Life	Being able to live a normal lifespan.
Bodily health	Access to nutrition, shelter, and healthcare.
Bodily Integrity	Protection from violence, freedom of movement, reproductive autonomy.
Senses, Imagination, and Thought	Education, freedom of expression, and aesthetic experience.
Emotions	Ability to form attachments and experience love, grief, anger without fear or repression.
Practical Reason	Capacity to form conceptions of the good, critical reflection and life planning.
Affiliation	Social interaction with dignity and protection from discrimination.
Other Species	Respect for animals, plants, and nature.
Play	Laughter, enjoyment, play, and recreation.
Control over one's Environment	Political participation, choices, freedom of speech and property rights.

T15.3 Nussbaum's list of capabilities NUSSBAUM[24]

Nussbaum's capabilities approach has influenced development economics, welfare analysis, and human rights:

- Paul Anand, Hunter & Smith found links between capabilities like health and affiliation and subjective well-being in British households.[25]
- Jamil affirmed the model's normative strength, while noting empirical challenges.[26]
- Coetzee proposed expanding the list to include digital literacy, privacy, and AI autonomy, adapting it to modern contexts.[27]

Overall, the model is most effective as a normative guide for policy, with selective empirical validation.

<u>An exploration - Creating a framework</u>

Maslow's hierarchy of needs and Nussbaum's capabilities offer distinct models of human development. Maslow's is psychological and motivational, progressing from survival to self-actualization. Nussbaum's is philosophical and normative, presenting ten capabilities for dignity and functioning.

Placing them side by side invites objections, especially from Nussbaum's perspective. Her framework rejects hierarchy, treating all capabilities as equally vital. Aligning it with Maslow risks misrepresenting her ethical commitments and methodological rigor.

This comparison, however, is not about equivalence. It's a heuristic exercise to identify shared domains: health, safety, affiliation, autonomy, that recur across both models. Mapping needs to capabilities helps illuminate how deprivation shapes worldviews. Patterns reveal how psychological and ethical deficits influence belief systems and political behaviour and can be applied to both history and the present.

Maslow and Nussbaum's models intersect in a many-to-many relationship. Starting with visceral experiences, hunger, fear, isolation, we can trace how deprivation ripples outward, constraining capabilities.

By beginning with Maslow's hierarchy, we can examine how unmet needs calcify into psychological barriers that prevent the flourishing Nussbaum envisions, revealing how deprivation reshapes not just opportunity but perception, aspiration, and identity.

A Maslow-to-Nussbaum mapping

Physiological Needs → Life, Bodily Health, Bodily Integrity

In Cox's Bazar, Bangladesh, over 1.1 million Rohingya refugees live in camps with poor sanitation and water access. Over 90% of women report urinary infections.[28] These conditions violate Maslow's physiological needs: food, water, shelter, and Nussbaum's capabilities: Life is endangered, Bodily Health compromised, and Bodily Integrity undermined by immobility and unsafe living.

Safety Needs → Life, Bodily Integrity, Control over One's Environment

During the Syrian civil war, millions were displaced. A 2021 UNHCR report found 70% of Syrian refugee women in Lebanon feared sexual violence, and over half lacked secure housing.[29] This reflects deprivation of safety: violence, instability, vulnerability, and violates Nussbaum's capabilities for Life, Bodily Integrity, and Control over One's Environment. Fear and powerlessness reshape worldview and erode dignity.

Love and Belonging → Affiliation, Emotions, Play

Youth vulnerable to Islamic radicalisation often face emotional isolation, alienated from family, excluded by peers. Extremist networks exploit this void, offering identity and community. A 2022 Europol report found radicalised youth felt "finally seen" in online forums.[30] This reflects breakdowns in Affiliation, Emotions, and Play, relationships replaced by dogma, emotional expression channelled into grievance, and recreation into indoctrination.

Esteem → Practical Reason, Affiliation, Control over One's Environment

Unemployment rates amongst the youth in North Africa exceed 30% in countries like Tunisia and Algeria. A 2023 ILO report found many feel "invisible" and disconnected.[31] This erodes Maslow's need for esteem and undermines Nussbaum's capabilities: Practical Reason is stifled by lack of purpose, Affiliation by social exclusion, and Control over One's Environment by economic disempowerment.

Self-Actualization → Senses, Imagination, and Thought; Practical Reason; Play

In Afghanistan, Taliban restrictions on education for girls has blocked intellectual growth. Since 2021, girls are banned from secondary and university education; boys face rigid, underfunded schooling.[32] This denies self-actualization and violates Nussbaum's capabilities: Senses, Imagination, and Thought are stifled; Practical Reason is obstructed; Play is diminished. Dreams are silenced, and identity reshaped around obedience.

Cognitive Needs → Senses, Imagination, and Thought; Practical Reason

China's internet censorship: blocking platforms like Wikipedia and news, restricts access to diverse ideas.[33] This suppresses Maslow's cognitive needs and Nussbaum's capabilities: Senses, Imagination, and Thought are narrowed; Practical Reason is constrained. Censorship reshapes cognition, limiting curiosity and worldview formation.

Aesthetic Needs → Senses, Imagination, and Thought; Play; Other Species

In Mumbai's slums, where over 40% of the population lives, aesthetic deprivation is reality. Cramped quarters, poor sanitation, and lack of green space dull sensory experience. A 2023 study found households lack access to parks or cultural spaces.[34] This violates Maslow's aesthetic needs and Nussbaum's capabilities: Senses, Imagination, and Thought are dulled; Play is constrained; Other Species is severed by urban sprawl.

Transcendence Needs → Affiliation, Practical Reason, Other Species, Emotions

Rising climate grief reflects a rupture in transcendence needs. A 2021 Lancet study of 10,000 youth across ten countries found 59% were "very or extremely worried" about climate change; 45% said it affected daily functioning.[35] This emotional distress maps to Nussbaum's capabilities: Affiliation is strained, Practical Reason clouded by despair, Other Species mourned, and Emotions overwhelmed by grief.

Deprivations and emergent worldviews

Effects of deprivation

T15.4 maps how the absence of needs and capabilities can shape reactive worldviews. Worldviews often emerge as adaptive responses to unmet needs, not cultural traits. Policy gaps that ignore these deficits risk reinforcing distorted beliefs, behaviours, and governance.

Need / Capability	Deprivation / Lack	Emergent worldview / disposition
Physiological / Life, Bodily Health, Integrity	Hunger, poor health, precarity, lack of shelter or sanitation	Survivalism, fatalism, short-termism, scarcity mindset
Safety / Bodily Integrity, Life, Control over environment	Exposure to violence, instability, precarity, displacement, legal insecurity	Authoritarianism, militarism, distrust of institutions
Belonging / Affiliation, Emotions, Play	Social exclusion, isolation, lack of community	Tribalism, sectarianism, hyper-individualism
Esteem / Control, Practical Reason, Affiliation	Disempowerment, lack of recognition, blocked autonomy	Resentment ideology, status obsession, nationalism, conspiracy thinking

Need / Capability	Deprivation / Lack	Emergent worldview / disposition
Self-Actualization / Senses, Imagination, Reason, Play	Suppressed creativity, lack of purpose, constrained expression	Anti-intellectualism, nihilism, despair, disengagement
Cognitive / Senses Imagination, Thought, Reason	Lack of education, restricted information, epistemic control	Dogmatism, conspiracy thinking, distrust of institutions
Aesthetic / Senses Imagination, Play, Other Species	Sensory monotony, degraded environments	Apathy, desensitization, aesthetic withdrawal
Transcendence / Emotions, Other Species	Loss of meaning, ecological alienation	Existential despair, anthropocentrism, moral disengagement

T15.4 Emergent worldviews due to deprivation

While not empirically unified, there is support for the idea that psychological deprivation and capability constraints shape ideological orientation and collective behaviour.

- Studies in self-determination show that basic psychological needs are intertwined with well-being and moral agency, and that their frustration can lead to compensatory belief systems, including conspiracy ideation and ideological extremism.[36]

- Empirical work on ideological obsession demonstrates how unmet existential and relational needs can drive moral disengagement and radicalization.[37]

These findings do not confirm the framework, but they provide backup to its heuristic value. The following sections draw on historical case studies and contemporary data, examining how unmet needs and constrained capabilities correlate with collective psychological and political shifts.

Historic and contemporary examples

T15.5 links areas of deprivation to historical examples and then explores the kinds of worldviews or ideologies that may emerge in response.

Historical Example	Contemporary Example	Emergent Worldviews
Deprived: Physiological / Life, Bodily Health, Integrity		
Bengal, (1943), Irish famines (1845–1852)	Rohingya displacement, Yemen crisis	Survivalism, short-termism, scarcity mindset
Deprived: Safety / Bodily Integrity, Life, Control over environment		
Stalin Purge (1936–1938), Khmer Rouge (1975–1979)	Syrian civil war / refugees, Drug violence Mexico	Authoritarianism, militarism, distrust of institutions
Deprived: Belonging / Affiliation, Emotions, Play		
Partition of India (1947)	Online Islamic radicalization	Tribalism, sectarianism,

Historical Example	Contemporary Example	Emergent Worldviews
Deprived: Esteem / Control, Practical Reason, Affiliation		
Post-WWI Treaty of Versailles	Youth unemployment in North Africa	Resentment, nationalism
Deprived: Self-Actualization / Senses, Imagination, Reason, Play		
Mao's Cultural Revolution (1966–76)	Educational repression in Afghanistan	Anti-intellectualism, nihilism, despair, disengagement
Deprived: Cognitive / Senses Imagination, Practical Reason		
Galileo's persecution	Internet censorship China	Dogmatism, conspiracy thinking
Deprived: Aesthetic / Senses Imagination, Play, Other Species		
U.S. slaves denied culture and leisure	Urban slums - no green public space	Desensitization, aesthetic withdrawal
Deprived: Transcendence / Emotions, Other Species		
Industrial revolution, 18/19C	Climate grief and ecological despair	Existential despair, disengagement

T15.5 Possible emergent worldviews – historic and contemporary examples

T15.6 presents a snapshot of global human deprivation. Each row reflects 2025 data on deficits ranging from hunger and safety to imagination and transcendence captured through key indicators sourced from international bodies.

Need / Capability	Key indicator	2025 global data
Physiological / Life, Bodily Health, Integrity	People facing hunger	673 million (8.3% of global population)
	Moderate or severe food insecurity	2.3 billion people
Safety / Bodily Integrity, Life, Control over environment	Countries with highest crime index (0 to 100). Global average 45.1	Venezuela (81.2), Papua New Guinea (79.7), Afghanistan (78.3)
	Global decline in safety	Eighth consecutive year of decline
Belonging / Affiliation, Emotions, Play	People affected by loneliness	1 in 6 globally (≈1.3 billion)
	Displaced persons lacking civic affiliation	117 million forcibly displaced (UNHCR, 2025)
Esteem / Control, Practical Reason, Affiliation	People with access to affordable healthcare (UHC)	500 million gained coverage (only half of WHO's target)
	Adults with low digital literacy	2.1 billion globally (UNESCO, 2025)
Self-Actualization / Senses, Imagination, Reason, Play	Crisis-affected children needing education support	234 million across 60 countries
	Out-of-school youth (upper secondary level)	138 million globally

Need / Capability	Key indicator	2025 global data
Cognitive / Senses Imagination, Thought, Practical Reason	Decline in global average IQ	Down to 87.8 (from 89.2 in 2000)
	Youth struggling with basic cognitive tasks	34% of global teens show deficits in focus and reasoning
Aesthetic / Senses Imagination, Thought, Play, Other Species	Urban populations lacking access to green/public space	1.1 billion people (UN-Habitat, 2025)
	Countries with lowest aesthetic infrastructure index	Haiti, Chad, Bangladesh
Transcendence / Emotions, Other Species	People reporting existential dissatisfaction	60% globally feel life is worse than 50 years ago
	Youth engaged in climate-related activism	1 in 4 globally (Fridays for Future, 2025)

T15.6 Global status of human needs / capabilities. Data sources include UN, WHO, OECD, UNESCO, and recent global surveys (2023–2025)

The data shows that despite global progress, over 2.3 billion people still face food insecurity, and 2.6 billion cannot afford a healthy diet. 234 million children in crisis contexts need educational support, and 138 million youth are out of school at the upper secondary level.

These figures represent not just lost potential but suppressed imagination. In regions like Afghanistan and Sudan, creative expression is repressed. In the "West" Graeber described bullshit jobs are pervasive as are beliefs in religious, commercial and nationalistic worldviews that cause individual and collective suffering.

A humanist worldview

<u>Grounding worldviews in humanist principles</u>

The humanist approach treats worldviews and life purposes as choices, informed by science, but not dictated by it. This interplay underpins the humanist worldview.

This compact does not collapse the Is-Ought distinction but recognizes that science clarifies assumptions behind moral decisions, while ethics helps determine what science should explore. Humanists revise beliefs and policies when evidence shows misalignment with reality, expanding moral consideration to include broader stakeholders and future generations. The bidirectional arrows in F15.2 represent this interplay, not a conflation. They illustrate a feedback loop where moral reasoning guides where to inquire, and empirical findings refine moral understanding.

Humanist morality is grounded not in fear, but in reason and well-being. It rejects dogma and absolutes, aligning ethics with scientific insights into our evolutionary roots, emotions, and creative drives. As primates, we evolved empathy, curiosity, and symbolic thought, traits humanism draws on to foster inquiry, expression, and ethical responsibility.

| Morality informed by assumptions that science clarifies | Moral decisions (Embracing and invigorated by emotions and imagination) | Morality helps identify which aspects to investigate |

| The findings of science (including evolution, cultural, and environmental phenomena) | Science as a test of knowledge (awareness of falsifiable, replicable feedback loop empirical processes) |

F15.2 The close compact between moral decisions and science

It values meaningful engagement over survival or pleasure, embracing pluralism, dignity, truth, and ecological integrity. Meaning is created, not given. Humanism evolves with evidence, adapts to context, and resists harm. Research like Vohs' on priming shows how cues shape behaviour, challenging notions of choice and conditioning. Platforms like Facebook, despite promises of connection, often fuel polarization and addiction. Humanism responds by revising behaviour and policy in light of evidence, expanding moral concern to long-term and collective impact.

The contours of a modern humanist worldview

A humanist worldview, centred on meeting needs and expanding capabilities, may appear compatible with other traditions. Yet beneath shared terms like dignity and flourishing lies a divergence. Humanism is neither dogma nor sentiment; it is a framework grounded in agency, tested by science, and committed to cultivating cognitive, emotional, aesthetic, and existential capacities.

Unlike theistic, nationalistic, ancestral, or economic worldviews, it does not subordinate the individual to divine command, lineage, market logic, or ideological purity. It treats imagination, play, and transcendence not as luxuries or instruments of power, but as essential to being human.

Humanism challenges systems that define worth through obedience or conformity, affirming each person's right to develop their capacities fully, not as a means, but as an end. It offers no resolution, only a direction: a way to navigate complexity without absolutism. Its strength lies in openness to revision and resistance to sanctified authority. In this way, it subverts dominant systems by re-centring the individual in awareness and responsibility for a better world.

Nor is it transcendental in the sense critiqued by Amartya Sen. It values partial resolutions and incremental progress, recognizing the mutual shaping of individuals and institutions in the pursuit of flourishing.

Across the final four domains of needs and capabilities, humanism diverges sharply from other worldviews reflecting fundamentally different assumptions (T15.7).

Needs, Purposes and a Humanist Worldview

Self-Actualization/ Senses, Imagination, Thought	Cognitive/ Senses, Imagination, Thought, Play	Aesthetic/ Senses, Imagination, Thought	Transcendence/ Emotions, Other Species
Modern Humanism			
Valued intrinsically fosters autonomy, creativity	Open-ended inquiry, pluralistic play, exploration	Celebrated as personal and cultural expression	Includes ecological empathy and emotional depth
Religious-Theistic			
Morally constrained; must align with divine purpose	Limited by theological orthodoxy; dissent often punished	Filtered through religions heretics punished	Divine salvation; Subordinated to moral purity
Mythic-Nationalistic			
Instrumentalized to reinforce heroic narratives and national identity	Subordinated to mythic destiny; dissenting play discouraged	Serves patriotic art and symbolic rituals	Martyrdom or sacrifice for nation; ecological empathy suppressed
Marxist–Leninist			
Suspicious unless serving revolutionary consciousness	Politicized; framed as dialectical materialism	Must expose exploitation	Illusion; redirected toward historical progress

Self-Actualization/ Senses, Imagination, Thought	Cognitive/ Senses, Imagination, Thought, Play	Aesthetic/ Senses, Imagination, Thought	Transcendence/ Emotions, Other Species
Market Fundamentalist			
Consumer preference and brands	Valued for innovation optimization	commodified art becomes product	Privatized and monetized; ecology sidelined

T15.7 Comparison of worldviews

While humanism seeks to expand all capabilities, alternative frameworks often elevate some while suppressing others, resulting in competing visions of flourishing, each with its own moral trade-offs.

Despite centuries of diverse worldviews, global deprivation persists across foundational needs and capabilities. Many frameworks accept this passively, while humanism places unmet human essentials, and planetary flourishing, at the heart of its ethical vision.

16

Activities in the Sciences, Arts, and Humanism

Cooperation and monopoly seeking competition

The big-tech view of competition

Are our worldviews driven by cooperation or the pursuit of monopoly-seeking competition? Peter Thiel, a confidant of Zuckerberg, believes that monopolies in low-regulation environments benefit everyone.[1]

This thinking draws on the model of 'dynamic monopolistic competition,' which focuses on markets where firms gain advantages through technology, branding, or local dominance. Joseph Schumpeter built on this with his idea of 'creative destruction,' where innovation replaces outdated systems, driving growth.[2,3] Monopolies, in this view, spark cycles of innovation and renewal.

These ideas have merged with models of 'disembodied technological change,'[4] productivity gains from ideas and methods (like AI), not physical machinery. But the risks are profound: such models obscure human, ecological, and ethical dimensions. They amplify inequality and prioritize efficiency over sustainability and human flourishing.

Adam Smith, empathy and the wealth of nations

Many proponents of unfettered markets misread Adam Smith, casting him as a prophet of laissez-faire capitalism. Yet Smith was foremost a philosopher. In the 'Theory of Moral Sentiments'[5] he emphasized:

- Sympathy (his term for empathy)
- Imagining how an impartial observer would judge actions (a precursor to Rawls' "veil of ignorance")
- Cooperation, social stability, and balancing self-interest with concern for others
- Justice and fairness

This moral foundation shaped 'The Wealth of Nations,'[6] where Smith argued:

- Specialization boosts productivity
- Markets self-regulate through non-monopolistic competition
- Government must intervene against exploitation and collusion
- While opposing excessive control, Smith supported public goods, fair taxation, and protections against monopolies

Thus, Smith's reasoning affirms that empathy and cooperation are compatible with genuine competition and practical governance. However, modern humanism breaks from both Thiel's distorted monopolistic model and Smith's market-centred vision. Drawing instead on Smith's moral sentiments to show that cooperation and social democracy effectively promote human flourishing.

Axelrod's experiments

Can empathetic societies thrive amid aggressive, monopolistic competition? In the 1980s, Robert Axelrod explored this through computer simulations based on the "Iterated Prisoner's Dilemma," a game where players repeatedly choose to cooperate or betray, learning from past interactions (AXELROD)[7].

In the classic dilemma, betrayal seems rational for self-interest, yet mutual cooperation yields better outcomes. Axelrod invited experts across disciplines to submit programs that played repeated rounds. The most successful strategy was "Tit for Tat": begin with cooperation, then mirror the opponent's previous move. Later versions introduced random errors to reflect real-world imperfections, revealing that cooperative strategies still thrived over long timeframes.[8] Two conditions proved essential:

- A small yet critical number of cooperative players help them survive and flourish. Below this number Tit for Tat collapses.
- Repeated interactions. Trust takes time. Without enough rounds, cooperation fails to take root.

Axelrod's key insight: even in competitive environments, simple cooperation strategies can succeed. In the real world, collaboration doesn't erase individuality or competition, it amplifies flourishing. Cooperation is not conformity but fair play. Each person brings imagination and drive; differences become tools for dialogue, dissent, and innovation.

Co-opetition

Brandenburger and Nalebuff have introduced the concept of 'co-opetition' – the bringing together of cooperation and competition.[9] Rather than viewing our interactions as a zero-sum war, co-opetition encourages strategic empathy and objectives based on shared benefits. It encourages fostering innovation through fair play competition creating win-win outcomes.

Berners-Lee exemplary work mixed cooperation and competition for everyone's benefit. Most of us pursue co-opetition much more often than dominance-based competition.

Axelrod was active in interdisciplinary collaborations and efforts at diplomacy in some of the most difficult environments (AXELROD)[10]. He wrote: "Opportunities for cooperation are much greater than generally thought, and understanding the conditions that foster it can help us navigate conflicts and build lasting partnerships."[11]

<u>Designs for coercion vs. designs for trust</u>

From monopolistic competition to punitive justice systems, coercive designs prioritize control and exclusion. Trust-based systems embrace solidarity, dialogue, and dignity. Our worldviews shape whether we design for domination or trust, a choice that shapes how we respond to conflict, dissent, and difference.

In technology, Tim Berners-Lee's decision to make the World Wide Web freely available was a landmark act of cooperative humanism. "You can't propose that something be a universal space and at the same time keep control of it," he wrote (BERNERS-LEE)[12]. His design trusted users to shape the web collectively.

By contrast, Facebook's evolution from social connector to manipulative platform exemplifies coercive design. Internal documents show content is amplified to boost engagement. Zuckerberg's goal to "rewire how people connect and what they see" reveals a chilling logic of engineered compulsion.[13] Berners-Lee's open web invited participation; Facebook's silo demands engagement. One builds trust, the other compels behaviour.

Isaac Asimov, longtime president of the American Humanist Association, wrote: "It is only when people feel free to think for themselves, using reason as their guide, that they are best capable of developing values that succeed in satisfying human needs" (Asimov)[14]. Coercive systems undermine freedom, replacing reflection with manipulation.

The same divide appears in justice systems. Restorative frameworks prioritize dialogue; punitive models enforce control. The U.S. incarceration system is dehumanizing. Norway's Halden prison, by contrast, emphasizes dignity, education, and reintegration.

Modern humanism is a moral design ethic. It chooses autonomy over compulsion, trust over manipulation. As Edward Said wrote, "Humanism is the only—I would go so far as saying the final—resistance we have against the inhuman practices and injustices that disfigure human history."[15]

Science and engineering

Berners-Lee and Zuckerberg

Berners-Lee invented the Web at CERN, a post-WWII institution devoted to scientific discovery and international collaboration. Born in 1955, he grew up in a society shaped by civic duty and shared responsibility. CERN publishes discoveries openly, using science as a method of inquiry grounded in falsifiability and replicable feedback.

Zuckerberg, born in 1984, was shaped by Silicon Valley's culture of individualism, deregulation, and venture capital. While its engineers have built the infrastructure of modern life, the model prioritizes profit, patents, and monopolies, often at the expense of privacy and collective welfare.

Moral purposes behind science and engineering

Science seeks truth through testing and admits no authority. Its moral purpose is discovery. Engineering applies scientific findings to build systems and products. It is amoral, driven by practical objectives, not truth.

Apple's ultra-slim iPhone 17 Air, for example, relies on scientific knowledge: electromagnetism, semiconductors, programming, but its engineering is shaped by commercial goals. Engineers work to meet specifications set by corporate authorities, not moral inquiry.

Technology, the product of engineering, demonstrates an additional aspect. Every engineered product (technology) has a built-in obsolescence, a design life shortened to suit market conditions, drive upgrades and revenue, not human needs. Thus, there is often an in-built failure and finality to the current technological 'solution.'[16]

Engineering has also served destructive ends. The Manhattan Project, hailed as a triumph of science, was an engineering feat that unleashed atomic devastation. Oppenheimer called the effort "technological enthusiasm."[17] Today, engineers build autonomous drones, hypersonic missiles, and cyberwarfare tools with similar fervour and without reflection.

Fritz Haber, who pioneered chemical warfare in WWI, believed science could serve national victory, even at the cost of mass suffering.[a] In contrast, scientists like Faraday and Rutherford resisted weaponization. Faraday refused to develop

---

a   A paradox of destruction and nourishment – he also synthesizes ammonia from nitrogen and hydrogen gases enabling the mass production of nitrogen-based fertilizers.

chemical weapons; Rutherford focused on submarine detection, not offensive arms.[18,19]

Understanding the meaning of technology

Technology is the physical and software instruments used and created by science (e.g. telescopes) and engineers (e.g. lithium-ion batteries). As "objects" technologies have no attribute related to morality, although they have been used for dual purposes – nuclear bombs and nuclear power.

Engineering as a creative activity

Engineering can also be a creative force that shapes and is shaped by culture. John Lienhard writes, "We are the machines we create," noting how technologies like farming, the printing press, and the Internet have transformed human thought and civilization.[20] Engineering, like science, is a manifestation of ingenuity.

Samuel Florman says that engineering is not just technical but an intensely creative pursuit that is driven by the human impulses to shape, build, and transform.[21] In this way engineering becomes a historical actor and a creator of meaning.

Engineering for humanist purposes

When engineers serve only governments, corporations, or militaries, they risk becoming instruments of domination. Lewis Mumford urges a reorientation toward humanist ideals: technologies that serve communities.[22] This was Berners-Lee's ethic in creating the World Wide Web.

Engineering has brought benefits: vaccines, electricity, medicine. But when reduced to utilitarian goals, it ignores ethical consequences. Leaded gasoline, nuclear weapons, and Facebook's engagement algorithms cause harm while chasing metrics.

Big tech algorithms create closed systems, shaping thought based on ad revenue. In 2024, major tech firms spent over USD 61.5 million lobbying the U.S. government to not legislate against their extractive actions. Meta spent USD 240,000 per day on lobbying when the U.S. Congress was in session and a total of USD 24.4 million in 2024.[23,24]

A humanist lens on civilizations and death

Humanism confronts more profound questions. Many engineers and religious ideologues seek permanence, one through technology, the other through myth.[25] Humanism embraces impermanence as clarifying truth, not defeat.

Epicurus wrote, "Death does not concern us… when it does come, we no longer exist."[26] He urged a life of understanding. Happiness, he said, comes from reasoned contemplation of nature and ethics.

As of mid-2025 we have accumulated approximately 4 million cubic meters of radioactive waste and over 104,000 metric tons of spent nuclear fuel[27]. Engineering looks to store this growing nuclear waste safely for 10,000 to over 1 million years. Humanism realizes that no empire or civilisation has lasted more than a few centuries. As Garrett Hardin cautioned, "We cannot safely say that civilization will persist long enough to guard its own radioactive legacy."[28]

Humanism, the creative arts, and science

Conjoined triplets

Jacob Bronowski argued that the conduct of science is a pursuit of creative acts. He compared this to artistic creation, emphasizing that both spring from and require imagination (BRONOWSKI)[29].

But Bronowski also explains that "the creative act is alike in art and in science; but it cannot be identical in the two; there must be a difference as well as a likeness. The practices of science and the creative arts are both fuelled by imagination, but science is self-correcting while the creative arts are unfettered by this constraint."[30]

The 'Creative Arts' is the expression of imagination through which ideas and stories of the universe are explored with an eye towards aesthetics. Einstein placed particular importance on creative imagination saying, "Imagination is more important than knowledge. For knowledge is limited, whereas imagination embraces the entire world, stimulating progress, giving birth to evolution."[31]

Humanism takes these common elements and binds the creative arts and science into a conjoined whole (see F1.1). It tempers the animating features of the arts using science, it synthesizes and makes moral commitments that result in environments within which both the creative arts and science can flourish. E. O. Wilson called this 'Consilience,'[32] the unity of otherwise isolated branches of knowledge and endeavours: science, the creative arts and humanist living.

Consequences of breaking the unity

When this synthesis is broken, neither of these three human activities can exist for long.[33] Examples include Lysenko under Stalin, Mao's anti-intellectual campaigns during the cultural revolution, and Joseph McCarthy's 1950s campaigns against the purported 'un-American' Oppenheimer, Leonard Bernstein, Arthur Miller and others. In the U.S. (2025) budget cuts have been made to science funding, a 50% cut to NASA's science division, USD 1.4 billion in National Science Funding grant cancellations, scientific integrity policies are controlled by political appointees, and funding threats to universities are pervasive.

The most egregious of these actions is the removal of climate change data from multiple federal agency websites.[34] This is akin to the burning of books as a tool of ideological 'purification.'

On May 10, 1933, Nazi-aligned university students across Germany incinerated over 25,000 books deemed 'un-German.' Works by Jewish authors including Albert Einstein were publicly torched while crowds saluted and Joseph Goebbels declared the end of "extreme Jewish intellectualism". Similar purges occurred when the Qin Dynasty destroyed Confucian texts, the Francoist regime suppressed Catalan literature, the jailing of writers in China (118 writers were in jail in 2024),[35] and in antiquity the destruction of the library of Alexandria.

The above are not mere historical events, they're warnings. When the bond between science, the arts, and humanism breaks, the fallout is severe. Knowledge is silenced, creativity crushed, and ethics bent to ideology. Under authoritarian, religious, and populist purges, science is politicized, art censored, and humanist values erased. What's lost is the foundation for human flourishing. Without this triad, science, and art lose meaning and cease to exist.

In today's fractured world, Jacob Bronowski may sound idealistic, yet his message carries a core humanist ethic:

> "Massacre is prevented by the scientist's ethic, and the poet's, and every creator's: that the end for which we work [our purpose] exists and is judged only by the means which we use to reach it. ... It is the basis of a society which scrupulously seeks knowledge to match and govern its power."[36]
> BRONOWSKI

Literature and philosophy: emotion and reflection

Great literature is an art that distils aspects of the human condition. Literature and philosophy give form to emotional truths, ethical dilemmas, and existential questions that resist easy explanation.

Philosophy, through argument and analysis, and literature, through narrative, metaphor, and voice, together help us navigate moral complexity. They reveal aspects of ourselves that remain hidden when expressed in scientific terms.

Yet these emotional and philosophical insights must be tempered as they are not immune to error. Emotions can mislead, arguments fallacious, and narratives harmful. Therefore, literature and philosophy must be held accountable to the scientific learning process: not in their form, but in their claims. This does not diminish literature's value, it deepens it.

Some years ago, I corralled a few friends into reading and discussing four impactful books: Adam Hochschild's "King Leopold's Ghost: A Story of Greed, Terror, and Heroism in Colonial Africa," Joseph Conrad's "Heart of Darkness," Chinua Achebe's "Things Fall Apart," and Sven Lindqvist's "Terra Nullius: A Journey Through No One's Land."[37,38,39,40]

Adam Hochschild's account of the Congo Free State under King Leopold II exposes the scale of colonial violence and Europe's moral blindness. His claims are supported by historical records: forced labour, mutilation quotas, and population collapse, all documented in missionary reports, diplomatic correspondence, and the 1904–1905 Commission of Inquiry. These affirm the core facts, including the use of severed hands as proof of military efficiency.

Joseph Conrad's novella explores the psychological breakdown of colonial agents, particularly through Kurtz's descent into madness. This literary portrayal aligns with reports of colonial environments that often induce moral disengagement and cognitive dissonance (see Harshini and Praveenkumar[41]).

Chinua Achebe's novel poses two key questions: How do colonial systems dismantle indigenous cultures? and What are the psychological consequences? Tewari shows that the Igbo society is historically accurate in its decentralized governance and social rituals.[42] The arrival of British missionaries and administrators, documented in colonial records, led to legal erasure, religious displacement, and generational rifts.

Lindqvist exposed the doctrine of 'terra nullius,' which declared Australia "empty" despite Indigenous habitation. This legal fiction was used to justify British sovereignty without treaty or consent. Historical records show that Indigenous Australians had complex systems of law, land stewardship, and oral traditions that were completely wiped out (see Nursoo[43]).

These brief paragraphs hardly do justice to these insightful and beautifully crafted works, but the objective is to show how each book is seen to be not just a literary work but a gateway to empirical inquiry. In this way, literature, philosophy, and science each contribute to human flourishing. Together, they offer a fuller account of what it means to live wisely, feel intensely, and think clearly.

The meaning of aesthetics

Aesthetics is the study and appreciation of beauty and artistic experience. It offers order, proportion, and emotional depth and permeates the arts, sciences, and ethics.

Aesthetics is a shared sensibility, a vital element in crafting theories and shaping moral structures, from artistic expression to acts of service. Aesthetic experience brings meaning, coherence, and emotional resonance, allowing each to speak a common language of depth and refinement.

Beauty in the sciences

Scientists often describe theories as elegant or harmonious, aesthetic judgments that guide inquiry. Symmetry, simplicity, and unity are prized not just for mathematical economy but for their beauty.

Physicist Frank Wilczek argues that the universe embodies "beautiful ideas," writing, "The world is a work of art."[44] Symmetry governs physical laws, particle interactions, and space-time. He likens atoms to instruments in a cosmic symphony, and supersymmetry to a poetic yin-yang of the quantum world. For Wilczek, beauty is a clue to reality.

Beauty as invention

Mathematics, once seen as discovered, is now viewed as invented, shaped by cognition and culture. It feels discovered because it's embedded in how we think (C31).

Anjan Chatterjee, a neuroscientist, and Ellen Dissanayake, an anthropologist, argue that beauty is shaped by biology, culture, and learning. Though not universal in content, it is universal in function. We invent aesthetic systems: musical scales, poetic forms, architectural styles, and refine them through generations.

Chatterjee writes, "Aesthetic experiences are shaped by evolutionary pressures, cultural contexts, and personal histories."[45] Preferences vary, but the neural architecture is shared.

Ellen Dissanayake proposes that aesthetics is a way humans have evolved to make things 'special' through ritual, art, and symbolic behaviour. She writes, "Aesthetic experience is not a luxury but a biological necessity, shaped by culture and refined through practice."[46]

The neuroscience of learned beauty

Neuroaesthetics explores how the brain processes beauty. Research by Anjan Chatterjee and colleagues shows aesthetic appreciation activates:[47]
- Sensory cortices for form and pattern
- Prefrontal areas for reasoning and judgment

- Reward centres like the orbitofrontal cortex and nucleus accumbens

Zaira Cattaneo's work suggests the brain prefers moderate complexity, neither too simple nor chaotic, aligning with art and science, which balance order and surprise.[48]

Cattaneo finds that long immersion in creative fields enhances neural responses to complexity. This implies:[49]

- Aesthetic appreciation is learned. Beauty in any discipline requires attunement.
- Mastery in one art form elevates appreciation in others. A physicist may see elegance in a painting. Aesthetic principles: symmetry, contrast, modulation, form a language across disciplines.

Aesthetics and Humanism

Aesthetics is a necessity. It helps us make sense of the world, express what matters, and connect across difference. Beauty invites contemplation and wonder. It celebrates our capacity to feel and imagine. Just as deprivation of basic needs shapes worldviews, so too does the absence of beauty. Without engaging senses and imagination, human flourishing is diminished. But when nurtured, aesthetics becomes a pathway to resonance, reflection, and meaning, enabling a humanist worldview grounded in empathy, dignity, and creative participation in life.

Humanist flourishing

Heritage of flourishing

Aristotle provided the most well-known explanation of flourishing (eudaimonia) as a lifelong process of living through

virtuous activity and rational reflection, guided by reason. Aristotle defined virtuous activity as the habitual practice of:
- Moral virtues (courage, temperance, generosity) by finding the Golden Mean—the balanced point between excess and deficiency. He assumed that such a balance point was obvious.
- Intellectual virtues (wisdom, understanding, prudence) developed through teachings and reflection.

Rational reflection involved the deliberate use of reason to guide ethical judgment, cultivate practical wisdom and align one's actions with a life of purposeful flourishing.

As discussed in the C1 S: Formative epochs of thought, there were other thoughtful, self-reflective conceptions, that resonate with humanist ideas and that are from other cultures and times. Each tradition offers a lens on what it means to live well, whether through rational inquiry, ethical cultivation, communal responsibility, or creative exploration.

It must be noted that humanist flourishing becomes truly comprehensive when a broader set of considerations for all life on Earth is taken into account. This leads to planetary flourishing, as discussed in C20.

Expressions of Flourishing

Modern humanism envisions flourishing as a continuation of earlier generative movements marked by expansive creativity, conceptual innovation, and a departure from inherited constraints. Yet it is also a radical revision, reimagining foundations and extending the humanist ethic in new directions. It draws upon science's practice of self-correcting learning, embraces the evolving findings of science, and integrates insights into emotion, imagination, and the creative arts.

This conception of flourishing centres on the active engagement of our cognitive, emotional, ethical, and creative capacities — not as ideals, but as lived practices. It is through such practices that modern humanism takes shape, guiding purposeful action and inviting reflection on which activities truly advance empathy, understanding, and collective well-being.

Pre-requisites and pleasures

Basic needs and capabilities must be satisfied before flourishing can be addressed. The first two foundational needs and capabilities: Physiological (Life) and Safety (Bodily Integrity) need to be met.

Once basic needs and capabilities are secured, humanism recognizes our evolutionary inheritance and drivers. It affirms the legitimacy of bodily pleasures such as good food, wine, and sex—pleasures that, especially in the case of sex, are not merely enjoyable but biologically essential and emotionally necessary when pursued consensually and without hurt or harm to others. These are not sinful indulgences but inherent evolutionary drivers, natural expressions, and celebrations of our human condition. Lucretius, reflecting Epicurus, wrote, "Pleasure is ours by nature's gift, and it is sweet to us; and we shrink from pain as something hostile to our senses."[50]

Such pleasures, however, do not constitute flourishing in themselves; they are foundational, not final ends. Epicurus argued precisely this, "The pleasant life is not produced by continual drinking and dancing, nor sexual intercourse, nor rare dishes of seafood and other delicacies of a luxurious table. On the contrary, it is produced by sober reasoning which examines the motives for every choice and avoidance..."[51] Pleasures, when moderated and ethically grounded, are part of a flourishing life but not sufficient.

<u>Relational and cognitive flourishing</u>

The third and fourth foundational categories as outlined in Chapter 15: Belonging (Affiliation) and Esteem (Control, Practical Reason), function not only as facilitators of flourishing but as integral processes through which flourishing proceeds.

These dimensions encompass love (as expressed by the classical Greeks), friendship, mutual recognition, and the capacity for reflection and action. Here philia (friendship love) is not only a facilitator of flourishing but a moral and cognitive achievement in itself, a relational structure through which flourishing is sustained and deepened.

Crucially, the above needs and capabilities are animated by kindness and compassion, affective dispositions that reinforce ethical engagement and relational depth. Far from being peripheral sentiments, kindness and compassion serve as connective tissue between affiliation and agency, enabling individuals to act with care, perceive others with empathy, and sustain cooperative, flourishing-oriented environments. This is exemplified in the Greek concept of agape, selfless, unconditional love rooted in moral commitment, which is activated through acts of care, empathy, and regard for others. It underpins the humanist ethic of treating others with dignity, regardless of difference or circumstance.

These enablers are not merely antecedents to flourishing but pathways: they shape and sustain the very possibility of flourishing by embedding ethical sensibility within relational and cognitive development.

Flourishing activities

Flourishing is realized through activities that enact it. These include:

- Scientific inquiry, which expands our understanding of the universe through tests of knowledge using empirical rigor and plausible reasoning.
- Artistic expression, including literary, visual, musical, and performance arts, which express beauty with emotional resonance.
- Self-reflection, which integrates philosophical reasoning with ethical and historical awareness to guide moral imagination.
- Service activities such as education, caregiving, and civic participation, which enhance dignity, equity, and shared purpose.
- Political, commercial, administrative, and engineering activities can be considered humanist only when pursued as forms of service — not for power, profit, or domination, but to enable coordinated action, equitable resource sharing, and the shaping of humane, sustainable systems. Their humanist value depends on ecological responsibility, social sensitivity, and a commitment to dignity and collective flourishing.

The activities listed under human flourishing are not a fixed or exhaustive catalogue. They are grounded in human agency, not collective uniformity, and guided by humanist principles aimed at enabling flourishing. Like Martha Nussbaum's ten central capabilities, this open-ended framework identifies domains where dignity, creativity, and ethical reasoning are actively cultivated. Too often, humanists hesitate to name these activities, even though they are broad, inclusive, and demonstrably vital. Just as importantly, they often avoid calling out activities that are clearly extractive, exploitative, or corrosive to human dignity, and therefore incompatible with humanist flourishing. Flourishing is not an abstraction; it gains meaning only through concrete action.

Commerce and politics

Commerce, politics, and administration are service functions, tools for coordinated action, not for profit or power. In a humanist model of flourishing, they are guided by ethical purpose and collective need, not market dominance or control. This is not forced collectivism but a principled path to individual and shared flourishing.

Plato's philosopher king governed by grasping the 'Form of the Good.' Modern humanism rejects this hierarchy, favouring distributed agency, participatory ethics, and moral imagination across all individuals. Flourishing means empowered to think, act, and collaborate with dignity and purpose. This is a shift from rule by the few to action among the ethically engaged many.

Commercial activity and governance are duties of flourishing — providing resources, capabilities, and pleasures for the many. This need not constrain innovation or creativity. Keynes wrote: "The love of money [and power] as a possession — as distinguished from the love of money [and service] as a means to the enjoyments and realities of life — will be recognised for what it is, a somewhat disgusting morbidity, one of those semicriminal, semi-pathological propensities which one hands over with a shudder to the specialists in mental disease" (bracketed additions mine) (Keynes)[52].

When oriented toward service, commerce and politics become domains of innovation, collaboration, and openness, contributing to collective well-being.

<u>Liberating Labour: Toward Meaningful Engagement</u>

There is an inherent dignity in every human being. Yet we must be honest about the nature of certain forms of work: manual labour in manufacturing, repetitive service roles, and many corporate tasks, that, while often valorised as dignified, can be as stultifying and harmful as commercial or political activities pursued solely for profit or power.

Humanism suggests that such roles be replaced by AI and machines, and, crucially, those engaged in them be supported with free and meaningful opportunities to build capabilities and meet their needs. This does not strip humans of dignity. Rather, it recognizes that true dignity lies not in enduring numbing routines, but in being empowered to flourish. Naming this truth is not dismissal, it is compassion. It calls for thoughtful economic, social, and educational frameworks.

<u>Self-actualization, transcendence, flourishing</u>

This outline answers the question 'What exactly are Maslow's self-actualization and transcendence?' and makes explicit the assumptions behind Nussbaum's capabilities framework.

Flourishing becomes a journey towards a better world.

Professional scientists – practices, pressures, and challenges

Science is an activity of human flourishing, a pursuit of truth, creativity, and service. Yet, like commerce and politics, it is often co-opted for personal gain, power, and prestige. The following section examines the state of professional science today through Clair Patterson's critique of its narrowing practices and his call for reform.

As of mid-2025 there are an estimated 8.1 million full-time professional scientists worldwide (T16.1).

Region	Number (mn)	Percentage
Asia and Pacific	3.2	39.5%
Europe	2.5	30.9%
North America	1.8	22.2%
Latin America and Caribbean	0.5	6.2%
Africa	0.4	4.9%
Middle East	0.3	3.7%

T16.1 Professional Scientists Worldwide. Source: UNESCO Institute for Statistics[53]

E. O. Wilson talks of two types of scientists, "The first go into science in order to make a living. The second do the reverse: they find a way to make a living in order to go into science" (E. O. WILSON)[54]. Patterson believed that the practice of science was being held hostage to the first type of Wilson's scientists resulting in perverse funding requirements and prioritization of utility, speed, and quantifiability over depth, ambiguity, and wonder.[55]

This results in a loss of a shared emotional and intellectual bond among scientists. Academic research is increasingly shaped by quantity-driven publishing, relentless funding pressures, undervalued replication, and restricted access — all of which undermine long-term inquiry and public benefit.

There is also a rush to enter into 'Big Science.' E. O. Wilson points out there is "no guarantee that the exploration of Earth's biosphere can be completed before the twenty-third century. The problem is a severe shortage of expert researchers. The solution to the problem is more naturalists."[56] Such naturalists need a tiny fraction of the resources the technology hungry big scientists require and they do work that is as much, if not more, important.

Also, while notable exceptions like Sen, Piketty, Stiglitz, Acemoglu, Banerjee, and Duflo, have advanced work on equality, justice, and humanist ethics, many social scientists, economists in particular, shy away from these themes, often retreating into arcane mathematical models that obscure rather than illuminate human realities.

This section does not aim to paint all scientists with the same brush. Yet the broader research environment often discourages humanist behaviour. Scientists are part of our global cultures — their challenges are our challenges. The vision of modern humanism outlined in this book applies as much to them as to the wider public.

N8 Last man on Earth – still dominating

"The trouble with the rat race is that even if you win, you're still a rat."[57] Lily Tomlin

Part 7

CONSEQUENCES of how the world is and ought to be

Planetary Flourishing

<u>Humanist approach to poverty, conservation and historic responsibilities</u>

The Shallow Pond

Collapse of the Aral Sea

The UN Durban Conference 2001

"Overcoming poverty is not a gesture of charity. It is an act of justice. It is the protection of a fundamental human right, the right to dignity and a decent life."[1]

Nelson Mandela

"The world is not given by his fathers, but borrowed from his children."[2]

Wendell Berry

"There are wrongs which even the grave does not bury."[3]

Harriet Jacobs

17

The Shallow Pond

In 1972, during the Bangladesh famine, Peter Singer wrote "Famine, Affluence, and Morality" (Singer)[1]. The article used the following 'Shallow Pond' allegory: "to reduce extreme poverty, not to make you guilty."[2] Its applications are as relevant now as in 1972.

"Imagine you're walking across a park. Somewhere in that park there's a pond. You know the pond is quite shallow, but you see something splashing in the pond. When you look closer, you're shocked to find that it's a small child who seems to have fallen into the pond and is flailing around because it's too deep for this small child to stand. So, you look around for the parents or the babysitter, but there's nobody. There seems to be only you and the child. Your next thought is, I better run down to the pond, jump into the pond, and grab the child. Not hard to do. No risk to me because the pond is shallow.

But then it does occur to you that [saving the child] is going to ruin your most expensive shoes. You'll be up for some hundreds of dollars to replace them and other clothes you might ruin. So, you think, why shouldn't I just walk away and not have to go to the expense of replacing my shoes? Now the question for everybody is: If somebody did that, would you think that was really the wrong thing to do? Would you think that you had done something seriously wrong in leaving the child very probably to drown? Most of the people who I ask this of say that would be an awful thing to do — it would be terrible to allow a child to drown because you didn't want to go to the expense of buying new shoes, even if they were expensive ones.

The point of the thought experiment is to then switch to the situation that we really are in. We live in an affluent society where we often have considerably more than we need to meet all our basic needs, enjoy life, and make reasonable provision for the future. We also are living in a world in which there are millions of children who die each year from preventable causes and there are effective organizations that would gladly accept a donation from you that would increase their ability to save some of these children. So, if you're not helping to save some of these children, then are you really all that different from the person who walks past the child in the pond?"[3]

PETER SINGER

Extract from *The Life You Can Save: How to Do Your Part to End World Poverty* pp 13 – 15 by Peter Singer, copyright © 2009. Reprinted by kind permission of Peter Singer.

The website at https://www.thelifeyoucansave.org/ provides details of charities through which you can make effective donations in the fight against extreme poverty.

Singer concludes by saying, "If it is in our power to prevent something bad from happening, without thereby sacrificing anything of comparable moral importance, we ought, morally, to do it."[4]

Singer gives a significant portion of his income to charity and lives modestly. In 2009 he founded a nonprofit organization, The Life You Can Save, that recommends vetted NGOs that help alleviate poverty.[5]

While the UN recognizes multidimensional poverty, which includes deprivation in education, health, and housing, the World Bank defines extreme poverty as living on less than USD 3.00 per day (2021 PPP). Estimates are that 9.9% of the global population (~792 mn) live in extreme poverty (mid 2025) (T17.1).

Region	Extreme poverty (population)	Comments
Sub-Saharan Africa	~560 mn	Highest global concentration
South Asia	~150 mn	Significant decline in India
East Asia & Pacific	~30 mn	Major progress in China and Indonesia
Latin America & Caribbean	~20 mn	Mixed trends: Honduras, Guatemala
Middle East & North Africa	~18 mn	Conflict zones (Yemen, Syria) drive higher rates
Central Europe & Central Asia	~10 mn	Concentrated in rural pockets
High-Income	~4 mn	Marginal but persistent poverty in urban fringes

T17.1 Global extreme poverty (less than USD 3 per day)[6,7]

A study by Likhar and Patil shows that malnutrition during the first 1,000 days—from conception to age two—can permanently impair brain development.[8] Deficiencies in nutrients like iron, iodine, and folate disrupt neural growth, leading to delays in cognition, motor skills, and language, with lasting effects on academic and life outcomes.

18

Collapse of the Aral Sea

The Aral Sea in 1960

In 1960, the Aral Sea in Central Asia was the world's fourth-largest inland lake, spanning 68,000 km² — larger than Sri Lanka and Lithuania, slightly smaller than Ireland. Fed by the Amu Darya and Syr Darya rivers, it had a depth of 40 m, salinity of 10 g/L, and an annual inflow of 56 km³. The region supported:[1]
- 638 plant species, including reed beds and riparian forests
- 24 native fish species, 310 bird species, and 70 mammals
- A thriving fishing industry employing 40,000 people, alongside agriculture

The Aral Sea in 2025

By 2025, the Aral Sea has shrunk to less than 10% of its original size. Over 40,000 km² of dried seabed has become the Aralkum Desert. Inflow from the rivers has dropped to just 3.28 km³/year. Each year, 45 million tons of salt- and pesticide-laden dust rise into the atmosphere.[2]
- Over half of plant species and 68 mammal species have vanished
- Native fish, amphibians, and reptiles are extinct
- The fishing industry collapsed by the late 1980s
- Salinity damaged 46% of Uzbekistan's irrigated land; crop yields fell by two-thirds

Dust storms now cause respiratory illness, birth defects, and high child mortality. In Karakalpakstan, child mortality rates range from 40 to 55 per 1,000 births (compared to 1.9 in Japan). In 2023, 99% of pregnant women suffered from anaemia, with no improvement projected. Multidrug-resistant tuberculosis spread widely between 2005–2020, with prevalence rates from 13% to 40%.

In 1989, the Aral split into North and South basins. The Kokaral Dam (2005) helped restore 3,065 km² of the North Aral Sea, reviving fisheries and reducing salinity. But the South Aral is desert or hypersaline (over 100 g/L). Regional warming is twice the global average.

Why the Aral Sea collapsed

Soviet planners diverted the Amu Darya and Syr Darya to irrigate cotton fields across Uzbekistan, Turkmenistan, and Kazakhstan. The Karakum Canal, built from 1954 to 1988, redirected water from the Amu Darya to transform the Karakum Desert. Poorly built canals leaked, wasting up to 50% of diverted water.

Annual inflow to the Aral Sea fell from 56 km³ in 1960 to under 10 km³ by the 1990s. By 2008, the eastern lobe had dried up. Salinity in some areas reached 130 g/L. Most of the former Aral Sea is now desert.[3]

Soviet planners justified this under Stalin's "Great Plan for the Transformation of Nature," imposing cotton monoculture across Central Asia. Cotton was a strategic export crop, and ecological concerns were ignored.[4] Nature was seen as subordinate to human will.

Despite signs of collapse, cotton acreage expanded into the 1980s. Between 1960 and 1985, irrigated land rose by 33% in Uzbekistan and 123% in Turkmenistan. Soviet engineers knew the risks but feared contradicting Politburo-approved plans. As Aleksandr Assarin said in 2002, "Nobody on a lower level would dare to say a word… even if it was the fate of the Aral Sea."[5]

Bioweapons and Vozrozhdeniye island

The disaster was worsened by a Soviet bioweapons facility on Vozrozhdeniye Island. Weapons tested included the viruses causing smallpox (Variola major), anthrax (Bacillus anthracis), plague (Yersinia pestis), and many others.[6]

- In 1971, a lab technician on a vessel near the island contracted smallpox, triggering an outbreak. Authorities quarantined a town, vaccinated 50,000 people, and incinerated 18 tons of goods.
- In 1988, 50,000 saiga antelope died within an hour near the island, likely due to bioweapons exposure.
- In the late 1980s, 200 tons of freeze-dried anthrax spores were buried on the island. Though neutralization efforts began in 2002, contamination remains.

With the Aral Sea dried up, Vozrozhdeniye is now connected to the mainland and accessible, posing ongoing risks.[7]

N9 Earth and Mars

"Only when the last tree has died and the last river been poisoned and the last fish been caught will we realize we cannot eat money."[8]

Cree Nation

19

The UN Durban Conference 2001

Background to the Durban conference

On 2 November 1973, the UN General Assembly adopted Resolution 3057, launching the "Decade for Action to Combat Racism and Racial Discrimination."[1] This was driven largely by African nations, through the Organization of African Unity (OAU), in response to apartheid and colonial legacies. The resolution reaffirmed the UN's commitment to eliminating racism, calling on all states to cooperate.[2]

Three major conferences followed:
- 1978, *Geneva*. Declared racial superiority "scientifically false, morally condemnable, socially unjust and dangerous," and called for sanctions against South Africa.[3]
- 1982, *Geneva*. Emphasized that eradicating racism required action at all levels—education, legislation, and cooperation.[4]
- 1993, *Vienna*. Coincided with the end of apartheid. It established the UN High Commissioner for Human Rights and supported a Special Rapporteur on Violence Against Women. U.S. Secretary of State Warren Christopher warned against cultural relativism as an excuse for human rights violations.[5]

On 12 September 1997, Mary Robinson, former President of Ireland, was appointed the first UN High Commissioner for Human Rights. The U.S. supported her nomination. Later that year, the UN adopted Resolution 52/111, calling for a global conference on racism. South Africa, having ended apartheid in 1994, was chosen as host.[6]

Preparatory conferences revealed divisions:[7]
- *Strasbourg* focused on European racism and migration, urging forward-looking dialogue without reparations.
- *Santiago* addressed indigenous rights and colonial legacies.
- *Dakar* called for reparations for slavery and recognition of historical injustices.
- *Tehran* excluded Jewish NGOs and Israeli delegates, centring on anti-Israel rhetoric.

The Durban conference

Held from 31 August to 8 September 2001, the Durban Conference was fraught with tension. Four major issues dominated:[8]
- *Slavery Reparations.* The OAU and Caribbean nations demanded reparations from Europe and the U.S., which rejected legal liability. The final declaration acknowledged slavery's harm but avoided compensation.
- *Israel and Palestine.* Arab states and Iran pushed to label Israel a racist state. The U.S. and Israel walked out on 3 September, with Secretary of State Colin Powell stating, "The conference has been hijacked by those who seek to use it to promote hatred."[9] The final text omitted references to Israel.
- *Indigenous Rights.* Latin American countries led efforts to include these, and they were recognized.
- *Islamophobia and Anti-Semitism.* Both were condemned in the final declaration, reflecting rare consensus between

Islamic nations and Jewish NGOs.

The declaration adopted recognized historical injustices: slavery, colonialism, apartheid, and called for education, legal reform, and international cooperation.[10]

The parallel NGO forum

Held from 28 August to 1 September, the NGO Forum drew 8,000 participants from 3,000 organizations. Its final declaration accused Israel of "apartheid and ethnic cleansing," and called for its isolation and sanctions.[11] Mary Robinson refused to endorse the document, saying, "I had urged the NGOs not to adopt it... I have a democratic right to reject that declaration."[12]

Aftermath

Three days later, the September 11 attacks occurred in the U.S., eclipsing media coverage of Durban. The U.S. withdrew support for Mary Robinson, who stepped down in 2002.

Follow-up conferences in 2009, 2011, and 2021 were boycotted by several countries. Issues of slavery reparations, colonial legacies, and the Israel–Palestine conflict remain contentious.

20

Poverty, conservation, and historical responsibility

An approach to global concerns

Circles of concern and circles of influence

Epictetus wrote, "Some things are within our power, while others are not."[1] This distinction between what we can shape, our choices, habits, and relationships, and what lies beyond immediate control, poverty, climate, history, remains relevant. However, while the Stoics emphasized inner mastery, there is a reciprocity: engaging with broader concerns can enrich our inner life, transforming how we lead, relate, and grow. Humanists move fluidly between both spheres, allowing personal development and moral action to reinforce each other.

Misuse of the ESG framework

Much lip service is paid to Environmental, Social, and Corporate Governance (ESG) issues and many businesses treat them as impractical or irrelevant. This disconnect may stem from ESG's origins in financial risk metrics, rather than in scientific or humanistic inquiry. The framework often oversimplifies complex, interwoven challenges: environmental degradation, social injustice, and governance failures, that require context-specific understanding.

Treating ESG as a checkbox exercise not only dilutes its potential but risks reinforcing the very problems it seeks to address. Each issue: poverty, conservation, historical responsibilities, must first be approached on its own terms. Only through individual analysis can shared strategies emerge with clarity and purpose.

Moral choices - poverty and precarity

Utilitarianism

Utilitarianism, shaped by Enlightenment thinkers Jeremy Bentham and John Stuart Mill, urges us to maximize happiness and minimize suffering, regardless of proximity or personal ties. Bentham's call to "remove all the misery you are able to remove,"[2] Mill's insistence on accountability for action and inaction,[3] and Peter Singer's Shallow Pond analogy all converge on a moral imperative: if we can prevent harm without sacrificing something of equal moral weight, we ought to do so. Singer's pond, where rescuing a drowning child is morally obvious, becomes a metaphor for aiding distant strangers, whether through poverty relief, pandemic response, or climate action (SINGER)[4].

Critiques of Singer's framework fall into two camps: those who accept utilitarian logic but question its demands, and those like Rawls and Sen who reframe the moral landscape.

Arguments against - within Singer's framework

Some critics say that Singer's utilitarian principle, that we must prevent harm unless doing so requires sacrificing something of equal moral importance, can feel overwhelming. It places a heavy moral burden on individuals when faced with countless "shallow ponds" of suffering across the world. Singer acknowledges this discomfort, initially advocating for giving up to the point of marginal utility—where further giving would cause as much hardship to the giver as it relieves for the recipient—but later suggesting that even donating 10% of one's income is a morally powerful and sustainable act.[5,6]

Other critics argue that Singer's framework oversimplifies the complexity of global poverty, which involves distant beneficiaries and uncertain outcomes. Yet Singer maintains that distance and scale do not diminish moral responsibility. He asks, "Should I consider that I am less obliged to pull the drowning child out of the pond if others nearby are doing nothing?"[7] His answer is no. He also points to improved NGO accountability and his initiative, "The Life You Can Save," which vets effective organizations.[8]

Precarity and inequality both also deserve moral attention. Singer might respond in two ways: by prioritizing immediate life-saving interventions, or by recognizing that improving lives at the margin is inseparable from saving them. In both cases, he would urge urgent action and generous giving.

Some resist Singer's conclusions by appealing to non-utilitarian values such as prioritizing family. Singer concedes this instinct but insists that moral concern must extend beyond personal ties. For Singer, preventing suffering is a universal imperative, grounded in moral clarity.

Re-framing Singer's allegory

John Rawls and Amartya Sen reframe Singer's urgency through broader lenses of justice. Rawls shifts the question from "how do we save the child?" to "why is the child drowning?" His theory of "justice as fairness," built under the veil of ignorance, compels us to design institutions that prevent such tragedies.

Charity, for Rawls, is secondary to systems that ensure no child is left behind.

Sen builds on this by arguing that no single metric, utility or suffering, can capture the full scope of moral obligation. His focus is on expanding freedoms and agency, advocating for democratic reasoning and collective responsibility. Sen's vision moves from rescue to empowerment: not just pulling children from ponds, but ensuring they can swim, flourish, and be heard. Martha Nussbaum extends Sen's capabilities approach, emphasizing that flourishing requires more than survival. While she agrees with Singer on the urgency of meeting basic human needs, she also includes all sentient animals and diverges on the philosophical path to get there.

In my view, these perspectives are not in conflict. A humanist must act now to alleviate suffering, while also working to build just institutions and nurture agency. Individual action and systemic reform are not opposing forces; they are complementary strands of moral commitment. Modern humanism draws from all these thinkers, combining urgency with reform.

Conservation

Lessons from the Aral Sea

The collapse of the Aral Sea, once the world's fourth-largest inland lake, exemplifies the Anthropocene worldview where human utility overrides ecological wisdom. Soviet planners diverted rivers to grow cotton, triggering salinization, toxic dust storms, mass migration, and extinction. Techno-optimists like Stewart Brand declare, "We are as gods," while Wilson counters, "We are not as gods… we're playing a global endgame."[9,10] E.O.

Wilson warns that utilitarian thinking erodes the "web of life,"[11] and Barry Commoner's axiom, "You can't do just one thing," is reflected in the cascade of unintended consequences.[12]

The Aral Sea's collapse is not isolated. Minamata's mercury poisoning, Lake Chad's shrinkage, Papua New Guinea's Ok Tedi mine disaster, global bee colony collapse, and China's Three Gorges Dam all reveal how technocratic-short-sighted solutions unleash ecological and human devastation. These cases show that ecosystems are interdependent networks, and their unravelling threatens biodiversity and human survival.

Wilson calls conservation a moral imperative. The Aral Sea disaster drove extinction rates a thousand times above prehuman levels, erasing species and knowledge. António Guterres called it "the biggest ecological catastrophe of our time," warning that without urgent climate action, such tragedies will multiply.[13]

Earth's systems are tightly interconnected. The Southern Ocean's currents regulate global climate, and their weakening disrupts rainfall, fisheries, and carbon absorption. The Global Tipping Points Report identifies 21 critical thresholds, from Amazon collapse to Arctic ice loss, while also offering hope through 10 positive tipping points like solar and wind energy.[14] Every fraction above 1.5°C increases the risk of irreversible change.

Restoration is possible but slow. Kazakhstan's Kokaral Dam revived parts of the North Aral Sea, restoring fish stocks and local economies. Yet recovery is minimal and even that is far harder than prevention. Conservation must be central, not an afterthought.

The Half-Earth project

The collapse of ecosystems reveals the urgency behind E.O. Wilson's "Half-Earth Project" (E. O. WILSON)[15,16]. Wilson argued that setting aside half the planet for nature is essential to prevent cascading extinctions and stabilize life systems. His research showed that protecting 50% of Earth's land and oceans could preserve 80% of species, offering a path to reverse our trajectory toward mass extinction.

Conservation can be embedded in business. Yvon Chouinard, founder of Patagonia, built his company around environmental responsibility. Inspired by the Iroquois "Seventh Generation principle," where one tribe member represented future generations during major decisions, Chouinard made long-term sustainability central to his business.

From pioneering low-impact climbing gear to switching to 100% organic cotton for his clothing lines, Patagonia prioritized ecological integrity over profit. In 2022, Chouinard transferred ownership to a trust and nonprofit, ensuring all profits support climate action and land protection (CHOUINARD)[17,18,19].

Chouinard insists this isn't philanthropy, it's the cost of living on Earth.[20] His model shows that ESG can be more than compliance; it can be a core business ethic. Other companies, Ørsted, IKEA, LanzaTech, Wasabi Technologies, and Seventh Generation, demonstrate that environmental responsibility is viable across industries. We don't need to replicate Patagonia exactly, but we can draw confidence from these examples: sustainability can work, and it must.

The Tragedy of the Commons

The Tragedy of the Commons, coined by Garrett Hardin in 1968, reveals how individual self-interest can deplete shared resources, causing collective harm.[21] A classic example is a hill pasture used by villagers. Each adds more cattle for personal gain, but overgrazing eventually depletes the grass, harming all.

Whether it's villagers overgrazing a communal pasture or nations exploiting the atmosphere, oceans, forests, or outer space, the damage is external to individual decisions and rarely priced into markets. Free riders, those who benefit from others' restraint without contributing, further undermine cooperation, especially in efforts like climate action.

Hardin argued there is "no technical solution" to this dilemma. His remedies included "mutual coercion, mutually agreed to" (rules to limit resource use), privatization, and government control. Free market proponents like Milton Friedman proposed taxation to internalize costs.

Joseph Stiglitz notes that due to externalities all these solutions involve some form of coercion (STIGLITZ)[22] and that pricing alone commodifies public goods and erodes civic responsibility. Privatization, while promising efficiency, often increases inequality and introduces moral hazard where private owners may overexploit resources, expecting others to bear costs. This was seen in post-Soviet Russia or satellite launches that ignore space debris. Stewardship cannot be left to profit motives alone.

Stiglitz offers a balanced framework combining regulations, prices, and public investment. Democratically designed rules protect rights, while investment aligns private incentives with public goods. Binding agreements and community enforcement counter free riding. This approach transforms shared vulnerability into shared responsibility, without resorting to authoritarian control.

Elinor Ostrom showed that voluntary governance can work. In Nepal, India, Guatemala, and Switzerland, small communities sustainably manage commons through norms and transparency. Yet scaling these models to global domains remains a challenge. Outer space, now crowded with satellites from players like SpaceX, has become a modern common under threat, our new pasture, grazed without oversight.

Corrective action is urgent. The 'One Health' approach, defined in 2021 by WHO, FAO, WOAH, and UNEP, provides a model that promotes cross-sector collaboration to optimize the health of people, animals, and ecosystems. The World Bank estimates that building such systems in low and middle-income countries would cost USD 3.4 billion annually, far less than the USD 570 billion impact of zoonotic outbreaks or the USD 13.8 trillion loss from COVID-19.[23] Such actions are not just morally sound; they are economically wise.

The power of individual actions

Can we truly shift regulations, investments, and resource management toward conservation? An answer lies in stories like Silent Valley, where individual and collective action averted ecological catastrophe.

In 1970, a proposed hydroelectric dam threatened to submerge 8 km² of pristine rainforest in the Nilgiri Hills of southern India. Romulus Whitaker, a herpetologist raised early alarms.[24] Steven Green, studying the endangered lion-tailed macaque, joined the call. Their findings reached the Bombay Natural History Society, sparking wider concern.[25]

Indian ecologists, Salim Ali, Madhav Gadgil, V.S. Vijayan, mobilized tribal communities and university students. Satheesh Chandran Nair and Shanthi Satheesh, students in zoology, led debates and data dissemination. The Kerala Sastra Sahitya Parishad, a collective of writers, rallied thousands.[26] Their unified voice compelled the government to review the project.[27] Silent Valley was declared a national park in 1984.[28]

This story, like Chouinard's, shows that individual actions, rooted in care, knowledge, and courage, can reshape policy. Each of us holds the power to contribute meaningfully, to protect what matters, and to imagine new directions for conservation.

Humanism and history

Humanist approach to history

History is the interpretive record of human choices and contingency. For Sven Lindqvist, it's a "moral excavation," not a lack of knowledge but a lack of courage to understand and act (LINDQVIST)[29]. He agreed with Erasmus: history is medicine for a sick mind, a vast archive of human conduct laid bare to guide us.[30]

Appreciate that getting at the 'facts' of history is a challenging task.

A humanist sees history not as static facts but as a complex, interpretive endeavour. Alfred North Whitehead reminds us that facts are never pure, they're steeped in theory, emotion, and purpose.[31] Yet humanists need to strive for clarity and triangulate evidence to build nuanced accounts. Objectivity may be a myth, but transparency and rigor are not.

Understand what happened, clarify why it happened, and what could have happened differently. Determine history's legacy and effects.

History's legacy is alive in the present. Colonialism, industrialization, and ideological conflict still shape inequality, ecological harm, and cultural identity. Humanists study these forces not to assign blame, but to understand how inherited narratives shape today's challenges. As Whitehead said, "A science which hesitates to forget its founders is lost," progress demands historical awareness and the courage to evolve.[32]

Learn lessons, think through and take actions based on the present situation.

Ultimately, history must lead to ethical reflection and action. Humanists learn from the past to shape the future, guided by empathy, critical thinking, and commitment to dignity. Buckminster Fuller put it best: "We are called to be architects of the future, not its victims."[33] Through education, dialogue, and civic responsibility, humanists take up that call.

A selection of examples whose consequences still affect us are provided below.

<u>The Indian caste system</u>

The Indian caste system, once a fluid framework around 1500 BCE dividing society into Brahmins, Kshatriyas, Vaishyas, and Shudras, ossified over centuries into hierarchies of purity and exclusion.[34] Though Article 15(1) of the Indian Constitution prohibits caste discrimination, it remains embedded in social life, only 5.4% of women in a 2011–12 survey had married outside their caste.[35]

Scheduled Castes and Tribes, historically marginalized, still suffer the highest poverty rates: 33.3% and 50.6% respectively, compared to 15.6% among higher castes.[36] Constitutional reservations in education, jobs, and legislatures aimed to correct this have now been expanded to other backward classes (a term used by the government Mandal commission) and poor upper castes. Yet, after 75 years, inequality persists.

Kaivan Munshi adds nuance to the understanding of caste in India by framing it not only as a social identity but as an economic institution.[37] His research shows that caste networks provide informal access to jobs, credit, insurance, and social support—especially in contexts where formal markets are weak or exclusionary.

Munshi also cautions that while these networks can be adaptive, they may reinforce segregation and limit broader integration. Even as constitutional reservations have expanded access to education and employment, caste-based networks continue to shape life outcomes in persistent ways. This dual role—both protective and restrictive—underscores the complexity of caste in modern India and the need for policies that strengthen inclusive public institutions.

<u>Colonialism and development in Africa</u>

By 1900, nearly all of Africa was colonized following the Berlin Conference of 1884–85 where 14 European powers established principles for "effective occupation."[38] Though formal decolonization began post–World War II, interference endures. The Congo's history is especially brutal: King Leopold II's personal colony saw 5 to 10 million deaths in the pursuit of rubber and ivory.[39]

After independence in 1960, Prime Minister Patrice Lumumba was assassinated with CIA backing, paving the way for Mobutu's 32-year dictatorship.[40] Today, the Democratic Republic of Congo remains in crisis, millions displaced and face food insecurity, while foreign companies continue extracting cobalt, coltan, and gold.[41] Heldring and Robinson's research links colonial brutality to enduring underdevelopment, weakened institutions, and entrenched inequality.[42]

Slavery in the U.S.

In the U.S., chattel slavery, where humans were owned, bought, and inherited, was codified by the late 1600s, with the Barbados Slave Code and Virginia laws making slavery hereditary and brutal. Over 12 million Africans were enslaved, fuelling southern plantations and northern industries.[43]

Though abolished in 1865, its legacy endured through Jim Crow laws, economic exclusion, and mass incarceration, effectively "slavery by another name."[44] Recently in the U.S. (2025), there is backsliding that has been described as a "deluge of new voter restriction laws" reminiscent of the Jim Crow era.[45]

Inequality across centuries

Thomas Piketty's work reveals how inequality persists across centuries: wealth's return outpaces economic growth, allowing inherited capital to snowball while labour income lags. From 1800 to 1914, Europe's top 10% held 90% of wealth; by 2020, they still held 58%, and up to 77% in Latin America. The bottom half of the world owns less than 5%, while the top 1% captures up to 30% of national income. Reductions in inequality came from war, redistribution, and welfare states, but neoliberalism since the 1980s reversed many gains, deepening disparities through privatization and regressive tax reforms.[46]

Patterns of collapse: five Lessons from history

Jared Diamond identifies five factors that contribute to societal collapse.[47] These are summarized below with reasons how they could apply to the Earth today.

Environmental Damage

Diamond's account of Easter Island is a stark warning: a society that deforested its land to build statues and expand agriculture eventually collapsed under the weight of its own ecological destruction. Without trees, they lost soil, food, and the ability to fish or flee, trapped in a dying ecosystem. Today, humanity is repeating this pattern on a global scale. Deforestation in the Amazon, coral collapse, and topsoil erosion threaten biodiversity and food security. If degradation continues unchecked, it could unravel agriculture, water systems, and habitability across continents, triggering widespread collapse.

Climate Change

The Ancestral Puebloans of the American Southwest flourished until a prolonged drought shattered their agricultural base, forcing mass abandonment. Now, rising seas, extreme weather, and shifting crop zones are already reshaping the planet. If temperatures breach critical thresholds, entire regions may become unliveable, sparking migrations, resource wars, and the breakdown of infrastructure and governance.

Hostile Neighbours

Hostile neighbours have played a role in collapse. The Norse in Greenland refused to adapt to Inuit survival techniques, and as their environment worsened, they lacked the means to coexist or defend themselves. Today's geopolitical tensions mirror that rigidity. Russia's invasion of Ukraine, Sudan's civil war, Congo's rebel violence, and the Israel-Palestine conflict all show how regional crises can spiral outward, drawing in global powers and threatening stability.

Loss of Trading Partners

The Maya civilization depended on trade networks for survival. When instability severed those ties, cities withered. Modern economies are as vulnerable. Global trade disruptions, from pandemics, war, or financial collapse, can cripple supply chains and trigger cascading shortages. The Great Depression and the 2008 financial crisis exposed this fragility. Reckless U.S. tariff policies in 2025 have reignited fears of trade wars and global fragmentation, threatening a new wave of upheaval.

Poor Decision-Making

Poor decision-making has always sealed civilizations' fates. Norse settlers in Greenland clung to European customs, refusing to eat fish or adopt Inuit clothing and hunting techniques. Their cultural rigidity prevented them from adapting to Greenland's harsh conditions, sealing their fate.

Today, humanity continues to ignore scientific warnings about inequality, climate change, and pandemics, paralyzed by denial, inertia, and short-term thinking. If we persist in this path, we risk irreversible damage to the foundations of global civilization.

The challenge of dialogues on history

Engaging with history is a humanist endeavour rooted not in rigid solutions or moral taboos, but in an ongoing effort to understand the past and its living legacy.

Debates around reservations in India or affirmative action in the U.S. are politically charged, often framed as unfair to those outside the favoured categories. Joan Williams notes how redistribution pits the "have-a-littles" against the "have-nots," eroding political support.[48]

Reparations provoke similar tensions, many Europeans question why they should bear the burden of crimes, while proponents argue that today's privileges are built on stolen lives and resources.

A failure to dialogue

The 2001 Durban Conference offers a cautionary tale. Intended to confront racism and intolerance, it collapsed under ideological clashes and walkouts. Arab states condemned Israeli policies, prompting U.S. and Israeli exits; African nations demanded reparations, Europe resisted; Indian delegates blocked caste discrimination language, sparking Dalit protests.

Mary Robinson, then UN High Commissioner, was criticized for lacking mediation skills, but the fundamental issue lay in the conference's collision of ideologies and moral fervour.

Key principles of humanist dialogue on historical responsibility

Humanist dialogue begins not with blame, but with curiosity. It seeks to understand the emotional and historical roots of why people resist reparations or feel burdened by inherited guilt, and why others demand justice for enduring harm. This ethic is grounded in compassion, humility, and engagement. It avoids moral absolutism, builds trust through empathy, and searches for shared values like dignity and fairness.

Justice, for the humanist, is a collective endeavour, not a zero-sum game but a project of healing. Practically, this means creating spaces for multi-perspective dialogue, using storytelling to humanize policy debates, and incorporating tools like restorative justice circles. It means thinking long-term, guided by principles like the Seventh-Generation ethic.

Once this groundwork is laid, humanists can articulate their commitments: accepting historical evidence, alleviating suffering, designing fair institutions, expanding capabilities and rights, and overcoming tribalism to embrace the broader web of life.

Ultimately, humanism grounded in rationality and compassion invites us to design institutions that are not only defensible but emotionally intelligent. It asks us to move beyond metrics of control and toward metrics of care such as the UN's Human Development Index that consider how people feel safe, seen, and supported. In this way, compassion and kindness become a form of ethical rigor, guiding us toward solutions that are both empirically grounded and morally expansive.

Planetary flourishing

A flourishing journey of life begins with the cultivation of individual capacities, but it cannot be a comprehensive flourishing without an expansion into wider moral and ecological dimensions. (See C15 S: Needs and capabilities)

Peter Singer is an exemplar modern humanist because he demonstrates that human flourishing is not a selfish or solely individual pursuit. Humanism entails concrete actions, as Singer shows by the examples of his lifestyle, his philanthropic commitments, and his efforts to guide others toward giving.

Singer, Chouinard, Berners-Lee embody a humanism that is both intellectually rigorous and transformative. Most of us are not at the level of these exemplars, but it is incumbent on us to keep up our efforts of service and giving. Humanist meaning is lost without actions towards planetary flourishing.

Part 8

CONSEQUENCES of is and ought

Structures and Frameworks

<u>Humanist ideas for economic, political, social structures, and education</u>

Keynes' vision for 2030

"Make the world work for 100% of humanity, in the shortest possible time, through spontaneous cooperation, without ecological offense or the disadvantage of anyone."[1]

BUCKMINSTER FULLER

Clarifying the Scope and Intent of Part 8

In considering the consequences of how the world is and ought to be, Amartya Sen's five instrumental freedoms have been broadly followed:
- Economic facilities (C23)
- Political freedoms (C4)
- Social opportunities (C25)
- Transparency guarantees (C23, 24, 27)
- Protective security (C23, 24)

C22 provides background to the above discussions.

The aim is to outline key features that reflect humanist principles and actively support the flourishing of life on Earth.

21

Keynes' Vision for the Future

John Maynard Keynes was an economist and philosopher. His ideas tangibly uplifted societies and remain profoundly relevant for equitable progress today.

In 1930 Keynes wrote an essay, "Economic Possibilities for our Grandchildren," painting a vision of the future in 2030 where:[1]

- Advances in science and technology would increase economic output, allowing people a standard of living four to eight times higher than in the 1930s
- There would be a transitional period where machines displace workers faster than new jobs can be created, a phenomenon Keynes called "technological unemployment."
- People would work 15 hours a week, not out of necessity but for fulfilment, as material needs become easier to meet.
- Humans would need to shift mindsets and priorities from working more efficiently and accumulating wealth to pursuing the art of living well.

Keynes wrote:

"Thus for the first time since his creation man will be faced with his real, his permanent problem-how to use his freedom from pressing economic cares, how to occupy the leisure, which science and compound interest will have won for him, to live wisely and agreeably and well.

The strenuous purposeful money-makers may carry all of us along with them into the lap of economic abundance. But it will be those peoples, who can keep alive, and cultivate into a fuller perfection, the art of life itself and do not sell themselves for the means of life, who will be able to enjoy the abundance when it comes.

There are changes in other spheres too which we must expect to come. When the accumulation of wealth is no longer of high social importance, there will be great changes in the code of morals. We shall be able to rid ourselves of many of the pseudo-moral principles which have hag-ridden us for two hundred years, by which we have exalted some of the most distasteful of human qualities into the position of the highest virtues. We shall be able to afford to dare to assess the money-motive at its true value. The love of money as a possession - as distinguished from the love of money as a means to the enjoyments and realities of life - will be recognised for what it is, a somewhat disgusting morbidity, one of those semicriminal, semi-pathological propensities which one hands over with a shudder to the specialists in mental disease. All kinds of social customs and economic practices, affecting the distribution of wealth and of economic rewards and penalties, which we now maintain at all costs, however distasteful and unjust they may be in themselves, because they are tremendously useful in promoting the accumulation of capital, we shall then be free, at last, to discard."[2]

JOHN MAYNARD KEYNES

22

Diagnosing the Crisis and Evaluating Possibilities

Humanism and economics

Economic theories shape the institutions and structures we inhabit. Keynes wrote, "The ideas of economists and political philosophers...are more powerful than is commonly understood."[1] Understanding the basics of economics, like any science and art, is a civic and ethical duty.

Unlike physics, macroeconomics with its reflexive systems proves more resistant to falsification and often yields conflicting prescriptions because it is partly normative. Economic assumptions are often reduced to Pareto efficiency (the point when no one can be made better off without making someone worse off) or utility (a measure of the satisfaction or benefit an individual derives from consuming goods and services). But this framing misses the often implicit, normative assumptions that shape economic theory and policy.

Different economic theories reflect distinct moral visions. Neoclassical models assume rational actors in efficient markets; Austrian theory champions individualism; Marxian theory focuses on class struggle; behavioural economics challenges rationality; ecological economics embeds environmental limits. None are neutral. Metrics like GDP, HDI, inflation, or unemployment reflect choices about what matters.

Zachary Carter writes, "Keynes ... came to see economic history as a fundamental political story... Economics... could not be a bloodless scientific investigation... but only a set of observations about trends in human arrangements."[2]

Humanists therefore must unpack economic ideas and uncover assumptions with care. This chapter begins with Keynes' vision, then compares it to today's world. Skidelsky called Keynes' essay an eloquent utopian vision.[3] But this chapter shows it is a 'conditional, rational utopia,' a vision grounded in economic trends and technological optimism yet tempered by moral and institutional realism.

Revolutions vs. dialogic justice

Building on Keynes's vision, this chapter turns to the humanist imperative for change—one grounded in reasoned engagement and dialogic justice, rather than coercion or revolution. Where coercion is necessary, in laws and regulations, it must arise from mutual agreement, made possible through inclusive and principled discourse.

Humanism embraces caution, applying the precautionary principle to prevent unintended harm. It recognizes that violence accompanies revolutions. Stanley Loomis' account of the French Revolution illustrates this: "The execution became the ritual of the new order... The guillotine... indifferent and inexorable."[4] The Russian Revolution promised equality but delivered gulags and purges. In both cases, violence became institutionalized, and the oppressed became new oppressors. What started as justice turned into terror.

Rather than radical upheaval, humanists advocate urgent yet feasible reforms: flexible work hours without income insecurity, expanded metrics like HDI (augmented with ecological health), and prioritizing Universal Basic Infrastructures: public health, education, and digital access. Cooperative and commons-based models deserve greater support.

Even in conditions of extreme poverty, the humanist imperative resists the notion that theft or violence is a legitimate form of justice. As Dostoevsky's "Crime and Punishment" illustrates, crime—even when rationalized by desperation or ideology—ultimately corrodes the self and fractures society. Raskolnikov's descent into guilt and isolation is not a religious parable, but a psychological truth: violence, even against the exploitative, dehumanizes the actor and perpetuates harm.

The extremely poor may ask, with painful clarity: "You speak of morality—but I'm starving. You defend law—but the law protects the powerful and rich." The modern humanist bears witness to these truths with fierce compassion, moral clarity, and unwavering solidarity. The response is not moral condemnation, but moral solidarity. Humanism insists that dignity must never be contingent on wealth, and that even under duress, the means we choose shapes the world we create. In contexts where institutional justice fails, Graeber and Roy's vision of everyday anarchism, where ethical order arises from voluntary cooperation and mutual aid, offer a path to dignity without domination—an alternative to both criminal desperation and authoritarian control.

These ideas are achievable within current systems and resonate with Keynes' humanist vision. But urgency must not lead us to embrace authoritarian solutions, however well-intentioned. Even benign Leviathans can become monstrous.

Yet even moderate proposals may appear radical to entrenched ideologies. As Keynes observed, "To suggest social action for the public good to the City of London is like discussing the Origin of Species with a bishop... The first reaction is not intellectual, but moral" (KEYNES)[5].

Status of Keynes' vision 2025

What is the status of Keynes' vision today?
- *Living standards.* Keynes predicted a 4–8x rise in living standards. U.S. real GDP per capita rose from ~USD 8,800 (1930) to USD 69,500 (2025), an eightfold increase.[6,7] Global GDP per capita (USD 13,000) exceeds basic needs and extreme poverty thresholds.[8]
- *Resource sufficiency.* The world produces enough food for over 10 billion people, despite livestock production using 80% of global agricultural land but providing less than 20% of the world's calories.[9] Resources and technology exist to meet human needs.
- *Automation and jobs.* AI and robotics are reshaping labour markets, increasingly performing many white-collar tasks. Yet developing countries remain labour-intensive. We're in the early, disruptive phase Keynes foresaw.
- *Energy.* Renewables dominate new power generation. Over USD 2 trillion invested in clean energy; 70% of new electricity is renewable. Efficiency gains come from digitization and decentralization.[10,11]
- *Wealth and Curiosity.* In the past, wealth enabled intellectual exploration: Darwin, Humboldt, Herschel, Cavendish. Today, wealth aligns with capital accumulation. While some wealthy fund conservation programs, they do not

focus on inquiry, the amateur scientist has nearly vanished.
- *Working Hours.* Keynes imagined 15-hour workweeks. Reality: India (45.8), China (44.8), Nigeria (39.6), U.S. (36.1). Economic pressures still drive long hours.[12,13]
- *Leisure.* Keynes hoped leisure would foster wisdom and fulfilment. Aristotle saw leisure as sustaining our souls and civic lives. Yet today, daily leisure is mostly febrile triviality of 3–6 hours spent on digital platforms (T22.1).

Country	Daily average (hours)	Top digital platforms (2025)
U.S.	3 to 4	YouTube, Instagram, TikTok, Facebook, and Netflix
India	3	YouTube, WhatsApp, Instagram, Netflix, and Disney+ Hotstar
China	5.5 to 6	WeChat, Douyin, Xiaohongshu, Kuaishou, Sina Weibo
Nigeria	3 to 4	WhatsApp, YouTube, Instagram, Facebook, and TikTok.

T22.1 Selected countries – Average daily hours spent on digital content[14,15,16,17]

Ortega y Gasset's 'Mass-Man'

Mass Man

In 1930, the same year Keynes wrote his essay, José Ortega y Gasset published "The Revolt of the Masses."[18] He warned that the rise of the 'mass man,'—who was unreflective, lacking ambition and historical awareness—threatened civilization's foundations.

Ortega y Gasset contrasted those striving for excellence with those content in mediocrity. He feared that as 'mass man' gained influence, intellectual elites would lose legitimacy. Though he supported liberal democracy, he cautioned against its slide into populism. Civilization, he argued, depends on a "select minority" defined by moral and intellectual character — not wealth, birth, or race. Entitlement-driven democracy risks shallow discourse, diminished creativity, and ethical confusion.

A chakra[a] - meaningless work to mindless time-wasting

Humanists acknowledge:
- Their own fallibility, shaped by biases, hubris, and evolutionary heritage.
- Affirm most people can flourish in creative fields.
- Accept many of Ortega y Gasset's observations, refusing to ignore uncomfortable truths or hide behind political correctness.

Yet, humanists also recognize that social, economic, and cultural systems shape human purpose. Stress from unmet basic and social needs often drives people into meaningless jobs, which then cascade into mind-numbing leisure. This cycle traps individuals in a loop of anxiety, purposeless work, and trivial distraction.

Thorstein Veblen's concept of "conspicuous consumption" described status-driven spending.[19] This is amplified by advertising and digital priming. Luan and You's 2025 study across 28 countries found that longer work hours, driven by pressure to match others' lifestyles, are associated with

---

a   Chakra – Sanskrit word for "wheel," used here to evoke a cycle of ceaseless, entrapping spin.

greater inequality. High earners benefit more because they can monetize their time effectively, while lower-income workers gain less and face greater strain.[20]

John Kenneth Galbraith's "dependence effect" explains how production creates the desires it claims to satisfy.[21] This illusion of affluence erodes the meaning of work and leisure, replacing intrinsic needs with externally manufactured wants.

Recent studies reinforce this:
- Woolley and Sharif found that low-autonomy jobs lead to more social media, a search for 'dopaminergic distractions' like scrolling and snacking.[22]
- A study of 1,200 university students showed that social media use increased stress and harmed academic performance and mental health.[23]

This chakra, formed by unmet needs, meaningless work, and escapist leisure, feeds on itself. In the absence of purpose or fulfilment, stress intensifies, and the cycle continues.

The fallacy of rankings and a question on change

In addressing Ortega y Gasset's mass-man observations, rather than the elevation of "noble" elites to a high rank, humanism elevates understanding. This means debates, discussion and dissent, an open conversation accessible to everyone.

Humanism does not deny differences in individual abilities, but is aware that (i) there are a multitude of vectors (traits), not one single trait, over which abilities manifest (ii) individual differences do not translate into group (racial, class) differences (iii) traits and abilities evolve and change as individuals learn and grow (Gould)[24]. Rather than a flat distribution, each individual embodies a 3-D bell (F10.4), composed of countless intersecting traits that together form a rich and varied personal landscape.

Humanism asks and addresses the question, 'What conditions suppress human potential?' and 'What might humanists do to engender change?'

Essential embodied well-being

A range of ideas and actions may be necessary to move towards a humanist worldview. Each offers distinct insights and practical suggestions.

Physical affection

James W. Prescott's research in the 1960s and 70s underscores the significance of physical affection and movement in early childhood (C6 S: Memes).[25] His findings suggest that societies that prioritize tactile nurturing, such as holding, carrying, and cuddling infants, tend to foster healthier emotional regulation and exhibit lower rates of violence across the lifespan.

To reframe affection as a public good, policies could extend paid parental leave to support early bonding, embed tactile engagement in childhood education and healthcare, normalize physical affection through developmental campaigns, and design public spaces that foster movement, touch, and caregiver interaction.

Celebration of the physical pleasures in life

An associated requirement is the celebration of the physical pleasures in life, always without harm to others or the environment. These are pursued without opprobrium or guilt.

To shift pleasure from taboo to celebration, recommendations might include advancing curriculum reform that integrates body-positive, pleasure-aware content alongside ethics, ecology, and consent; designing sensory-rich public spaces, from parks to dance venues, that foster embodied experience; supporting uncommodified media that celebrates joy; and moving beyond destigmatization toward affirming safe, consensual adult pleasures.

Kindness and lived compassion

The two bedrocks of humanism, kindness and lived compassion need to be embedded in institutions. Without kindness and recognition of another's vulnerability, ideals like dignity, justice, and freedom remain inert, technically admirable but emotionally uninhabitable.

To institutionalize these values, train frontline professionals in empathetic care, redesign welfare around dignity and relational support, foster inclusive gatherings that build mutual recognition, and elevate narratives that frame kindness as strength.

Environmental and nutritional foundations of mental health

A humanist worldview must also confront the material conditions that shape mental and emotional development:
- Exposure to environmental toxins—such as lead, in Jessica Reyes's research linking lead exposure to increased violent crime—can have lasting neurological and behavioural consequences (C7 S: The use of probabilities and statistics).

- Likhar and Patil's study showing that malnutrition during the first 1,000 days of life leads to irreversible neurodevelopmental disorders (C17)
- Street and Bernasconi's research highlights growing concerns about the impact of microplastics on children's development.[26] It explains that children are especially vulnerable due to their smaller body size, developing organs, and behaviours that increase exposure, such as hand-to-mouth activity. Microplastics impair neurodevelopment and contribute to cognitive and behavioural issues in children.

Policies must prioritize environmental cleanup, regulate neurotoxic pollutants, and ensure universal access to nutritious food. Unless these conditions are addressed, other policies aimed at human flourishing will remain aspirational and incomplete.

23

Approaches to Economic and Social Structures

Summary of humanist-economic themes

While many scholars have and continue to contribute to humanist economics, this chapter highlights a group of leading voices whose work exemplifies distinct and influential perspectives. Their contributions are summarized around key humanist themes that provide an overview of practical insights and serve as a foundation from which humanists can build actionable frameworks for economic reform grounded in ethics and values.

1: Inequality and redistribution (T23.1)

- *Systemic* thinkers seek structural change (e.g., tax reform, institutional overhaul). *Pragmatic* thinkers focus on actionable solutions.
- *Ethical* framing emphasizes justice, dignity, and fairness. *Economic* framing focuses on market dynamics, efficiency, and fiscal outcomes.

	Ethical	Economic
Systemic	**Sen**: Inequality as capability deprivation; justice demands expanding freedoms.	**Piketty**: Capital accumulation drives inequality; calls for global wealth tax. **Acemoglu & Robinson**: Inequality stems from extractive institutions; reform must be structural.

	Ethical	Economic
Pragmatic	**Banerjee & Duflo**: Justice through context-sensitive interventions; ethics guide experimentation.	**Stiglitz**: Market failures create inequality; policy must correct distortions. **Keynes**: Redistribution stabilizes demand; inequality is a macroeconomic concern.

T23.1 Inequality and Redistribution

2: Role of institutions (T23.2)

- *Normative* views ask what institutions ought to do morally or democratically. *Functional* views assess how institutions operate and what outcomes they produce.
- *Macro-level* thinkers analyse institutions as national or global systems. *Micro-level* thinkers examine how institutions affect individuals and communities in practice.

	Normative	Functional
Macro-Level	**Sen**: Institutions must foster public reasoning and democratic participation. **Acemoglu & Robinson**: Institutions shape prosperity; inclusive ones prevent elite capture.	**Piketty**: Institutions entrench inequality; calls for democratic control over capital. **Keynes**: Institutions stabilize demand and employment.
Micro-Level	—	**Banerjee & Duflo**: Reform institutions based on field evidence. **Stiglitz**: Institutions often fail due to capture; need targeted reform.

T23.2 Role of institutions

Most thinkers operate at the macro level, but Banerjee, Duflo and Stiglitz bring institutional analysis down to the granular level, testing and fixing mechanisms rather than theorizing ideals.

3: Justice and fairness (T23.3)
- *Procedural* justice focuses on fair processes (e.g., public reasoning, participation). *Distributive* justice concerns the allocation of resources and outcomes.
- *Idealists* propose visionary frameworks for justice. *Realists* emphasize what is feasible or effective within existing constraints.

	Procedural	Distributive
Idealist	**Sen**: Justice is expanding freedoms through public reasoning.	**Piketty**: Justice means dismantling inherited privilege and democratizing capital.
Realist	**Banerjee & Duflo**: Justice is what works in context; guided by evidence.	**Stiglitz**: Justice requires correcting distortions and ensuring opportunity. **Keynes**: Justice is tied to employment and macroeconomic stability. **Acemoglu & Robinson**: Justice emerges from inclusive institutions.

T23.3 Justice and fairness

Sen and Piketty offer visionary ideals, while others focus on justice—a practical outcome, whether through policy, experimentation, or institutional design.

4: Growth and development (T23.4)

- *Human-centred* approaches prioritize well-being, freedom, and capabilities. *Output-centred* views: GDP, investment, and productivity.
- *Structural* thinkers propose systemic reforms. *Experimentalists* test interventions through data and fieldwork.

	Human-centred	Output-centred
Structural	**Sen**: Development is freedom; must expand capabilities. **Acemoglu & Robinson**: Growth depends on inclusive institutions.	**Piketty**: Growth without redistribution leads to oligarchy. **Keynes**: Growth driven by demand and investment.
Experimental	**Banerjee & Duflo**: Focus on improving health, education, and savings through trials.	**Stiglitz**: Growth must be inclusive and sustainable; critiques GDP-centric models.

T23.4 Growth and development

5: Globalization and capital mobility (T23.5)

- *Sovereignty-focused* thinkers defend national autonomy and democratic control. *Integration-focused* thinkers explore global systems.
- *Critical* stances highlight risks and injustices in globalization. *Cooperative* stances seek reform or design mechanisms to make globalization work.

Approaches to Economic and Social Structures

	Sovereignty-focused	Integration-focused
Critical	**Piketty**: Global capital flows undermine sovereignty; calls for global wealth registry. **Sen**: Globalization must respect local capabilities.	**Stiglitz**: Globalization's uneven benefits require reform; supports fair trade.
Cooperative	**Acemoglu & Robinson**: Globalization works if institutions are strong.	**Keynes**: Supported Bretton Woods; wary of destabilizing flows. **Banerjee & Duflo**: Prefer local experimentation over global models.

T23.5 Inequality and Redistribution

Both Piketty and Sen critique globalization from the standpoint of its impact on national or local autonomy, rather than embracing global systems as the primary framework, while Stiglitz urges global reforms.

6: Measurement of well-being (T23.6)
- *Qualitative* approaches assess well-being through freedom, dignity, and capabilities. *Quantitative* approaches rely on metrics like GDP, HDI, or randomized trials.
- *Historical / systemic* thinkers analyse long-term patterns and institutional effects. *Empirical / field-based* thinkers measure through data and experimentation.

	Qualitative Metrics	Quantitative Metrics
Historical/Systemic	**Sen**: Called for HDI and capability as measures of freedom. **Acemoglu & Robinson**: Emphasize institutional persistence over short-term metrics.	**Piketty**: Uses historical data to track inequality; sceptical of GDP. **Sen**: Animated by Sen others quantified HDI and capability metrics to assess freedom.
Empirical/Field-Based	—	**Banerjee & Duflo**: Randomized trials to assess impact. **Stiglitz**: Called for multidimensional indicators; critiques GDP. **Keynes**: Relied on employment and output aggregates.

T23.6 Measurement of well-being

7: Technology and automation

Humanist economics must confront the impact of AI, platform economies, and automation on labour and inequality. While some argue that automation will democratize productivity, others warn it may exacerbate inequality by amplifying capital returns and displacing workers.

Banerjee and Duflo caution against simplistic narratives. They emphasize that technological adoption must be context-sensitive: policies must account for behavioural responses, local constraints, and institutional readiness. Automation is not neutral; it encodes values and priorities. (See C25 S: Education, technology, and AI)

8: Migration and borders

Humanist economics must address refugee rights and border regimes. Migration is driven by inequality, conflict, and climate stress. Policies must balance national interests with moral obligations to safety, opportunity, and dignity.

Banerjee and Duflo have studied immigration's effects on labour markets and social cohesion.[1] Their findings challenge xenophobic assumptions: immigrants complement native workers, contribute to growth, and enrich civic life. They argue for policies fostering trust, simplifying access to services, clarifying legal pathways, and investing in integration.

Other humanist thinkers

Other thinkers who provide ideas to humanist economic frameworks include David Graeber, John Rawls, and Michael Sandel. They are focused on particular aspects or philosophical in method. Their ideas are incorporated in the relevant following sections

Economic institutional designs

Sen on capabilities, justice, and the ethics of flourishing

Amartya Sen's *capabilities approach* urges policy to look beyond economic metrics and focus on expanding real freedoms, what people are actually able to do and be. His framework responds to utilitarianism and Rawlsian resourcism (justice as fairness via resource distribution) by redefining progress not through GDP or resource distribution, but through human capabilities: the freedom to live with dignity, meaning, and ethical agency. Autonomy is not merely instrumental; it is intrinsically valuable. (C20 S: Moral choices – poverty and precarity)

Democratic reasoning. Rather than prescribing a fixed list of capabilities, Sen invites societies to define priorities through democratic reasoning and empirical scrutiny. Nussbaum's list (C15 S: Needs and capabilities) offers a practical alternative, but Sen's flexibility makes the model adaptable across contexts.

In *"Development as Freedom,"* Sen argues that development is the expansion of freedoms: from hunger, access to education, political participation, and meaningful relationships (SEN)². Flourishing is not a utopian endpoint but a dynamic unfolding of individual and collective agency.

In "The Idea of Justice," he critiques transcendental ideals and advocates *comparative justice* rooted in pluralistic reasoning and public deliberation.³ Societies must openly decide which capabilities matter and how they're shared.

Humanist resonance. Modern Humanism strengthens Sen's emphasis on scrutiny by grounding justice first in what IS, before moving to what OUGHT to be (C27 S: Humanist discourse as a layered process). Justice becomes a direction—a negotiation between agency and institutions.

Sen's influence reaches far beyond academia. His ideas shaped Nussbaum's list, the Human Development Index, and global debates on poverty, education, and gender justice. By redefining poverty as capability deprivation, he shifted development's focus from material accumulation to human flourishing.

Thomas Piketty on equality and reform

Ethical foundations of participatory socialism. Thomas Piketty's participatory socialism is not mere redistribution; it's a humanist framework rooted in the ethical imperative to restore dignity through democratic control of capital. His proposals reconfigure the economy for flourishing, not entrench inequality, which threatens access to basic goods and political equality. Progressive wealth taxation, up to 90 percent, is restorative, dismantling rent-seeking structures and redirecting idle wealth toward lives of meaning, creativity, and civic engagement.

The *universal capital endowment* at age 25 is radical: it democratizes access to capital, freeing individuals from ancestral privilege and affirming the right to develop capacities as an end in itself.

A humanist economy must balance *redistribution with decommodification*. Taxing commodified sectors can fund equity, but essentials like education and healthcare should be removed from market logic. Only then can taxation support ethical outcomes without contradiction.

Democratic agency and institutional reform. Piketty's vision aligns with humanism's emphasis on agency and pluralism. Worker voting rights in governance challenge shareholder dominance and reimagine the workplace as a space of democratic expression. His push for transparent ownership registries and reform of trade and monetary rules reflects a commitment to epistemic integrity—moral instruments for confronting entrenched power and designing systems grounded in justice and sustainability.

Feasibility and moral urgency. Critics call the proposals infeasible. But humanism measures feasibility by moral urgency, not inertia. There are solutions we must look at. For example, Piketty's wealth tax and universal capital endowment proposals are criticized as leading to intergenerational conflict. Younger generations benefit from public spending funded by taxes on older individuals' accumulated wealth.

Piketty counters this by reframing wealth as a collective inheritance, not a purely individual achievement. His proposed tax measures are aimed at billionaires and large estates, rather than middle-class retirees. The taxes are intended to fund ambitious public projects that enhance social and cultural capital—recycling wealth into public goods, and thus creating a virtuous cycle of investment and inclusion.

Automation and AI are central to this vision, sustaining high-tax economies by lowering the cost of public ambition and expanding the scope of what's possible without relying on private capital returns.

Without such changes, Piketty warns, when return on capital outpaces growth, inequality becomes arbitrary and corrosive, concentrating media and lobbying power and undermining democracy (Piketty)[4].

Stiglitz on reforming capitalism

Shared prosperity and ethical agency. Joseph Stiglitz reimagines prosperity not as accumulation but as shared flourishing. "The only true and sustainable prosperity is shared prosperity," he writes, a vision that centres dignity and ethical agency at the heart of economic life (Stiglitz)[5]. He reframes freedom not as deregulation but as the ability to live a dignified life, shifting economics from market ideology to the lived realities of meaning, security, and connection.

Fiscal justice and economic participation. His proposals: fair taxation, closing loopholes, taxing capital gains, and setting corporate tax floors, are moral correctives, not mere fiscal tools. They serve the common good (C20 S: Conservation). Antitrust enforcement and labour protections affirm autonomy, breaking monopolies and empowering unions to enable meaningful participation in economic life.

Universal social security and investment in infrastructure, education, and healthcare are moral commitments. His focus on externalities, especially climate change, elevates ecological empathy: humanism sees flourishing as planetary, not just personal.

Critics argue he overestimates state capacity. But humanism doesn't sanctify authority, it calls for reform grounded in evidence and imagination. Stiglitz advocates smart regulation, not blanket intervention, pointing to Nordic models and stimulus programs as proof of possibility.

Community and existential erosion. His concern for identity, community, and solidarity is intrinsic. "Perhaps the greatest cost imposed by the top 1 percent," he writes, "is the erosion of our sense of identity in which fair play, equality of opportunity, and a sense of community are so important."[6] This erosion is existential, undermining the relational and cognitive structures through which we flourish.

Sandel on market morality and meritocratic justice

Markets and moral foundations. Michael Sandel's economic critique centres not on policy prescriptions but on the moral foundations of economic life. He links justice to civic renewal.

Sandel interrogates the critical question of what kind of society markets help shape. He argues that markets, while efficient in allocating resources, are not neutral spaces; they reflect and reinforce societal values.

Commodification and civic erosion. In "What Money Can't Buy," Sandel warns that "the more things money can buy, the more the market thinking comes to dominate our lives" (SANDEL)[7]. This commodification of goods, from education and healthcare to civic duties, erodes the moral fabric of society and undermines solidarity and hence equity.

Meritocracy and dignity. In "The Tyranny of Merit," Sandel critiques the assumption that economic success reflects moral worth. He writes, "The meritocratic ideal is corrosive. It encourages the successful to inhale too deeply of their own success and to look down on those less fortunate" (SANDEL)[8]. This critique has implications for institutional design: economic systems must not only redistribute wealth but also redistribute dignity. Sandel calls for structures that recognize the value of all forms of labour and resist the tendency to equate income with virtue.

Graeber on debt

Debt as moral and political structure. Graeber's critique of debt adds a moral lens to economic structures. In "Debt: The First 5,000 Years," David Graeber reframes debt as a moral and political institution, not just an economic transaction.[9] He shows how debt has enforced hierarchy and coercion, often backed by state violence.

Jubilees and alternative economies. In ancient Israel, jubilees, debt cancellations every 50 years, freed slaves and restored land, preventing permanent concentration of wealth and power. Graeber advocates similar debt forgiveness, critiques exploitative global finance, promotes mutual aid systems like time banks and gift economies, and prioritizes socially meaningful labour over market metrics.

Forgiveness and moral hazard. He rejects that forgiveness creates moral hazard. The real hazard, he argues, lies in indefinite repayment enforced by violence or shame. Historical jubilees were stabilizing, not reckless. Most modern debt, especially among the poor, stems from systemic exploitation, not irresponsibility.

Global activism and grassroots solidarity. Graeber's vision of alternative debt structures finds expression in sovereign debt cancellation campaigns in Zambia, Ghana, and Kenya; informal COVID-era aid in India; worker cooperatives in Brazil; and community savings groups in South Africa. Activists in New York abolished over USD 31 million in debt, reframing cancellation as political action. During COVID-19, cities such as Oakland and New York saw grassroots networks rooted in reciprocity and solidarity.

Crisis and institutional hypocrisy. The 2007–2009 financial crisis revealed how institutions, often critics of Graeber's proposals, operated under privatized profits and socialized losses, resisting reform even as they received bailouts. "There's no better way to justify unequal power than by reframing it as a moral obligation" (GRAEBER)[10].

Acemoglu and Robinson - inclusive or extractive institutions

In "Why Nations Fail,"[11] Daron Acemoglu and James Robinson argue that the determinant of a nation's long-term prosperity is the nature of its institutions, specifically, whether they are inclusive or extractive. Inclusive institutions distribute political and economic power broadly, protect property rights, and encourage innovation and participation. Extractive institutions concentrate power and wealth in the hands of a few, suppress opportunity, and stifle growth.

The authors reject explanations based on geography, culture, or natural resources, instead pointing to institutional design as the key variable. They illustrate this with contrasts: North and South Korea share the same geography but diverge radically in institutional structure; colonial Latin America was built on extractive systems that enriched European elites; and countries like Zimbabwe and Congo continue to suffer under elite capture and institutional decay (C24 S: The social contract and boundaries of power).

Cautionary tales - India

Stuck at 4%: India's lost decades

India's post-independence model was rooted in ideals of self-reliance and equity: centralized planning, nationalization, marginal tax rates above 90%, and protectionism. Yet by 1990, it had yielded stagnation, inefficiency, and corruption. The state controlled banking, industry, and trade, but lacked transparency, participation, and accountability. The result: a bloated public sector, entrenched patronage, and the stifling 'License Raj' that suppressed innovation.

While top income shares declined modestly, inequality endured in subtler forms. Access to capital, education, and opportunity stratified. Public services were underfunded, and redistributive intent was thwarted by bureaucracy and elite capture. Without transparent ownership registries or universal capital access, redistribution lacked depth and durability.

Piketty, Stiglitz, Galbraith – redistribution and democracy

Piketty might view this as redistribution without real democratization. His participatory socialism calls for worker voting rights, universal capital endowments, and public wealth registries, none of which India's centralized model offered. Stiglitz would highlight poor institutional design: government must be active, smart, accountable, and incentive-compatible. India exemplified regulatory capture and distorted market signals.

Galbraith's "countervailing power" envisioned institutional checks between the state, corporations, organized labour, and other interest groups, believing that each could offset the dominance of the others. Yet he underestimated the influence of Indian business families that shaped policy and discourse beyond civic counterweights. India plans concentrated power in the state, directing lives without empowering citizens.

Sen and Rawls – human development and justice

Amartya Sen critiqued India's neglect of social infrastructure. Unlike South Korea or China, India failed to invest in education, healthcare, and nutrition, leaving its HDI abysmally low. Bureaucratic opacity and limited civic engagement diluted political freedoms, corruption and exclusion undermined transparency and security.

Rawls's principles, which include equal liberties and allowing inequalities only when they benefit the least advantaged, were often violated. Constitutional guarantees were hollowed out by elite capture. Scholars argue Rawlsian justice remains vital: affirmative action and legal safeguards reflect his 'difference principle' but implementation lags. His 'veil of ignorance' (designing justice as if you didn't know your own place in society) offers a powerful critique of state-driven policymaking that failed to deliver fair outcomes.

Sandel – meritocracy and moral discourse

Michael Sandel has called for an Indian meritocracy rooted in dignity and inclusivity, not competitive success. He warned that hyper-nationalism and liberalism clash in a vacuum of moral discourse (SANDEL)[12]. When ethical debates vanish from public life, fundamentalists rush in. Sandel questioned whether secular politics can endure without engaging culture, suggesting India is a nation for reimagining democracy through pluralistic moral reasoning.

Undertrials and the hollow promise of equity

India's justice system reveals institutional failure. Undertrials—presumed innocent—make up 77% of the prison population, often jailed for years due to delays, lack of representation, or inability to afford bail. This injustice reflects a governance model lacking humanist foundations.

Despite gains in poverty reduction, over 70 million Indians remain below the poverty line, with many more in precarity. Liberalization lifted millions, and the free food grain program begun during COVID-19, PMGKAY, now continues, benefiting over 800 million people. This is welcome welfare expansion, but structural transformation remains absent.

Cautionary tales - China

China is praised for lifting hundreds of millions out of poverty, yet this achievement masks a more precarious reality. Its rapid ascent has come at immense environmental, political, and epistemic cost.

China's infrastructure boom, including over 87,000 dams, has caused ecological disruption, community displacement, and hydrological instability (C20 S: Conservation). The Three Gorges Dam alone has triggered landslides and seismic concerns. These projects reflect a top-down approach that values control and spectacle over sustainability and consent. Sobering precedents like the Aral Sea disaster go unacknowledged.

The COVID-19 pandemic exposed the dangers of opacity: suppressed data, silenced whistleblowers, and controlled scientific discourse delayed global response and prolonged the crisis. As Stiglitz warns, regimes that subordinate truth to expediency undermine democratic function.

Despite its economic gains, China faces youth unemployment, inequality, and a brittle real estate sector. Without democratic feedback, course correction is elusive.

Sen calls development without freedom a hollow achievement. China's literacy and health coverage exist, but inequality in voice and liberty remains stark. Piketty critiques its lack of transparent wealth registries and democratic oversight. Stiglitz warns that innovation and resilience require open institutions.

China must also be viewed through a historical lens. Its autocratic system is vulnerable to the 'bad emperor' problem, unchecked power concentrated in a single figure. Mao Zedong's rule, marked by the Great Leap Forward and the Cultural Revolution, led to an estimated 70 to 80 million deaths.[13] Deng Xiaoping's famous dictum, "It doesn't matter whether a cat is black or white, as long as it catches mice," justified the Tiananmen Square crackdown, revealed the regime's readiness to pursue any means to support the supremacy of the Communist party. This is the risk of China where truth and liberty remain subordinate to control.

Experimentation in the social sciences

Abhijit Banerjee and Esther Duflo use randomized controlled trials and behavioural insights to reshape development economics by focusing on how poor people actually live and decide. Small interventions, like tutoring or health incentives, have improved millions of lives. Their method is a vital bridge between humanist ideals and empirical evaluation.

MIT's J-PAL, founded by Banerjee, Duflo, and Sendhil Mullainathan, has conducted over 1,000 evaluations in 90+ countries, influencing policies across education, health, gender, labour, and finance. Its innovation was to bring falsifiable, replicable methods, once confined to the natural sciences, into the realm of social policy.

In "Poor Economics," Banerjee and Duflo dismantle the myth of a universal 'poverty trap.'[14] Poverty is not monolithic, it's shaped by local constraints, psychological burdens, and policy failures. Their experiments reveal how small, context-sensitive interventions, like offering lentils to increase child vaccinations, can outperform grand reforms. These insights challenge ideological rigidity and highlight the need for humility in policy design.

They emphasize that poverty is not just material, it's cognitive and emotional. Scarcity taxes the brain, leading to short-term thinking and decision fatigue. Their proposal is to design policies that reduce cognitive load by simplifying access and using behavioural nudges.

J-PAL's approach is human-centred: scale up proven interventions, invest in experimentation, avoid silver bullets, and treat the poor as partners, not passive recipients.

On climate, Banerjee and Duflo stress that solutions must address behavioural and political dimensions, not just technical fixes. The Clean Cookstoves initiative failed because many recipients didn't use them—designs ignored cooking habits, fuel access, and trust. Technologies must be affordable and culturally accepted, or they remain idle.

Banerjee and Duflo remind us that progress often begins not with ideology, but with grounded, human-centred experimentation.

Universal Basic Income

Among the boldest of experiments are trials of universal basic income—unconditional cash transfers that meet Maslow's D-type needs (C15 S: Needs and capabilities). The results challenge the stereotype of the poor as free riders (people who benefit from a service or resource without contributing to its cost or upkeep).

In Finland, unemployed recipients of €560/month showed improved mental health and worked more. In the U.S., OpenResearch's USD1,000/month led to better housing, healthcare access, and trust in institutions. Ireland's basic income for artists increased creative output and reduced anxiety. Barcelona's B-MINCOME improved housing stability and community participation. Kenya's GiveDirectly reached 20,000 people, boosting entrepreneurship, health, and social cohesion.

Other experiments

Other experiments show uneven reform. Brazil's Bolsa Família affirms protection but falls short of Piketty's universal capital or Stiglitz's labour protections. South Africa has progressive taxation but lacks ownership transparency. France and Nordic states embrace wealth taxes and labour rights, yet global coordination remains elusive.

These trials suggest the need for coherent policy architectures that integrates redistribution, regulation, and democratic participation as a unified strategy of self-correction. Yet theoretical models of social choice remind us: widely satisfactory policies have inherent trade-offs.

Libertarian economics vs. humanist counterpoints

Ludwig von Mises's "Human Action," and Murray Rothbard's "Man, Economy and State," are libertarian texts.[15,16] They argue that economics is driven by rational choice (praxeology), and that free markets, unfettered by government, are the most efficient and moral way to organize society. Rothbard, in particular, condemned the New Deal and claimed that government intervention prolonged the Great Depression, asserting that only the massive spending of WWII restored full employment.

Humanist economists challenge this worldview on several fronts. They argue that libertarianism overestimates the rationality and autonomy of individuals while ignoring structural inequalities, market failures, and the social foundations of economic life. Keynes showed that markets don't self-correct quickly and that government spending is essential during downturns. Stiglitz demonstrated how real-world markets suffer from information asymmetries and monopolies, requiring regulation. Sen emphasized that freedom must include access to education, health, and dignity, not just market choice. Piketty warned that unchecked capitalism concentrates wealth and undermines democracy, advocating for redistribution.

Together, these humanist-economists reject the libertarian ideal of a self-regulating market. They argue that human welfare, justice, and stability demand thoughtful governance, public investment, and institutions that serve people, not just profits.

Despite their ideological rifts, Keynes and Mises maintained a respectful rapport—an enduring testament to the power of humanist dialogue across intellectual divides. Their civility reminds us that even the fiercest debates can be grounded in respect, where ideas clash but dignity remains intact

Social Choice theory and Arrow's Theorem

Social choice models examine how to fairly combine individual choices (for example, preferences for different candidates in an election) with outcomes (such as granting one political party the power to govern) that are fair, equitable, and efficient. Can such a system of governance be designed?

Kenneth Joseph Arrow's 'Impossibility Theorem,' in the book "Social Choice and Individual Values," showed that no voting system can produce a decision that satisfies five reasonable criteria:[17]

- Unrestricted Domain: Voters can have any preference order
- Non-Dictatorship: No single voter should always determine the outcome
- Pareto Efficiency: if there's no other option that makes one person better off without making any of the others worse off.
- Independence of Irrelevant Alternatives: The choice between A and B shouldn't be affected by a third option C
- Transitivity, or Consistency: If A is preferred over B, and B over C, then A should be preferred over C.

It reveals that democratic decision-making cannot be both fair and logically consistent. (Amartya Sen has provided an excellent, brief proof of the theorem in the book, "The Arrow Impossibility Theorem"[18])

The Theorem reveals the inherent tension in democratic decision-making: that no voting system can satisfy all fairness criteria simultaneously. Amartya Sen writes that "the subject of rational democratic decisions seemed to be inescapably doomed... [but] it also emerged that while impossibilities and impasses of his kind can arise with considerable frequency and amazing reach, they can also be, in most cases, largely resolved"[19] through information and, therefore, education.

Examples of change

The New Deal

Nearly a century ago, John Maynard Keynes proposed an economic philosophy rooted in counter-cyclical investment and state responsibility for full employment. His ideas defied the orthodoxy of balanced budgets and market self-correction, provoking fierce resistance. Yet with extraordinary political will, Franklin D. Roosevelt embraced Keynesianism as the backbone of the New Deal. Infrastructure projects, safety nets, and regulatory reforms helped lift the U.S. from the Great Depression.

Sweden post WWI

Sweden's transformation, as Thomas Piketty notes, shows that inequality is not a cultural constant but a political construction. Before WWI, voting power was scaled by wealth, some municipalities were legal dictatorships. Through deliberate mobilization, this system was dismantled. The rise of the Social Democrats brought to power leaders who reimagined the state and introduced progressive taxation to fund public systems beyond profit logic.

A humanist kinship

What's striking in these examples is not just the boldness of the change, but its speed and peaceful nature. Within years, each country redefined the state and laid the foundations for a new social contract.

Keynes memorably wrote, "There is no reason why we should not feel ourselves free to be bold, to be open, to experiment, to take action, to try the possibilities of things."[20]

Despite differences, humanist social scientists and philosophers share a kinship with a commitment to confronting inequality and rethinking the role of the state. They advocate for inclusive, dignity-centred economics. Their belief in the human power to correct systemic failures, whether through enhancing human capabilities, progressive taxation, universal basic services, or climate justice, places them firmly within a humanist tradition.

24

Approaches to Democracy - Power and Politics

Mapping humanist thinkers

Humanist ideas of IS and OUGHT call for a reconstructed social contract, rational, inclusive, and morally accountable. Yet, humanist politics produces a spectrum of frameworks. T24.1 highlights leading voices along two key dimensions in political organization:

These differences show that humanist democracy is a journey and an experiment with many paths but a shared goal: collective flourishing.

The social contract and boundaries of power

Leviathan, liberty, and the humanist struggle

Hobbes's "Leviathan" (1651) imagined a chaotic state of nature, justifying surrender to absolute authority for peace.[1] Locke's "Two Treatises of Government" (1689) defended natural rights and limited government, with rebellion justified when rulers violate them.[2] Rousseau's "The Social Contract" (1762) added a vital dimension: the "general will" as the collective moral purpose of people.[3] For Rousseau, legitimate authority arises from participatory self-rule – an idea that resonates with M. N. Roy's ethical citizenship.

Their tension persists in today's democracies—liberal, participatory, deliberative, and illiberal—balancing freedom, equality, and authority. Coercion, as Joseph Stiglitz notes, is sometimes necessary; even traffic rules require enforcement.

Such necessity often arises in situations involving externalities or complex interactions among many individuals—where coordinated action cannot rely solely on voluntary compliance. The challenge lies in its justification, limits, and accountability.

		Approach to coercion and consent		
		Authoritarian	Democratic	Deliberation
Political orientation	Centralized	**Hobbes** (Leviathan logic)	**Piketty** (redistributive state)	
	Decentralized		**Acemoglu & Robinson** (strong society) **Locke** (natural rights, limited government)	**M. N. Roy** (radical humanism, rotational leadership)
	Technocratic		**Stiglitz** (smart regulation)	
	Ethical		**Sen** (public reasoning, plural justice) **Montesquieu** (separation of powers)	**Sandel** (civic virtue), **Popper** (critical rationalism), **Rousseau** (general will)

F24.1 Coercion/Consent vs Political Orientation

Leviathan in disguise

Hobbes's Leviathan, born of fear, risks becoming the monster it seeks to prevent. History shows protectors often morph into oppressors. Humanism rejects this trade-off: security must never cost dignity or democratic participation.

In "Twilight of Democracy," Anne Applebaum explores how democratic societies can slide into authoritarianism—not through violent coups, but through the subtle betrayal of democratic norms by intellectuals and elites.[4] Applebaum's "clercs"—intellectuals who abandon truth to serve power—offers a chilling reminder that the erosion of democracy often begins with the corrosion of integrity among those in positions of influence.

Post-9/11 emergency powers: surveillance, detention, war mandates, hardened into permanent infrastructure. The 'state of exception' became the rule. The U.S. model of emergency governance remains intact and emulated globally. The challenge is early recognition and redressal of Leviathan's disguises.

The Narrow Corridor

Acemoglu and Robinson's "The Narrow Corridor" argues liberty emerges when strong states are balanced by strong societies.[5] Their "Red Queen effect" captures this dynamic: both must evolve to check each other (C10 S: Evolution by natural selection).

Montesquieu's theory of separation of powers – legislative, executive, and judicial – remains key to this balance.[6] His insight that concentrated power breeds tyranny underpins modern constitutional design and reinforces the need for institutional checks. In the corridor, democratic resilience is built upon the designs of Montesquieu's architecture.

- Britain entered the corridor through centuries of struggle.
- India's caste system fragmented resistance and weakened societal checks; democratic institutions remain uneven.
- China is a 'Despotic Leviathan'—growth without liberty.
- Russia's brief democratic opening collapsed under centralized control.

The fragility of consent

Democracy depends on institutions and consent. When citizens disengage or feel unheard, even democratic systems risk sliding into technocracy or populism. As Michael Sandel argues, civic virtue must be cultivated, not assumed. Without shared moral reasoning, democracy becomes procedural, not purposeful.

Technology and power

In "Power and Progress," Acemoglu and Johnson show how elites use technology to entrench dominance.[7] AI and automation risk repeating this pattern, amplifying inequality and undermining democratic agency.

Varoufakis's "Technofeudalism" warns: tech giants now own the digital terrain, extracting rent and shaping behaviour through algorithms.[8] Yet technology is not liberating or oppressive; its democratic potential depends on design and governance.

Civic tech platforms like Decidim (Barcelona) and vTaiwan (Taiwan) show how digital assemblies can foster participatory policymaking. Blockchain voting, open-source deliberation tools, and algorithmic transparency offer avenues for democratic renewal. But without public oversight, these tools risk becoming instruments of surveillance and manipulation.

Humanist democracy must reclaim the digital terrain, not reject it. This requires confronting digital exclusion: algorithmic bias, connectivity gaps, and literacy barriers, especially for marginalized groups across caste, race, and disability.

Humanism's path

Humanism charts a path between radical libertarianism and authoritarian control, grounded in reason, compassion, and democratic experimentation. It accepts coercion only when guided by science, reasoning, and the common good.

Far-right critics of mandates ignored how unchecked autonomy endangers others; authoritarian regimes buried dissent beneath metrics. Across humanist thinkers, a shared principle emerges: coercion may be justified, but only when rooted in democratic legitimacy and moral reasoning. It must be principled, inclusive, and compassionate, never arbitrary or absolute.

COVID-19: ethical tensions

- Stiglitz backed public health mandates, arguing that individual choices, such as drinking bleach as a supposed COVID cure, impose costs on others. The medical burden falls on society, and the harm ripples through families and communities.[9] Mandates are justified to address these externalities. Yet, he criticized India's lockdown for overlooking poverty, noting that "markets won't provide the right level of precaution" without inclusive policy design. (STIGLITZ)[10].

- Sen praised Kerala's response but condemned India's lockdown: "Precaution must be justified through public discourse" (SEN)[11].
- Piketty focused on inequality: "Who decides what risks justify coercion—and whose interests are protected?" (PIKETTY)[12].

Science guides humanist ethics. In March 2020, epidemiologist Marc Lipsitch warned: "Waiting and hoping for a miracle… is not an option."[13] Distancing bought time, but that time demanded a massive societal effort. Humanism begins with science, then informs moral choices. Epistemic dimensions are explored further in C27 S: Humanist discourse as a layered process.

Environmental democracy

Externalities are not only economic, but ecological. Climate change, biodiversity loss, and resource depletion demand democratic responses that transcend borders. Ecological democracy insists that stewardship must be participatory, just, and inclusive.

From indigenous land rights to youth climate movements, resistance to extractive models reflects a humanist ethic: the Earth is not a commodity, and future generations are stakeholders. Confucian governance, with its emphasis on harmony and moral duty, and African communalism, which centres interdependence and stewardship, offer philosophical anchors for ecological democracy.

Humanist politics must evolve to meet challenges, not through technocratic fixes alone, but through moral imagination and collective responsibility.

Historical and alternative models

Political structures – examples from history

We often assume that humans are bound to a narrow set of 'natural' social structures. This belief stems from historical narratives and the rigidity of current institutions.

The ancient Greek city-states experimented with diverse arrangements: Athens practiced direct democracy among male citizens, Sparta combined dual kingship with oligarchy and militarism, Corinth favoured mercantile oligarchy, Thebes oscillated between democracy and authoritarian rule, Delphi was governed by priestly authority.[14] Yet all shared coercive hierarchies: male citizens held rights while women and slaves were excluded.

Mary Wollstonecraft's "A Vindication of the Rights of Woman" (1792) challenged this exclusion.[15] She argued that reason and citizenship must extend to women. Her humanism expands the boundaries of civic imagination, demanding that participatory democracy be truly inclusive. In this light, historical models reveal not just innovation but also injustice

David Graeber and David Wengrow, in "The Dawn of Everything," challenge the myth of linear political evolution (GRAEBER and WENGROW)[16]. They document egalitarian hunter-gatherer networks, governance alternating between kingship and councils, and consensus-based models like the Haudenosaunee Confederacy, where women held veto power and chiefs were accountable.[17]

Graeber asks: "Are we supposed to believe that before the Athenians, it never really occurred to anyone, anywhere, to gather all the members of their community in order to make joint decisions in a way that gave everyone equal say?"[18] He described pirate societies in Madagascar, particularly the Zana-Malata, who developed democratic and egalitarian governance (GRAEBER)[19].

James Scott's work on Zomia, a highland region across Southeast Asia, shows how societies deliberately avoided state formation and centralized authority.[20]

History reveals that our current systems are contingent, not inevitable or permanent. They show that non-hierarchical and rotational systems are not fantasies but realities. Seasonal leadership, federated councils, and consensus models challenge the assumption that centralized authority is necessary for order.

Comparative democratic experiments

Beyond ancient and anthropological cases, modern experiments offer practical insights:
- Kerala's participatory planning integrates local councils with state policy.
- Porto Alegre's budgeting empowers citizens to allocate public funds.
- Zapatista governance in Chiapas blends autonomy with collective decision-making.
- Swiss referenda provide input on national policy.
- African communalist traditions emphasize consensus-building, shared responsibility, and the moral authority of elders, offering a model of deliberation rooted in relational ethics and collective well-being.[21]

- Confucian governance, with its emphasis on virtue, merit, and moral cultivation, offers a contrasting vision of leadership grounded in ethical example rather than procedural legitimacy.[22]

These models show that democracy is not a fixed form but a flexible practice, shaped by culture, history, and imagination.

M. N. Roy – Radical Humanism

M. N. Roy (1887–1954) evolved from Marxism to Radical Humanism, a philosophy rooted in reason, ethical individualism, and secular democracy. He rejected both authoritarian communism and party politics, viewing politics as a moral vocation. He dismissed political parties as engines of conformity, where individuals surrender judgment to collective dogma. He envisioned a party-less democracy of ethical citizens guided by conscience, not partisan loyalty.

Democracy's failure, Roy argued, stems not from constitutions but from a passive public mindset. Representation, in his view, replaced direct democracy resulting in "The most skilful demagogue [becoming] the most successful democrat." (Roy)[23].

Decentralization and community control

Roy championed decentralized self-rule via democratic councils, distinguishing administrative devolution from true political empowerment. Communities should legislate and execute decisions, with higher bodies coordinating, not dominating, local governance. Essential services, he believed, must be run by boards of workers, consumers, and citizens. This aligns with Piketty's participatory socialism, though rooted in local agency.

Voluntary federation and rotational leadership

Roy's model federates units: village councils, cooperatives, assemblies, through voluntary coordination, forming a layered yet non-hierarchical system. To prevent elite entrenchment, he proposed rotational leadership: "No one should hold power long enough to forget that it is a trust, not a privilege."[24]

India's Panchayati Raj system offers a parallel. Constitutionally mandated in 1992, it enables temporary leadership by ordinary citizens: farmers, artisans, teachers, who serve and return to their trades. With over 1.8 million development plans uploaded since 2019–20, and over 40% of seats reserved for women and lower castes, it reflects Roy's ideals of participation and rotation. Yet challenges persist: financial dependency, bureaucratic interference, elite capture, and capacity gaps. Its promise depends on reforms in autonomy, equity, and education – especially for historically marginalized groups.

Switzerland's civic model offers another glimpse. Politics, here, becomes an activity of service—not domination (C16 S: Humanist flourishing).

Campaign finance and the economics of power

Roy warned against the corrosive influence of money in politics. He called for strict campaign finance controls and free access to public records. Today, opaque funding, from India's Electoral Bonds to U.S. Super PACs, has entrenched elite influence. Even in Europe, public funding coexists with corporate lobbying and media capture.

Graeber and Roy: everyday anarchism

David Graeber's anarchist anthropology shares Roy's ideals. Both saw ethical order emerging from voluntary association, not coercion. Graeber argued that mutual aid—waiting in line, resolving disputes informally—is the basis of social order: "Anarchism is just the way people act when they are free… and deal with others who are equally free."[25] Like Roy, he saw egalitarianism as a recurring choice, not a historical anomaly.

Piketty and Roy: two humanist lenses

Roy and Piketty offer complementary visions for justice and democracy. Both critique concentrated power—Roy targets centralized states; Piketty, capital's grip. Piketty's participatory socialism relies on legal reform and state-led redistribution. Roy's radical humanism calls for a revolution in consciousness.

One builds from the outside in; the other, from the inside out. Together, they offer a dual lens: Piketty reshapes institutions; Roy reimagines the citizen. Yet tensions remain—Piketty's model demands a robust state apparatus, precisely the kind of institutional concentration Roy resists.

Other thinkers and the limits of localism

Humanists like Sen stress the need for organized structures to articulate collective interests. Critics of Roy argue parties aggregate views and mobilize change—Scandinavian democracies thrive on this. Roy counters that ethical autonomy must remain central, warning against party-induced conformity. He acknowledges challenges like voter fatigue, proposing rotation and education as partial remedies.

Stiglitz insists global crises demand central authority and enforceable norms. Roy prefers voluntary coordination, but goodwill alone often fails. The Paris Agreement and persistent tax havens expose the limits of principled localism. In today's interconnected world, parts of Roy's vision, though attractive, struggle to meet the scale and urgency of global challenges, shaped by externalities that transcend borders, institutions, and ideologies.

Democracy and the ethics of resistance

Democracy is not just consensus, it is also conflict. Civil disobedience, protest, and dissent are vital correctives to institutional inertia. From Gandhi's satyagraha to Black Lives Matter, resistance has redefined democratic norms. Humanist democracy must protect dissent, recognizing that justice often begins with refusal. Institutions must be strong enough to absorb critique without collapsing into repression.

Civic foundations and democratic renewal

Humanist and democratic renewal

Aspects of Roy's radical humanism may no longer be entirely practical in a hyper-connected world, where externalities—economic, environmental, technological—transcend borders and demand coordinated, often centralized responses. The pursuit of equity and fairness today frequently requires collective action at national and global scales, challenging Roy's vision of decentralized governance.

Still, his insistence on moral autonomy, civic responsibility, public funding of elections, a pluralistic media landscape, and politics as service remain strikingly relevant in an environment of rising authoritarianism and democratic fatigue

Sandel and the moral foundations of democracy

Michael Sandel's political philosophy aligns with Roy: democracy must be rooted in moral reasoning, not managerial neutrality. In "Democracy's Discontent," Sandel warns that "a politics emptied of moral argument leads to a public life that is impoverished and alienating."[26]

He calls for institutions that cultivate civic virtue and shared purpose, not merely aggregate preferences. He critiques economic elites for eroding democratic legitimacy, but goes further, insisting democracy must restore dignity, not just redistribute resources.

Popper and the Open Society

Karl Popper's vision of government was rooted in his defence of the "open society," a political order defined by its capacity for self-correction (POPPER)[27]. For Popper, strength lay in tolerating dissent, revising policies, and protecting freedoms without collapsing into authoritarianism. His "piecemeal social engineering" reflected a humanist commitment: progress must be experimental, inclusive, and grounded in humility.

Civic imagination and democratic literacy

Popper saw education as foundational to the 'open society,' a political order defined by its capacity for self-correction. He rejected authoritarian instruction, advocating a pedagogy rooted in critical thinking, intellectual humility, and the freedom to revise beliefs. Education, for him, was not about obedience or truths, but about cultivating rational inquiry and ethical independence.

Roy reflecting Popper's concerns, wrote: "Education must aim at producing rational human beings, not merely skilled workers or obedient citizens."[28]

Humanist democracy depends not only on institutions but on citizens capable of ethical participation. Civic imagination, the ability to envision shared futures, must be cultivated through education. Democratic literacy involves understanding rights, responsibilities, and the mechanics of governance. Without it, even the best-designed systems falter.

Closing reflection

Humanist democracy is not a destination but a discipline: a commitment to reasoned disagreement, moral imagination, and ethical experimentation. It asks not only how power is organized, but how it should be justified, shared, and transformed. The journey of democracy is one of constant revision.

25

A Framework for a Humanist Education

Education – A vital lever

Humanists often forget they hold a vital lever: sustained engagement with humanist education. Education drives reform and equips individuals for collective action. Yet it can masquerade as propaganda and entrench dogma—prioritized conformity over inquiry and suppressing dissent and justify obedience.

Humanist education, by contrast, applies universally and lifelong. It is a means and an end: subversive, engaging, persuasive, and liberating. It is a weapon of liberation. It cuts through entrenched power structures with wisdom, challenging orthodoxy and awakening critical thought.

The printing press shattered monopolies on knowledge, enabling ordinary people to read, reflect, and resist. Literacy movements, from Renaissance pamphleteers to postcolonial educators, have repeatedly shown how access to ideas can destabilize empires and dogmas. When deployed, humanist education becomes insurgent: it equips minds to question authority, imagine alternatives, and act with conscience.

If humanist education is to counter dogma and inertia, its agenda must be consciously crafted, its principles clear, and its methods aligned with human flourishing.

Importantly, foundational conditions such as environmental safety and early childhood nutrition are inseparable from educational equity (C22 S: Essential embodied well-being).

Each depends on the other; progress demands a shift across foundations and systems.

A modern humanist framework for education

Efforts have been made to implement humanist educational principles across diverse contexts, from democratic schooling models to project-based learning environments, but these often remain partial, context-bound, or inconsistently aligned with the philosophical scope of humanist thought.

The following summary framework offers a more precise and systematically integrated approach, not just in pedagogical style, but in curricular structure and long-term developmental aims.

Quotes have been deliberately included to serve not only as anchors but also to stay connected to the lived values and enduring insights that humanist education aspires to cultivate.

One of the most inspiring illustrations of this spirit is found in Sarah Bakewell's book "Humanly Possible" (Bakewell)[1]. With warmth and generosity, Bakewell traces the lives of humanists across centuries, showing how their commitments to curiosity, reason, and compassion were lived practices. Her portraits reveal that humanism is a lifelong adventure of learning and questioning. It offers a reminder that education, at its best, is inseparable from the art of living.

Purposeful and lifelong

Education is a means of human flourishing and an integral part of life's journey. As part of life's journey, learning continues across all stages of life.

John Dewey, the humanist-educator wrote, "Education is a process of living and not a preparation for future living" (Dewey)[2].

Wilhelm Humboldt (1767–1835), an architect of modern liberal education, shared Dewey's conviction that education is inseparable from life itself. For Humboldt, the purpose of education was the cultivation of the individual's inner freedom and moral character through engagement with the world. He believed that "the ultimate task of our existence is to shape ourselves into a harmonious whole," emphasizing that lifelong learning must foster autonomy, curiosity, and the development of one's full humanity.[3]

Universal and free

Education is a public good and has to be accessible freely to all, irrespective of geography, identity, or socioeconomic status. This is a moral and civic imperative.

Thomas Piketty reinforces the point: "Over a long period of time, the main force in favour of greater equality has been the diffusion of knowledge and skills" (Piketty)[4].

Understanding and demonstrating the unity of knowledge

Knowledge acquisition integrates the sciences, humanities (pluralistic and global), creative arts, ecology, and design. Two important aspects related to this recommendation are:
- First, this is not an argument for generalities across a multitude of subjects. As Whitehead writes, "Let the main ideas which are introduced into a child's education be few and important, and let them be thrown into every

combination possible. The child should make them his own and should understand their applicability here and now in the circumstances of his actual life" (WHITEHEAD)[5].

- Second, this broad understanding of the unity of knowledge belongs less to the accumulation of details and more to the moral imagination. It is about perceiving the connections—ecological, cosmic, human—that bind the sciences, humanities, and service.

This need to understanding the unity of knowledge cannot be emphasized enough. Buckminster Fuller coined the phrase Spaceship Earth to describe the planet as an interdependent system with finite resources and no instruction manual. "We are not going to be able to operate our Spaceship Earth successfully nor for much longer unless we see it as a whole spaceship and our fate as common. It has to be everybody or nobody" (FULLER)[6].

Deep: cultivating mastery through focus

An explanatory and cautionary note should be made about the above broad aspects of education. There are a set of core aspects in the unity of knowledge that give rise to an appreciation of a broad knowledge base. In addition to this we should follow Whitehead's additional suggestion: "What you teach, teach thoroughly."[7] Such learning to master a few subjects with rigor and intensity cultivates intellectual discipline and fosters the ability to think critically and creatively.

Importantly, Whitehead's emphasis on depth does not contradict the ideal of broad cultural appreciation. On the contrary, he argues that "the art of a teacher is to stimulate imagination and to develop the power of analysis."[8] By mastering the skill of inquiry in one domain, students become equipped to engage meaningfully with diverse fields. The capacity to learn deeply enhances one's ability to appreciate the interconnectedness of ideas across disciplines, making future explorations accessible and enriching.

This vision of deep and broad learning aligns with Humboldt's ideal of 'Bildung,' — a holistic process of self-cultivation through intellectual and aesthetic engagement. Humboldt argued that true education arises when individuals encounter the world in its richness and complexity, and then reflect upon it to develop their own capacities. In this sense, mastery of a subject is not an end in itself, but a gateway to understanding the unity of knowledge and the moral dimensions of inquiry.[9]

Numerate and analytically literate

Numeracy is not just a requirement for the sciences. It is a vital skill to address questions of life that, as John Allen Paulos writes, "arise naturally when one transcends one's self, family, and friends; How many? How long ago? How far away? How fast? What links this to that? Which is more likely? How do you integrate your projects with local, national, and international events? with historical, biological, geological, and astronomical time scales?"[10]

To be a full participating citizen of Earth, "Remembering this formula or that theorem is less important ... than is the ability to look at a situation quantitatively, to note logical, probabilistic, and spatial relationships, and to muse mathematically."[11]

This includes the broad understanding of plausible reasoning, such as that behind Jesica Reyes' statistical analysis showing that "changes in childhood lead exposure are responsible for a 56% drop in violent crime."[12] Similarly, to appreciate the latest WHO estimate that as of 2025 air pollution is responsible for approximately 7 million premature deaths annually worldwide one needs to understand the methodology used.[13] Otherwise, this is easily dismissed, remains a number or is accepted without understanding – all of which hamper public discourse and decision making on existential issues.

Barbara Oakley provides a resource for anyone, of any age, to begin the journey of numeracy and to explore new learning techniques.[14]

Experiential and with nature

Learning proceeds through direct engagement in nature, within the community, and through observations and experimentation. There are many aspects to experiential learning.

One part comes from direct engagement with reality via science's self-correcting process that is empirically based and that fosters both intellectual rigor and democratic sensibility. Denis Diderot, the Enlightenment philosopher and co-editor of the Encyclopédie, emphasized this model of education and knowledge acquisition. He wrote, "There are three principal means of acquiring knowledge… observation of nature, reflection, and experimentation."[15]

Another part involves cultivating awareness of ecological interdependence, fostering stewardship through conservation, and embedding experiential learning in natural environments. Outdoor engagement, whether through gardens or forest walks activate sensory perception, emotional resonance, and systems thinking. As Rachel Carson wrote, "The more clearly we can focus our attention on the wonders and realities of the universe about us, the less taste we shall have for destruction" (CARLSON)[16].

Lucy Jones, in "Losing Eden," argues that our psychological well-being is entwined with the natural world, and that disconnection from nature contributes to rising mental health issues.[17] Reclaiming this bond is not a luxury but a necessity for emotional resilience and societal healing.

The social sciences, history and philosophy are also subjects for experiential learning. Sven Lindqvist's "Dig Where You Stand" advocates for direct engagement with archives, oral testimonies, and workplace records, asserting that "the person who has lived the history is often best placed to uncover it."[18]

Mentoring, collaborative and intergenerational

Learning is a shared activity across the globe and between generations. In Rachel Carlson's reflections on childhood, she emphasized that "If a child is to keep alive his inborn sense of wonder, he needs the companionship of at least one adult who can share it" (CARLSON)[19]. Equally, the act of teaching gives adults insight into the subjects being taught and forces them to clarify their own understanding. Richard Feynman wrote, "If you can't explain something to a first-year student, then you haven't really understood it."[20]

Ethically and civically grounded

Martha Nussbaum recommendation is for a two-year national service program. This service would involve work in areas such as elder care, environmental restoration, and community health. These domains can cultivate empathy, responsibility, and solidarity. In this way, by integrating service into education,

Nussbaum envisions a system where students not only learn about justice but actively participate in its realization.

An ethically and civically grounded learner understands systemic trade-offs, appreciates the designs of decision systems, and engages with justice.

Practical and skilful

Education cultivates skills that serve ethical design, ecological stewardship, and sustainable organization, activities that should not be replaced by automation or machine learning (AI). Even for those activities that AI and automation can and will cover, practical skills and the knowledge of crafts of engineering and design will still have to be learned and retained. John Lienhard wrote, "Engineering is not only the application of science, but also the exercise of imagination, judgment, and skill in the service of human need."[21]

Difference from a liberal arts education

The explanations above clarify how a humanist conception of education diverges from familiar liberal education frameworks. While liberal education emphasizes breadth, critical thinking, and civic engagement, it often remains tethered to institutional traditions, and disciplinary silos. Humanist education, by contrast, is grounded in a curriculum of ethical universality, lifelong experiential learning, and free access.

Contemporary liberal arts institutions focus on young students, often demand exorbitant fees, saddle students with debt, and channel them toward employment, frequently in roles that resemble David Graeber's notion of "bullshit jobs."

In contrast, humanist education is inclusive of all age groups and centres on individual and planetary flourishing. It encourages lifelong engagement in creative endeavours akin to Mihaly Csikszentmihalyi's concept of creative flow.

Education, technology, and AI

Michael Dertouzos warned in 2001 that computing had stalled, not from lack of innovation, but from failing to serve human needs. He envisioned tech shaped by five principles: natural interaction, automation, personalized access, collaboration, and customization. Much of this exists today: voice assistants, algorithms, remote tools, but the greater goal remains unmet. Usability rose, yet dignity, empathy, and empowerment lag.

Tech leaders like Musk and Karp chase scale and control. Humanist critics—Turkle, Benjamin, Harris—ask how tech affects emotion, justice, and attention. They call for design rooted in empathy and autonomy, not distraction or bias.

Most AI is machine learning—pattern recognition—not understanding. It excels in narrow tasks: protein folding, text generation, diagnostics, but lacks moral reasoning or ethics. True AGI would grasp nuance and moral judgment. That frontier is still ahead—technically and philosophically.

Khan Academy broke elitism with free, innovative learning.[22] Salman Khan sees AI as transformative, personalizing education and expanding access, but warns it must augment, not replace, the ethical core of teaching. AI can scale, but not mentor, empathize, or guide.

Digital learning risks dehumanization: avatars erode empathy (Medusa Effect), and anonymity fosters hostility

(Online Disinhibition). Without philosophical grounding, AI may amplify bias and inequality.

Automation displaces millions; re-skilling often serves profit over purpose. Instead, humanism would redirect automation's wealth to fund creative, scientific, and service-oriented learning, fostering meaning, identity, and community.

This vision reimagines universities. In an AI-shaped world, they must evolve from elite, profit-driven institutions into public sanctuaries of learning and service, moving beyond credentialing to active outreach for displaced workers, underserved communities, and curious minds.

Through such efforts, universities can regain trust and shift from knowledge production to human flourishing, advancing Keynes' vision for the future.

N10 Febrile triviality

"Bread and circuses."[23] Juvenal. Ancient Rome's formula for pacifying the masses

Part 9

VISIONS OF LIFE in a better world

Directions Towards a Better World

From cleverness to wisdom

An Ape looks at a Blue Dot

"The visions we offer our children shape the future. It matters what those visions are. Often they become self-fulling prophecies. Dreams are maps.

I do not think it irresponsible to portray even the direst futures; if we are to avoid them, we must understand that they are possible. But what are the alternatives? Where are the dreams that motivate and inspire? We long for realistic maps of a world we can be proud to give our children. Where are the cartographers of human purpose? Where are the visions of hopeful futures of technology as a tool for human betterment and not a gun on hair trigger pointed at our heads?"[1]

SAGAN

26

An Ape Looks at a Pale Blue Dot

The Pale Blue Dot and Carl Sagan's vision for the future

Worldviews shape our visions

Our visions for the future are rooted in the worldviews we have about the universe. These perspectives orient us toward certain values and shape what we consider possible. Sagan advised that "broadening our perspective, even if we do not find what we are looking for, gives us a framework in which to understand ourselves far better" (SAGAN)[1].

The first photograph of Earth taken from space

The first photograph of Earth was taken at an altitude of about 105 kilometres by a V-2 rocket launched into space on October 24, 1946 (F26.1). The camera then fell back to Earth but the film survived, protected in a steel case.

Fred Rulli, a young army man who retrieved the steel case and film when it crashed back to Earth, said, "when they first projected [the photos] onto the screen, the scientists just went nuts."[2] The scientists were astonished to see the curvature of Earth set against the blackness of space, a view no human had ever seen before.

The biosphere (all of life on Earth at a given time) exists on a "razor-thin" surface "like a membrane, it cannot be seen sideways with unaided vision orbiting outside Earth's atmosphere" (E. O. WILSON)[3].

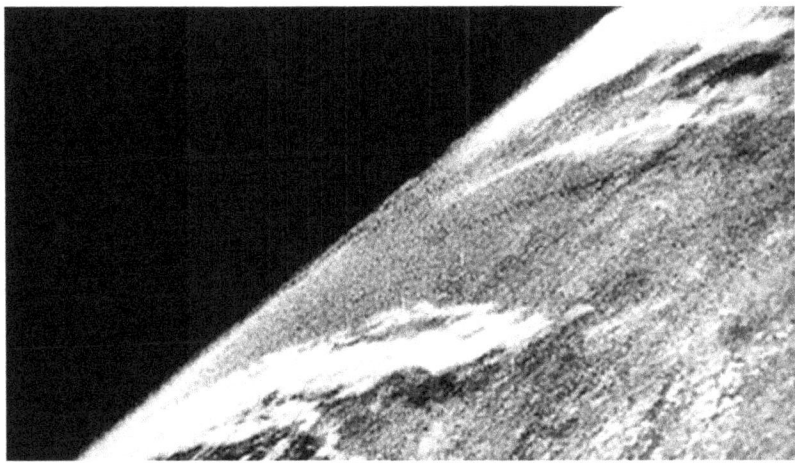

F26.1 First photograph of Earth from space. Image courtesy of White Sands Missile Range/Johns Hopkins University Applied Physics Laboratory.[4]

A transformative photograph of the universe saturated with galaxies

In 1995, astronomers pointed the Hubble Space Telescope at a seemingly empty patch of sky. The result was astonishing with nearly 3,000 galaxies appearing within that tiny spot in the sky. The night sky, even in its darkest corners, is teeming with entire galaxies, each galaxy containing billions of stars and planets[5] (F26.2).

F26.2 First Hubble Deep Field Image. Image courtesy of NASA, ESA, and R. Williams (STScI); the Hubble Deep Field Team.[6]

Image of the Pale Blue Dot

The 'Pale Blue Dot' photographs of Earth were taken on February 14, 1990 by Voyager 1 at the suggestion of Carl Sagan. Voyager 1 was around 6 billion kilometres from Earth, making it the most distant image of our planet ever captured. The final image (F26.3), a montage of photographs, is neither too close (where the Earth dominates the picture frame) nor too distant (showing an unimaginably large universe with no sight of Earth).[7]

Sagan writes that "From this distance the planets seem only points of light…Because of the reflection of sunlight off the spacecraft, the Earth seems to be sitting in a beam of light, as if there were some special significance to this small world. But it's just an accident of geometry and optics. …Had the picture been taken a little earlier or later, there would have been no sunbeam highlighting the Earth."

F26.3 The Pale Blue Dot. Image courtesy of NASA/JPL-Caltech. Processed by Kevin M. Gill with input from Candy Hansen and William Kosmann.[8]

Sagan's vision of the future

Carl Sagan was an exemplar humanist. His eloquent, poetic description of the pale blue dot photograph and the vision for our future that this photograph of "a mote of dust suspended in a sunbeam," engendered is inspiring.[9]

Sagan's vision calls for stewardship of Earth. The photograph shows how vulnerable Earth is. Sagan urged us to "preserve and cherish the pale blue dot, the only home we've ever known." The image provides us with a unique perspective on our successes and follies. Sagan writes about how the image highlights the absurdity of human arrogance — our wars, our imagined self-importance, our divisions. From the perspective of the image, all our dogmas and separate tribal ambitions disappear.

Sagan extends this perspective into a hopeful future. The adventures of science and the creative arts are what truly unites humanity, not as conquerors of nature, but as curious, compassionate stewards of life on Earth. Sagan's vision is that we should live with "responsibility to deal more kindly with one another, and to preserve and cherish the pale blue dot, the only home we've ever known."[10]

The Voyager missions[a]

The story of Voyager begins with Walter Hohmann, who in 1925 calculated a spacecraft could move between planetary orbits using minimal fuel, a crucial insight, since fuel comprises up to 90% of launch mass. In 1954, Derek Lawden proposed gravity assist: using planetary flybys to alter trajectory and velocity.

---

a   Ben Evans has written a comprehensive, yet relatively unknown, history of the Voyager missions: Ben Evans, *NASA's Voyager Missions: Exploring the Outer Solar System and Beyond*, 2nd ed. (Cham, Switzerland: Springer, 2022)

By 1961, Michael Minovich refined this with numerical methods, enabling spacecraft to gain speed via planetary slingshots. His work guided Mariner 10's journey from Venus to Mercury.

In 1965, Gary Flando identified rare trajectory windows, enabled by a once-in-176-years planetary alignment, that could allow a single spacecraft to visit Jupiter, Saturn, Uranus, and Neptune using gravity assists. This became the dream of the 'Grand Tour.' By 1969, James Long at JPL designed missions to pursue it. But with Nixon's shift away from space competition, NASA dropped the Grand Tour in 1973, funding two smaller Mariner-type missions instead. On March 4, 1977, they were renamed Voyager. Though the full tour was no longer official, scientists hoped for an extended mission.

Voyager 2 launched on August 20, 1977, followed by Voyager 1 on September 5. Each spacecraft measured 6.4 by 3.7 by 4.0 meters, with a 13-meter magnetometer boom and weighed 825 kg. Titan-Centaur rockets launched them into space at speeds up to 17 km/s (50 times faster than a Boeing 747).

For millennia, planets were seen as divine lights. Galileo's discovery of Jupiter's moons shattered the view, revealing planets as objects like the Earth. The second revolution came with spacecraft flybys: Mariner 4's images of cratered Mars in 1965, Pioneer 10's journey through the asteroid belt to Jupiter in 1972, and Pioneer 11's slingshot to Saturn in 1973.

In 1979, Voyager 1 reached Jupiter, followed by Voyager 2. They flew by the gas giants and their moons, revealing dynamic, geologically active worlds. Io had volcanoes, Europa a subsurface ocean. Saturn's rings showed intricate structures;

Titan had a dense atmosphere. Voyager 2 later uncovered Uranus's tilted magnetic field and Neptune's supersonic winds.

Though only approved for Jupiter and Saturn, Voyager 2's trajectory preserved the option for more. After Voyager 1's successful Titan flyby in 1980 and Voyager 2's Saturn flyby in 1981, NASA extended the mission. By 1989, the Grand Tour was achieved.

In 1990, Carl Sagan proposed a photograph of Earth from deep space. On February 14, Voyager 1, 6 billion km away and 32 degrees above the ecliptic plane, turned back and captured the "Pale Blue Dot." Sagan believed this image could inspire humility, a visual confirmation of our place in the universe.

Voyager 1 became the most distant human-made object in 1998. It entered interstellar space in 2012; Voyager 2 followed in 2018. As of mid-2025, Voyager 1 is over 24 billion km away (22.5-hour signal delay), and Voyager 2 is 20 billion km away (18.5-hour delay). Though aging, both still transmit data.

Each Voyager carries a 12-inch gold-plated copper disc, a message to any extraterrestrial intelligence. Designed by a team led by Sagan, the record includes 116 images, natural sounds, greetings in 55 languages, and 90 minutes of music. A bottle cast into the cosmic ocean, it reflects humanity's diversity, creativity, and hope.

27

Augmenting Reasoned Discourse

A landscape of knowledge

Psychologists Joseph Luft and Harrington Ingham introduced a model for interpersonal awareness, now known as the "Johari Window"—a blend of their first names.[1] This framework can be adapted into a broader 'landscape of knowledge,' as illustrated in T27.1.

	Known	Unknown
Known	Known knowns *Shared knowledge, culturally embedded* *Mercury, Venus, Mars, Jupiter, Saturn* *Known since antiquity*	Known unknowns *Hypothesized, inferred, but not confirmed* *Ceres, Kuiper Belt (Hypothesized much before confirmed)*
Unknown	Unknown knowns *Known (observed) but not truly understood* *Uranus pre-1781 CE (In 1690 CE Babylonian's saw Uranus but misclassified it.)*	Unknown unknowns *Invisible until instruments and imagination catch up* *Cosmic Microwave Background Radiation (discovered 1965)* *Fast Radio Bursts (discovered 2007)*

T27.1 The Johari Window Landscape of knowledge[2]

The adventure of knowing about our solar system has a long history. The Babylonians tracked the planets Mercury, Venus, Mars, Jupiter, and Saturn across the sky. Uranus, Neptune, and Pluto remained unknown until discoveries between 1781 and 1930.

The Kuiper Belt is home to over 100,000 icy objects larger than 100 km in diameter, with a handful comparable in size to Pluto. These bodies, stretching from Neptune's orbit to 8 billion kilometres from the Sun—over 50 times the Earth–Sun distance—were first hypothesized by Kenneth Edgeworth in the 1940s and independently by Gerard Kuiper in 1951. Direct observational evidence came in 1992, when David Jewitt and Jane Luu discovered (15760) Albion, the first Kuiper Belt Object beyond Pluto and Charon.

There are also many unknown–unknowns. Some we now recognize include:

- Cosmic Microwave Background Radiation, discovered in 1965 by accident. It revealed the Big Bang's afterglow—an unknown unknown that transformed cosmology.
- Fast Radio Bursts (FRBs), discovered in 2007. These signals were unpredicted and remain poorly understood.

E. O. WILSON describes our journey of understanding as, "The cost of scientific advance is the humbling recognition that reality was not constructed to be easily grasped by the human mind. This is the cardinal tenet of scientific understanding. Our species and its ways of thinking are a product of evolution, not the purpose of evolution."[3]

The Voyager missions exemplify the humanist pursuit, deepening our known-knowns, confirming known-unknowns, illuminating unknown-knowns, and revealing new unknown-unknowns.

Humanism vs Idols of the Mind

How science changes the landscape of knowledge

Knowledge evolves through science's self-correcting lens. Ancient cosmologies offered mythical origins, not numerical timelines—Earth's age was unknowable. Genesis imagined a six-day creation; Bishop Ussher dated it to 4004 BCE. In the 18th century, Hutton introduced 'deep time,' and Lyell laid geology's foundations. Estimates grew: Phillips (96 million), Kelvin (20–400 million), Rutherford (500 million). In 1956, Clair Patterson finally measured it: 4.54 billion years. The unknown became known.

Why the age of the Earth matters

Does Earth's age matter? The modern humanist says yes. Theologian Douglas Wilson disagrees: "The issue is always what God said, and not how old something is."[4] But replacing fact with fantasy weakens policy, education, and public trust. Earth's age, 4.54 billion years, is confirmed by geology and evolution. Ignoring it undermines the very methods that revealed it.

This denial spreads: eroding vaccine trust, climate action, and scientific literacy. Creationist theories sever our link to evolution, which explains traits like empathy, cooperation, and moral reasoning. Rejecting deep time also breaks our ecological awareness, without it, stewardship becomes sentimental, not informed.

Truth matters. Science isn't optional, it's the foundation of justice and responsible action. As myths fade, like Jupiter as a god, we uncover realities more awe-inspiring than fiction: a swirling gas giant with immense magnetic fields. Reality, not fantasy, is our most astonishing inheritance.

Humanism rejects the sacred bias

Why should humanists privilege certain ideologies or be less sceptical? Scientists, drawn to the Dalai Lama's openness, have long engaged Tibetan monks in scientific dialogue. The Dalai Lama himself has shown interest in neuroscience, quantum physics, and cosmology, hosting 'Mind and Life' conferences since 1987.

Yet this embrace of science coexists with metaphysical beliefs. Tibetan Buddhism, often seen as benign, had a feudal past marked by harsh punishments—mutilation for theft—no different from Sharia law. This history does not justify China's brutal occupation, which continues to inflict cultural erasure and human rights abuses.

The Dalai Lama concedes that beliefs like reincarnation are unfalsifiable and lie outside empirical scrutiny. In the documentary, "Wisdom of Happiness," he acknowledges that some beliefs are "personal and subjective" and "cannot be studied scientifically".[5] His principal English translator, Thupten Jinpa, clarifies: absence of evidence is not evidence of absence.[6]

This argument is often used across ideologies, effectively granting religious and spiritual convictions an 'opt-out' from the rigorous demands of evidence and falsifiability. But without a test of knowledge, how do we know whether these beliefs are delusional and imaginary?

Meditation without myth

Meditation has measurable effects on brain function and emotional regulation. Sam Harris writes in "Waking Up: A Guide to Spirituality Without Religion," "Being mindful is not thinking more clearly about experience; it is the act of experiencing more clearly."[7] Benefits appear across traditions— Tai Chi to yoga.

Yet the utility of meditation does not validate the truth claims of the broader belief system they emerge from. Harris frames meditation as a skill that can be cultivated without religious belief. Reincarnation, like resurrection or karma, remains a myth, not a model.

Beware of the Idols of the Mind

Francis Bacon (1561–1626) who laid the theoretical foundations for the modern scientific method warned, "Beware of the idols of the mind."[8] Dogmas, even when accepted in some areas of life, not only promote predominantly anthropocentric worldviews that conflict with rational inquiry but also shirk responsibility for evaluating and enacting the changes that we must undertake.

From persecution to exclusion to tolerance to humanism

Martin Rees warns that challenging religion may alienate believers from science.[9] But humanism engages and persuades through reason, not coercion. It demands no tests of belief—only dialogue. To grasp its moral depth, we must trace the journey from persecution to exclusion, tolerance, and humanism.

Persecution treats dissent as danger. From Mansur al-Hallaj's execution for mysticism to Europe's burning of heretics, history shows regimes suppress thought. Nazi Germany revived this pattern, and it persists today—from Iran's oppression of Bahais to the fatwa against Salman Rushdie, culminating in his 2022 attack.

Exclusion marginalizes without overt violence. Athenian democracy excluded women and slaves; Rome denied rights to non-citizens. Jim Crow laws segregated Black Americans after slavery ended. Today, LGBTQ+ individuals and immigrants still face legal and social barriers.

Tolerance is conditional and is granted reluctantly. Ritchie Robertson: "What is tolerated is also disapproved of."[10] Rees's call for coexistence avoids engagement. Fauci, during COVID-19, tolerated vaccine denial rather than confronting it with compassion and evidence. Amartya Sen calls disengaged tolerance a "lazy resolution" of the form: "you are right in your community, and I am right in mine."[11]

Humanism seeks understanding and compels engagement and persuasive dialogue, recognizing our solidarity as humans - shared biology and emotional capacities. Kant's ethics demand we treat others as ends, not means. Jefferson saw dialogue and civility as moral respect, though his relationship with Sally Hemings reveals the limits of his ideal. Genuine humanist discourse requires equity—no party should be subject to subjugation.

This journey from heresy to humanism is not just historical—it is moral. It reflects a commitment to dialogue, fought for by reformers who saw difference as opportunity to engage and learn.

Humanists must challenge dogma. Unquestioned beliefs distort reality and defer to imagined authorities. BERTRAND RUSSELL[12,13] defined religions as "a set of beliefs held as dogmas, dominating the conduct of life, going beyond or contrary to evidence."[14] He wrote that he was as "convinced that religions do [as much] harm as I am that they are untrue."[15] Examples can be found in the Catholic Church's opposition to artificial birth control and the Taliban's barring of girls from schools.

M.N. Roy warned that religion implants habits of obedience, undermining democratic thought. Belief in divine predetermination fosters servitude, not just spiritual, but political and economic. Today, people seek salvation in demagogues in a manner similar to theological deference.

Without active engagement and persuasive dialogue, humanists risk allowing pseudoscience, authoritarianism, and inequality to flourish. Silent toleration is not neutrality; it is moral failure.

Distinctions between intelligence and rationality

Keith Stanovich, in "What Intelligence Tests Miss: The Psychology of Rational Thought,"[16] draws a distinction between intelligence (as measured by IQ tests) and rationality, which he defines as cognitive competencies involving judgment, decision-making, probabilistic reasoning, and epistemic self-regulation (the ability to think carefully about what one believes and why, and being willing to change one's mind when the evidence says one should).

Highly intelligent individuals are often just as vulnerable to irrational thinking as anyone else. Intelligence may enhance one's ability to process information, but it does not necessarily lead to sound reasoning or objective decision-making.

Keith Stanovich introduced 'dysrationalia' to describe this phenomenon (STANOVICH)[17]. Stanovich explains:[18]

- Intelligence measures how well a person can hold beliefs in short term memory and efficiently process information.
- Rationality assesses if beliefs are worth computing with. It is the holding of beliefs and analysing goals that are commensurate with available evidence. Beliefs are calibrated appropriately to evidence, and one accepts that ambiguous evidence leads to tentative beliefs.[19]

Characteristics of irrational and rational individuals are shown in T27.2

Irrational	Rational
Cognitive inflexibility Belief perseverance	Cognitive flexibility Beliefs change based on evidence
Need for closure	Lives with uncertainties
Over confidence	Aware of ignorance / knowledge gaps
Confirmation bias	Looks for falsifications
Insensitive to inconsistencies	Investigates inconsistencies
Taboos and lack of intellectual engagement	No taboos – intellectually curious and engaged

T27.2 Characteristics of irrational and rational individuals. Adapted from STANOVICH[20]

Humanist discourse as a layered process

Rationality, justice, and the humanist worldview

C15 S: A humanist worldview highlighted how the close compact between science and moral decision making serves as a cornerstone for the emergence of the humanist worldview. This section provides the epistemic foundation underpinning that compact. The result is a layered conception of discourse, one that integrates science and ethical engagement.

The 'Is-Ought' divide and enlightenment rationalism

David Hume's 'is-ought' problem warned against deriving moral imperatives from factual premises. He wrote that "the rules of morality,…, are not conclusions of our reasons"[21] and cannot be logically derived from purely factual premises. Yet Enlightenment thinkers like Kant and Rousseau conflated epistemic and normative domains, believing reason alone could yield universal morality. Modern philosophers like Rawls and Parfit continued this pursuit. (See Appendix C31 for a critique of Parfit's mathematical formulation.)

Sen's rationality: justice through public dialogue

Sen's framework emphasizes rationality as ethically engaged, pluralistic, impartial, inclusive, and comparative (i.e. instead of seeking perfect justice, we aim to reduce injustice). Sen's rationality is not just about what is true, it is about what is fair. He relies on "positional objectivity" and "open impartiality," which invite global perspectives into moral reasoning leading to "government by discussion."

Scientific rationality: A self-correcting epistemic foundation

Scientific rationality is falsifiable, replicable, peer-reviewed, and Bayesian. It does not conflate "is" with "ought." It often yields partial knowledge when findings are ambiguous and it cannot be applied to normative questions — those that ask 'what ought to be.'

Case studies in epistemic breakdown

Consider the global challenge of human-engendered climate change where the application of scientific rationality on the 'is' question gives clear results (DESSLER):[a,22] anthropogenic emissions are destabilizing the planet. Sen's inclusive model invites all voices:

- Religious conservatives. Calvin Beisner, founder of the Cornwall Alliance has said, "We believe Earth and its ecosystems—created by God's intelligent design and infinite power and sustained by His faithful providence—are robust, resilient, self-regulating, and self-correcting."[23]
- CEOs. Darren Woods, CEO of ExxonMobil, has stated, "We do not support a carbon tax that would disadvantage U.S. industry and consumers."[24]
- National leaders. Trump while announcing the U.S. withdrawal from the Paris Agreement in 2017, said, "I was elected to represent the citizens of Pittsburgh, not Paris."[25]

Without a shared epistemic baseline, dialogue breaks down. The 2001 UN Durban Conference, during the leaded gasoline debate, and throughout the COVID-19 pandemic—where

---

[a] An outstandingly clear and authoritative resource on climate change science and policy is Andrew E. Dessler, *Introduction to Modern Climate Change*, 3rd ed. (Cambridge: Cambridge University Press, 2021)

misinformation and ideology overran evidence—illustrate this failure.

These examples show where multiple moral claims coexist and underlying assumptions are not challenged, then none can be disproven. This weakens the epistemic clarity needed for decisive moral action.

The layered process

Layer One

The first layer of humanist discourse is understanding—drawing on psychology, neuroscience, sociology, history, and anthropology—to explore biases, motivations, priming effects, and cultural memes that shape both the humanist and their interlocutors.

By acknowledging these factors, the humanist cultivates a more authentic compassion, recognizing that all participants of dialogue are shaped by complex, often invisible forces. This empathetic awareness fosters a genuine openness to understanding others. Only then can dialogue begin: not as a contest of ideas, but as a shared journey toward mutual insight and human flourishing.

Layer Two

Humanism begins its dialogue by focusing public engagement on the 'IS'—what we know and how we know it—through data, observation, and critical scrutiny of the issues at hand.

How might scientists engage in such a discussion? Through engagement and persuasion. This layer involves wider public dissemination, debate, and discussion on all aspects of science.

This layer is rarely pursued with vigour. Even well-meaning protesters who join marches cannot cogently explain what drives climate change. No wonder many 'grow up' and begin to resemble the 'authorities' they once opposed.

Layer Three

By first encouraging public scrutiny and inclusive reasoning about empirical insights, the third layer addressing human costs and context, becomes easier (though of course not easy) to achieve.

This layer turns to the 'OUGHT' questions and fully engages with Sen's ideals of "giving serious consideration to distinct and contrary arguments and analysis coming from different quarters"[26] as part of a participatory process. Having established a common base of discussion on what 'is' during *Layer Two*, better attempts can be made to be "impartial" and conduct "non-parochial scrutiny" of the 'ought' questions.

Applying the process – climate change discussions

International climate change discussions might seem to follow the layered process. The Intergovernmental Panel on Climate Change (IPCC), an independent scientific body, first examines the evidence and produces authoritative assessments on climate science, impacts, and mitigation options. Then the Conference of the Parties (COP) — the decision-making body of the UNFCCC — convenes to make the moral and political decisions about what actions countries will commit to. Together, the IPCC and COP form the science–moral–policy interface.

In practice, however, this interface is strained. IPCC reports are dense and technical, and the public typically receives only filtered summaries via media outlets. These summaries are shaped by local and global lobbies and interests, meaning that public understanding is often partial or distorted. Political leaders, who respond to public opinion, end up advocating positions that reflect lobby-driven narratives rather than scientific urgency.

Moreover, while COP sessions formally "note" or "welcome" IPCC reports, delegates rarely engage deeply with the findings. Negotiations proceed with the reports as background documents, not as central anchors. As a result, COPs are driven more by political expediency than by scientific understanding.

The layered process suggests a need for much wider and deeper public engagement with climate science. A more informed citizenry would raise pressure for ambitious action. Likewise, COP sessions that begin and end with structured, detailed reviews of IPCC findings — rather than merely acknowledging them — would strengthen the science–moral–policy connection.

Political decisions would still be shaped by national interests, economic concerns, and short-term geopolitics. However, the suggested layered process would help democratize climate science and embed it more deeply in political negotiation. In doing so, these steps may increase the likelihood of the accelerated climate action that is urgently needed.

Bridging the empirical and the ethical

None of the layers is easy. The point however is that while moral philosophers, social scientists and economists offer valuable frameworks, when the discourse is detached from a rigorous scientific foundation, they cannot reliably yield justice, fairness, or any meaningful form of equity.

We owe an immense debt to Sen and other humanist-social scientists for focusing on essential, core questions of life, but they rarely discuss the actual mechanics of doing science or the findings of science in their works. Modern humanism expands their ideas into an integrated framework. It results in a modern, transformative response to the challenges of our time.

Flourishing without fixed moral truths

Many traditions assume moral truths exist independently of human minds, fixed, eternal, and discoverable through reason or revelation. Humanism rejects this - we do not need fixed ideologies to live meaningful, ethical lives.

Sharon Street's "Darwinian Dilemma" challenges moral realism by arguing that our moral intuitions are shaped by evolutionary pressures, not access to objective truths.[27] If beliefs are evolutionarily contingent, they either coincidentally align with objective truths (implausible) or they don't—and are unreliable.

For humanists, this is not a crisis but a liberation from dogma. Humanism acknowledges the absence of fixed moral truths yet avoids nihilism. It makes explicit moral choices grounded first in using science—a test of reality—and then in a broad framework for human and planetary flourishing. These choices are reasoned, revisable, and responsive to evidence and experience.

28

From Cleverness to Wisdom

Homo sapiens – A clever animal

We disguise our basic drives with trapping of intelligence. We say, 'humans and animals,' as if we're not animals. We label other species as 'primitive,' reinforcing a hierarchy. This linguistic sleight of hand disguises our continuity with life and reinforces a feeling of separateness.

We are a clever animal. Language has allowed us to transfer our inventions to following generations. What once took centuries, like the transition from the wheel to the printing press, now unfolds in mere decades or even years. Innovations in one area gave rise to another. The result is a world where technological change is not just fast, it seems to be self-propelling, transformative, and unending (F28).

The S Curve

In his book, "Critical Path,"[1] Buckminster Fuller visualized this accelerating momentum as an 'S-curve' of human endeavours. He described how humanity's knowledge and capability follow a three-stage trajectory: slow growth at first, then a steep exponential rise, and finally a plateau or break-down. Fuller argued that we are now in the steep part of this curve (F28.1) where change is not only rapid but global in scope.

Modern Humanism

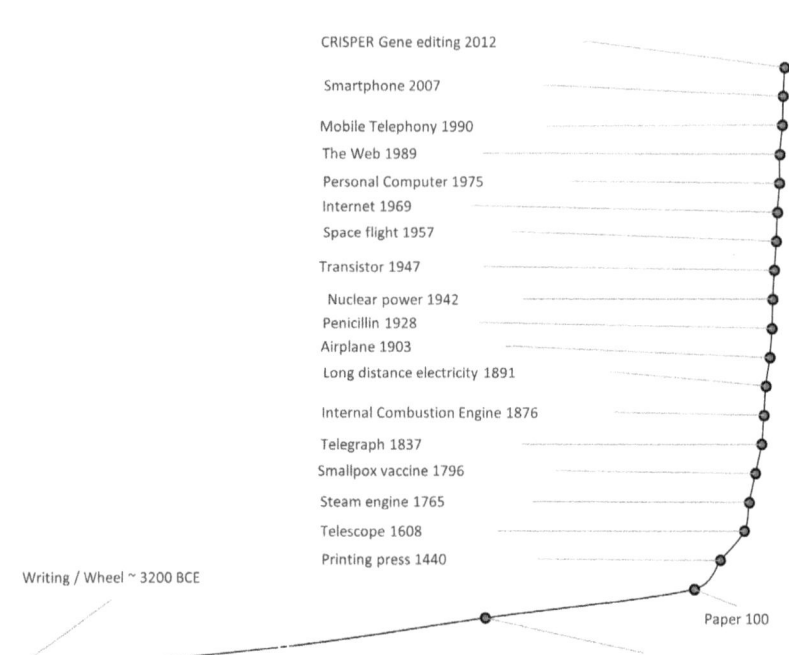

F28.1 Increased rate of technological inventions

However, this unending process of compound interest progress is shown to be a dangerous fallacy. F28.2 is based on ideas from GARRETT HARDIN[2], and succinctly illustrates this point. The quantity, for example, of population, pollution or technological 'progress' if it keeps growing in a compound interest manner goes off the chart, off the page, off our measuring scales and is not sustainable.

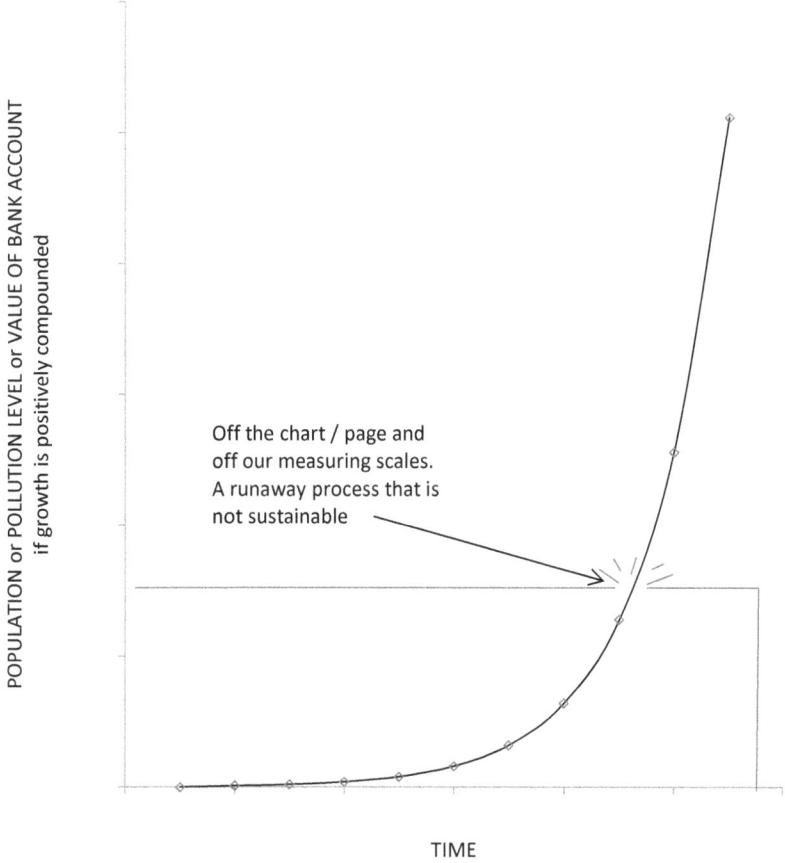

F28.2 The fallacy of non-stop compounding. Adapted from Garrett Hardin, "Living Within Limits: Ecology, Economics, and Population Taboos"[3]

Eventually compounding ends and a limit, called the 'carrying capacity of the environment,' is reached (see C20 S: Conservation). Fuller warned that without conscious design and ethical foresight, the exponential phase of the S-curve (F28.3) could lead to collapse rather than equilibrium. He saw this "critical path" as a moment of decision for humanity. Either we harness our ingenuity to create systems that serve all life on Earth, or we overshoot the limits and face ecological and social breakdown.

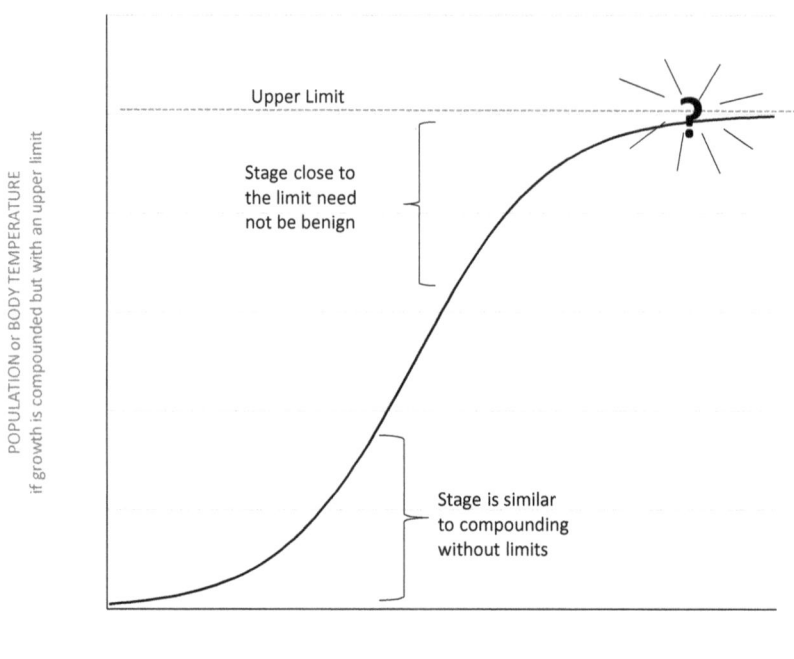

F28.3 Limits to compound growth – S curve / logistics function

Tipping points

Disasters often strike near thresholds, which may rise but not infinitely. A human analogy helps: our body maintains 37°C via homeostasis—a negative feedback loop. Within 36.5°C to 37.5°C, internal sensors adjust temperature. But extreme heat disrupts this, triggering positive feedback. Above 40°C, heatstroke and organ failure lead to death. We are not exempt from nature's limits. Yet "limits to growth" warnings are dismissed as premature, partly because phenomena unfold slower than our attention spans. Climate change has unfolded over centuries and decades, but is only now accelerating within years.

Earth regulates temperature through cloud formation, carbon sinks, and blackbody radiation. But our actions trigger positive feedback: melting ice reduces reflectivity, accelerating warming; thawing permafrost releases methane; warmer oceans absorb less CO_2 and emit more water vapor, trapping heat. If tipping points are breached - ice sheet collapse, rising seas, disrupted currents, rainforest loss—Earth's climate regulation fails, ushering in a new regime.

Elixirs of death - Testaments of hope

In the mid-20th century, organochlorines like DDT were hailed as miracle solutions, repurposed from wartime chemistry and sprayed en masse with little scrutiny. Rachel Carson's "Silent Spring," especially Chapter 3, "Elixirs of Death," exposed these as poisons with long-term impact on soil, water, wildlife, and human health (CARLSON)[4].

Carson's warning now applies beyond pesticides—to emerging threats like AI bias, gene editing, digital platforms, PFAS (per- and polyfluoroalkyl substances, persistent 'forever chemicals'), and microplastics. Her question haunts us: Are we creating tools that serve life, or unleashing forces we barely grasp?

We are a Johnny-come-lately. Dinosaurs ruled for 165 million years; Supersaurus endured 8 million.[5] Bacteria and archaea still dominate life. *Homo sapiens* emerged just 300,000 years ago; writing, a mere 5,000. On the 24-hour clock, we arrive seconds before midnight.

This scale humbles us. Cleverness — innovation, manipulation, short-term fixes — is not wisdom. Without humility and ethics, intelligence breeds hubris. From lead poisoning to ecological collapse, we've seen the cost of unmoored cleverness. What we need is not more power, but humanist wisdom: to reason ethically, live with uncertainty, revise beliefs, and prioritize life's flourishing.

Carson's warning stands. Yet history offers hope: smallpox eradication, the genome project, the free Web, Voyager, the Universal Declaration of Human Rights, the Montreal Protocol, Apollo 11. These are our testaments of hope.

We can unleash "elixirs of death" — or choose to build with foresight toward a better world.

A humanist scale of civilizations

The Kardashev scale

In 1964, astrophysicist Nikolai Semenovich Kardashev proposed a framework for imagining the future of civilizations: the Kardashev Scale (See C30 S: Kardashev scale for quantitative details).[6] It classifies civilizations based on their ability to harness and use energy — a proxy for technological advancement:

- *Type I:* A planetary civilization capable of utilizing all the energy available on its home planet. Earth is not yet Type I, but we are inching closer. We are still dwarfed by the planetary energy potential, highlighting how far we are from harnessing energy on a truly global scale.
- *Type II*: A stellar civilization that can capture the total energy output of its star.

- *Type III*: A galactic civilization that can tap into the energy of an entire galaxy.

This scale is utilitarian in nature. It doesn't ask what we become as (and if) we ascend—only what we consume. It invites reflection: will we focus on becoming more efficient extractors, or wiser stewards of the cosmos?

The Sagan scale

A humanist direction asks whether civilisational advances cultivate empathy, art, and self-awareness. Of greater, existential concern is Carl Sagan's warning about technological adolescence — our current moment in history — where our tools are outpacing our wisdom. The Kardashev Scale assumes progress means more technology and energy. Modern humanism says that progress means wiser understanding and broader compassion. The two are not mutually exclusive, but the creation and use of an alternative scale is vitally important as humanism guides the advancement and use of technologies.

Imagine a 'Humanist Scale of Civilization' that parallels Kardashev's focus on energy but instead traces scientific understanding and ethical evolution, creative endeavours, and stewardship of Earth. Such a scale can be named in honour of Carl Sagan—the Sagan (S) scale and can be used to trace the moral arc of a civilization based not on domination but on rational thought, empathy and planetary stewardship. Instead of asking how much power we can control (the Kardashev scale), it asks how wisely we live in relation to what we have (the Sagan scale). Using this S-scale civilisations are categorized as:
- *Type S-0* Tribal, competitive, survivalist civilisations based on myths and narratives of human supremacy

- *Type S-1* Utilitarian efficiency where progress is imagined as technological control. Such civilisations optimize primarily for efficiency, productivity, and resource extraction. Ethical reasoning exists, but it is used to justify optimization rather than question it.
- *Type S-2* Reflective and inclusive. Progress is imagined as harmony and planetary flourishing. Civilizations that recognize the limits of extreme resource extraction and begin reorienting toward coexistence, kindness, and compassion expanded to include non-human life. A global civilization that adopts the care of the Earth as a responsibility.
- *Type S-3* Cosmic stewardship where peace is a planetary norm. Such a civilization puts into practice a 'universal humanism' common to wise life across the universe – this understands that the journey of life is one of further explorations of the mystery of "atoms with consciousness...matter with curiosity" and that all life is "a universe of atoms...an atom in the universe" (FEYNMAN)[7]. Progress in scientific, creative and ethical endeavours will be seen not just over generations but across spacetime. These civilizations would approach Sagan's maturity not by conquering galaxies, but by internalizing a cosmically universal modern humanist ethic.

When Voyager 1 turned its camera back toward Earth it captured an image of our planet as a tiny speck, "a mote of dust suspended in a sunbeam,"[8] as Sagan famously put it. From that vantage point Earth's borders, conflicts, and divisions vanish. What remains is a fragile world, shared by all life we know.

Sagan used the image to ask, "What kind of species do we want to be in this vast cosmos?"[9]

Sagan pointed out how the image underscores the absurdity of human arrogance, our wars, our imagined self-importance, our divisions. From these planetary distances, all human history collapses into a single pixel. Sagan saw this as a chance to embrace a shared identity as Earthlings and move towards a direction of planetary flourishing.

A civilizational rite of passage

The odds of finding habitable worlds are promising. Voyager and follow-on missions have opened minds to several within our Solar System. Beyond that, over 5,900 confirmed exoplanets (as of mid-2025) orbit stars beyond our Sun, with thousands more awaiting verification. These span gas giants with scorching winds, rocky Earth-sized worlds, and planets with comet-like tails or silicate clouds, each expanding our understanding of planetary formation and the potential for life.

Yet, as Enrico Fermi famously asked, "Where is everybody?" There is the "Great Silence" – despite decades of SETI, we have no direct evidence or communication with extra-terrestrial life.[10] Robin Hansen proposed the "Great Filter," a bottleneck in the evolutionary path from simple matter to advanced, spacefaring civilizations.[11] It is possible that one or more steps toward intelligent life may be so improbable that few, if any, succeed. The formation of Earth—and the emergence of life itself—was shaped by a cascade of contingencies: the planet's stable orbit, the presence of the moon, the protective magnetic field, and the chance impacts and climate shifts that influenced evolution.

Another possibility: a Type S-3 civilization may choose techno-restraint for ethical reasons. Such civilizations might not broadcast, not because they don't exist, but because they've embraced Keynes' "art of living well." They may value fidelity to science and the creative arts over expansion, using technology to understand and celebrate the universe's beauty.

Perhaps the Great Silence reflects the rarity, and instability, of Kardashev Type II civilizations. Without ethical maturity (S-2 and above), they may self-destruct through nuclear war, ecological collapse, runaway genetic engineering, or AI. These civilizations might ignore existential risks and fall to them. Only those that cross the humanist threshold may survive. As Sagan warned, "A nuclear war...would be evidence of a technological civilization that never became mature."[12] The S-2 threshold may be a civilizational rite of passage.

It's also possible that when our destructive wars subside, and we put aside our political, economic, and technological hubris—outgrowing the zeal and destructiveness of religions and ideological dogma—we may finally begin to live wisely on our beautiful-fragile world. Then, perhaps, we will develop the clarity to hear kindred cosmic-humanist voices from afar.

Supreme existential challenges –reflections of a humanist

<u>The first challenge – confronting the existential moment</u>

Humanism begins with a sobering truth: we are in existential danger—individually and globally. Is transformation within reach, or a romantic hope against our destructive tendencies? Can engagement and persuasion scale in a world of inertia and technocratic arrogance? And can we change fast enough?

Rachel Carson warned that nature adjusts over millennia, but "now in the modern world, there is no time" (Carlson)[13]. Annie Jacobsen, in her chilling book "Nuclear War – A Scenario," quotes General Robert Kehler: "The world could end in the next couple of hours" (Annie Jacobsen)[14]. This is not scaremongering; it is a plausible reality.

Rawls and Sen believe "free and equal persons" possess 'moral powers,' and "by and large all of us are capable of being reasonable" (Sen)[15]. Our evolutionary heritage is double-edged: empathy from bonobos, aggression from chimpanzees. Stephen Jay Gould noted our "capacities for destruction" are side effects of complexity (Gould)[16]. Brains evolved for survival, but their complexity enables both creation and catastrophe.

We must confront the present crisis with hope: that if we change—if we reach the modern humanist's vision of universal health, education, and the cultivation of human capabilities, grounded not in endless desire but in the sufficiency of what truly sustains us—then we can begin to build a path toward planetary flourishing. The choices we make now will shape not only our survival, but the character of generations to come.

The second challenge – personal humanist living

Daily life is shaped by systems that desensitize us to injury and make disruption routine. This erosion of values does more than dull our moral senses, it normalizes harm. The link between routine and ruin is no accident; it reflects our failure to resist harm.

It's easy to blame others. We imagine ourselves as more aware, reasonable, humanist. But the truth is we must examine our own lives. The fault, dear humanist-*Homo sapiens*, lies not in our stars, nor in others, but in ourselves.

We must reflect on our extractive consumption and complicity in the issues we condemn. Gandhiji's dictum remains vital: "Be the change you want to see." The question is not just humanist ideals, but whether we embody them daily.

Readers of this book, likely not in extreme poverty, must ask: what does it mean to advocate justice and flourishing while living in extractive ways that quietly perpetuate harm?

The third challenge – creative works and speaking up

Are we spectators, disapproving in private? Or do we pause our creative work to engage, persuade, mentor, educate? We must not abandon our creative pursuits. As Mihaly Csikszentmihalyi reminds us, deep creative engagement is essential to human flourishing. To forsake our core projects in favour of constant proselytizing risks turning humanism into a rigid ideology, mirroring the doctrinal zeal preached from pulpits and platforms—the very fervour it aims to resist.

But sustained creative labour must be paired with speaking up. Silence sustains epistemic inequality, leaving humanism's transformative power unused. Earlier epochs faltered because their insights stayed confined to elites—philosophers, court scholars, academies. Without mass literacy, strong institutions, or empirical frameworks like evolution, cosmology, and neuroscience, humanist ideals were eclipsed by religious, mythic or authoritarian worldviews.

Today, we stand at a new threshold. With universal education, digital access, and global science, we can revive, modernize, and democratize humanism. The challenge is no longer access but the will to sustain open, pluralistic dialogue.

It is through both expression and action, imagination and dialogue, that humanism lives. If we don't, our future—and humankind's—is in peril.

<u>The fourth challenge – maintaining courage</u>

We are born, live, and die feeling that we are discrete individuals. Yet we are biologically, socially, and cosmologically integrated with each other and the universe. There is so much to be sorrowful about—personal losses, failures and struggles, the fragility of our planet, the weight of our collective destruction, and the deep ache of our own irreversible mistakes.

But it is imperative that we stay the course. There is no other option. To paraphrase James Baldwin, "Despair is a luxury we can't afford."[17]. We have begun an incredibly exciting journey to understand the universe—from atoms to evolution by natural selection, to self-knowledge, and to the vast structure of spacetime. We carry the privilege of this heritage of understanding: a foundation to build on, to explore further, and to give us courage.

Having gone through the adventure of this book, I hope readers can see that we are not alone. Patterson emphasized how true scientific discovery and understanding is a process where one joins a community of seekers stretching back millennia. Einstein wrote, "Although I am a typical loner in daily life, my consciousness of belonging to the invisible community of those who strive for truth, beauty, and justice keeps me from feeling isolated" (EINSTEIN)[18].

What sustains us is the courage we draw from the philia of friends, the quiet grace of those who forgive, and the creative artists, scientists, and service-humanists—past and present—who walk beside us as we strive after our works and seek to change direction toward a better world.

N11 The last wager

"In the midst of winter, I found there was, within me, an invincible summer" (ALBERT CAMUS)[19].

"It is not the strongest of the species that survives, but the one most responsive to change."[20] A paraphrasing of Darwin's theory of natural selection by Leon C. Megginson.

"What we do now echoes in eternity."[21] Marcus Aurelius

Appendices

C29 Aspects from the Philosophy of Science

C30 Plausible reasoning and a probability calculation

C31 Mathematics – Its effectiveness and the risks of its allure

"Science is not only a disciple of reason but, also, one of romance and passion."[1]

Stephen Hawking

"Probability theory is nothing but common sense reduced to calculation."[2]

Pierre-Simon Laplace

"Mathematics is not a careful march down a well-cleared highway, but a journey into a strange wilderness, where the explorers often get lost."[3]

W.S. Anglin

"Either man will abolish war, or war will abolish man."[4]

Bertrand Russell

29

Aspects from the Philosophy of Science

Scientific revolutions and method

Thomas Kuhn observed that science evolves through paradigm shifts. Scientists often work within dominant models until anomalies force a revolution, as with Einstein's overhaul of Newtonian gravity. Despite human bias and a messy process of discovery, science remains our most reliable tool for truth, self-correcting through empirical testing and falsification. Periods of stability (normal science) alternate with upheaval (paradigm shifts), but the method endures.

Feyerabend equates methodological diversity with epistemic equality, suggesting "anything goes." This overlooks the fact that not all methods yield equally reliable knowledge—science earns its epistemic privilege through rigorous testing, reproducibility, and unmatched track record of explanatory and predictive success.

Interdisciplinary safeguards

Feedback loops can mislead if initial assumptions are flawed. If the starting point of a feedback loop is flawed, whether due to conceptual error, limited data, or bias, the process may reinforce its own inaccuracies rather than correct them. Each iteration builds upon prior distortions, creating a self-validating system that drifts further from truth, resulting in errors that are mistaken as insights.

Interdisciplinary input helps break these traps, as seen when geology embraced seismology and oceanography to validate plate tectonics.

The problem of induction

Hume's famous pig analogy illustrates induction's limits. A pig is fed every day by a farmer. Based on its past experiences, the pig reasons all is well and that the farmer will continue feeding it indefinitely. One day, however, the pig is slaughtered. The pig's inductive reasoning fails. Thus, past patterns don't guarantee future outcomes. Science mitigates this through probability, falsification, re-evaluation, and plausible reasoning. Models remain tentative, not absolute.

The Gettier problem

This problem reveals that justified true belief may not be sufficient for knowledge, as beliefs can be true by coincidence rather than reliable reasoning. Modern humanism and science address this challenge through feedback-based learning: falsification and replication ensure that knowledge is not just justified and true, but also robust and repeatable. By constantly testing claims and correcting errors, science builds understanding that goes beyond lucky guesses — grounding knowledge in evidence, transparency, and collective scrutiny.

Conceptions of reality

Underlying scientific frameworks is a central tension: do scientific models and theories reflect actual truths about the universe (a position known as scientific realism), or are they merely tools for prediction and coherence (instrumentalism), useful but not necessarily representative of reality? This question cuts across all four categories, shaping how each interprets the nature of knowledge, truth, and human understanding. Science-informed humanism offers four metaphysical views.

- Mechanistic Naturalism. Reality is matter and energy governed by laws. Thinkers include Democritus, Newton, Einstein, Dennett, Deutsch, Dawkins, Carroll, and Feynman — who saw understanding as creation.
- Relational Processes. Reality is dynamic, defined by relations and change. Whitehead, Rovelli, Bohr, and Anaximander emphasized interdependence and emergence.
- Experiential Humanism. Reality is perception and meaning. Sartre, de Beauvoir, Camus, James, and Dewey focused on ethics, choice, and lived experience.
- Poetic Naturalism. Reality is natural yet awe-inspiring. Lucretius, Sagan, Carroll, Einstein, Whitehead, and Rovelli blended materialism with reverence and beauty.

These frameworks span domains, affirming that meaning is made, not given, a core humanist idea. They promote awareness, scientific humility, and existential depth without supernaturalism.

Theory-Ladenness of observation

This asks whether observations are objective or influenced by prior knowledge, beliefs, and theoretical frameworks. To reduce bias:

- Use interdisciplinary teams.
- Rely on instruments for standardized data.

- Apply peer review and replication.
- Define terms and methods clearly.
- Conduct blind studies and double-blind studies to prevent expectations from influencing results.

This brief appendix has only pointed out areas where interested readers can investigate further. For a more detailed exploration, CURD, COVER, and PINCOCK's book is highly recommended.[1]

30

Plausible Reasoning and A Probability Calculation

Understanding risks

Risks of accidental nuclear war

The Federation of American Scientists (FAS) is a group who provide "analysis of and solutions to protect against catastrophic threats to national and international security."[1] According to the FAS, there are 13,410 nuclear warheads (2023), of which approximately 3,720 are categorized as "Deployed strategic warheads." These are nuclear warheads on "intercontinental missiles and at heavy bomber bases" ready to be fired at a moment's notice.

Assume nuclear weapons won't be launched intentionally, and we've achieved 99.99% reliability against accidental launch. What then is the risk of nuclear annihilation by mistake or miscalculation?

Let's model this using a biased coin: heads means all is well, tails means disaster. With a 0.01% chance of tails (1 in 10,000), a single toss is unlikely to end the world. But what if we toss more?

- 2 coins: ~0.02% chance of at least one tail.
- 100 coins: ~1% chance.
- 1,000 coins: ~9.5% chance.
- 3,720 coins (matching the number of deployed strategic warheads in 2023): ~31.1% chance of disaster.

Plausible Reasoning and A Probability Calculation

If reliability rises to 99.999% (the error rate drops to 0.001%), the risk for 3,720 coins is ~3.65% (approximately 4 out of 100). The binomial distribution shows how even tiny probabilities compound into real threats when scaled.

<u>Two types of errors</u>

The following example is taken from L. Rastrigin.[2] Consider a detective trying to catch a criminal. The detective has a photograph of the criminal and a photograph of a suspect. The question to answer is: do the two photographs show the same person or different people?

- *Null hypothesis:* They are different people.
- *Alternative hypothesis:* They are the same person.

Two mistakes can occur (T32.1):

Type II error (false negative). They're the same, but the detective lets the suspect go.

Type I error (false positive). They're different, but the detective wrongly concludes they're the same.

		Prediction	
		+ve	-ve
Actual condition	+ve	True +ve	False -ve Type II error
	-ve	False +ve Type I error	True -ve

T30.1 Illustration of Type I and Type II errors

Plausible reasoning

Inference

Science thrives on inference, drawing conclusions from indirect data. Cunningham and Herr use Pluto: we accept its size and orbit not from direct measurement, but from observations, models, and calculations.[3] Even simple tasks, like measuring the thickness of a book's page can be estimated dividing a book's width by its page count – thus relying on inference.

Harold Jeffreys, scientist and philosopher of science, argued that science is fundamentally probabilistic, and that beliefs should be updated through Bayesian reasoning. He laid the groundwork for a systematic, probabilistic approach to scientific reasoning and remains a cornerstone of Bayesian philosophy.[4]

CERN and the Higgs Boson

This is explained using an example of CERN's search for the Higgs boson. CERN physicists see a signal from the Large Hadron Collider (LHC). Is it a new particle (the Higgs boson) or noise?

H_0= Null hypothesis – it is noise

H_1= Alternative hypothesis – it is the new particle

CERN physicists then weigh the likelihood of each explanation—noise or new particle—against their prior expectations. Even with a tiny initial belief in a new particle, strong data can dramatically shift that belief. This is the essence of Bayesian reasoning: updating what we believe based on how well new evidence fits each possibility. In this case, a faint signal becomes a compelling discovery.

Readers may wish to consult Jordan Ellenberg (more accessible to non-mathematicians) and the book by Howson and Urbach, which is better suited to readers with some mathematical background.[5,6]

Environmental lead and violent crime

C7 S: The use of probabilities and statistics, described research by Jessica Reyes that found "changes in childhood lead exposure are responsible for a 56% drop in violent crime in the 1990s."[7]

Building on studies by Dr. Needleman, she tracked lead levels in gasoline across U.S. states and compared them to crime rates 22 years later. To isolate lead's effect, Reyes built a statistical model that adjusted for other factors: income, policing, abortion access, geography, and national events.

Her analysis revealed a strong link between reduced lead exposure and falling crime, with a 22-year lag, and estimated that 56% of the crime drop (1992–2002) was due to removing lead from gasoline and 29% was due to legalized abortion.

She estimated that the "elasticity of violent crime" with respect to childhood lead exposure was approximately 0.8. A 1% reduction in lead exposure led to a 0.8% reduction in violent crime. For legalized abortion she found an elasticity of 0.23, meaning a 1% increase in abortion access led to a 0.23% decrease in violent crime.

Reyes' study demonstrates the power of using plausible reasoning, probabilistic calculations, and rigorous statistical modelling, akin to Bayesian thinking, to isolate the impact of lead exposure on violent crime. Reader may wish to consult the book by Howson and Urbach.[8]

The normal distribution

Probability distributions describe how outcomes are spread. A fair die gives a uniform distribution. The normal distribution, first described by de Moivre, expanded by Laplace and Gauss, is defined by its mean and standard deviation.

Thanks to the Central Limit Theorem, when many small, independent factors shape a result, the outcome tends to follow a normal distribution, even if the individual factors don't. Human traits like height or chest size often fit this pattern.

Gabriel Lippmann quipped: "Everyone believes in the law of errors [the normal distribution]—the experimenters because they imagine it is a mathematical theorem, and the mathematicians because they think it is an experimental fact."[9]

This reminds us that despite this amazing distribution occurring in so many unrelated areas, it is worth keeping in mind that it is a model of phenomena in the universe and may not be a universal truth.

A binomial distribution calculation

Coming back to the risks of nuclear annihilation, there is no easy way forward. Our social and political structures have not kept pace with the increased power of our technologies. Is a 3.7% risk of the first missile going off and then leading to massive retaliation and another mass extinction acceptable?

The probability calculation, shown below, is relatively simple.

Plausible Reasoning and A Probability Calculation

$X \sim B(n, p)$

$n = 3{,}720$ (nuclear missiles or coins tossed)

$p(0)$
= probability that in a collection of 3,720 nuclear missiles not even one will be fired by mistake

$1 - p(0)$
= prbability that in a collection of 3,720 nuclear missiles at least one nuclear missile being fired by mistake

$= 1 - \left(\binom{n}{0} p^0 (1-p)^n\right)$

Scenario 1 99.99% reliability, 0.01% probability of a missile fired by mistake

$1 - p(0) = 1 - \left(\binom{3720}{0}(0.0001)^0 (1-0.0001)^{3720}\right)$

$\cong 0.311$ (31.1% chance of nuclear war)

Scenario 2 99.999% reliability, 0.001% probability of a missile fired by mistake

$1 - p(0) = 1 - \left(\binom{3720}{0}(0.00001)^0 (1-0.00001)^{3720}\right)$

$\cong 0.0365$ (3.65% chance of nuclear war)

Kardashev scale – energy estimates and calculations

Civilization type	Power (W)	Energy per day (Wh/d)
I: *Planetary*	$\sim 10^{17}$ W	2.4×10^{18}
II: *Stellar*	Output of the Sun: $\sim 10^{26}$ W	2.4×10^{27} Wh/d
A Dyson Sphere megastructure would consist of orbiting satellites designed to capture solar energy and beam it back to the civilization's home world.		
III. *Galactic*	$\sim 10^{37}$ W	2.4×10^{38} Wh/d

T30.2 Kardashev scale-energy estimates

Human energy consumption comparison:
- LED bulb: ~240 Wh/day
- Human body: ~2,400 Wh/day (just enough to keep us alive)
- Average American: ~211,000 Wh/day
- Global daily energy use (if all humans consumed like Americans): ~1.7×10^{15} Wh/day

Readers can refer to an article by Robert Gray for further details.[10]

31

The Allure of Mathematics and Moral Certainty

<u>The urge for certainty</u>

Humans have a psychological urge for moral certainty, a longing to anchor their values in something unchanging and universal. This need often stems from the fear of moral chaos: the dread that without fixed principles, societies might descend into relativism, conflict, or nihilism. Many turn to systems that promise clarity and permanence. The fervour with which religious and other ideological people defend their universal moral visions is replicated by some moral philosophers who attempt mathematics type proofs for certainty.

Derek Parfit is just one example of a long tradition of moral philosophers, including Jeremy Bentham and John Rawls, who seek certainty in ethics. Parfit is a representative case to highlight the folly of such attempts.

<u>Parfit's attempt</u>

Derek Parfit sought to establish a universal, fixed morality by synthesizing:[1]

- Kantian ethics. Morality is grounded in duty and reason, the "categorical imperative," regardless of consequences.
- Rule consequentialism. Moral rules are justified if following them generally leads to the best outcomes.
- Contractualism. Morality arises from fair principles that no one could reasonably reject.

Parfit likened this 'moral truth' to mathematical truths: abstract, objective, and discoverable through reason. His vision was that ideal rational agents, free from bias, would arrive at the same moral conclusions, revealing a shared moral reality.

The mathematical analogy: A flawed foundation

Parfit's analogy falters on two fronts:

- *Mathematics as invention.* Mathematics enchants with its elegance and reach. Platonists like Penrose see mathematics as revealing timeless truths.[2] There is also the "unreasonable effectiveness of mathematics" in describing the physical world.[3] Yet, as E. Brian Davies explains, mathematics reflects how we model the world, shaped by symmetry and structure.[4]

  Mathematics is not discovered like natural laws but invented as a formal system of symbols, axioms, and rules of inference, independent of meaning or empirical truth. It is a human abstraction that reflects human cognition and utility, not objective metaphysical truths.[5]

- *Gödel's Incompleteness Theorems.* Kurt Gödel proved that any sufficiently complex formal system (meaning it can express basic arithmetic) is either incomplete or inconsistent.

Parfit's attempt to "prove" morality based on a mathematics type system rests on an epistemic error: like mathematics it is (i) invented and (ii) cannot be both comprehensive and coherent. This undermines efforts for universal fixed moral systems. Similar to religious dogmas, Parfit's framework imposes a rigid moral structure to a domain that resists such closure.

Humanism: A richer moral landscape

Humanism rejects the notion of a fixed, universal morality. Instead, it embraces moral pluralism—an openness to diverse ethical perspectives—grounded in empirical reality. This is the realm of what Hobbes referred to as the "IS": the observable facts of the world. These facts inform, but do not dictate, our moral decisions about what "OUGHT" to be. The "OUGHT" remains a human choice, shaped by values, context, and consequence.

While the IS and the OUGHT are distinct, they are not disconnected. The IS offers the terrain—background, evidence, and perspective—on which moral deliberation takes place. The OUGHT, in turn, charts the direction of our moral striving, often influencing which aspects of the IS we choose to investigate or prioritize. Yet the outcomes of empirical inquiry—the IS—remain separate from the normative judgments we make.

Humanism accepts this complexity as a strength, not a failure. It reflects the depth of human experience, not the illusion of moral certainty.

T31.1 summarizes two 'assured' moral philosophies—hedonist utilitarianism and Parfit's universal morality—contrasted with Modern Humanism's acceptance of the world as it IS and that morality must be grounded in humanist choices of what OUGHT to be - compassion, dignity, and flourishing.

Hedonist utilitarianism and Parfit both reach for the comfort of moral certainty, while Humanism resists that urge—choosing instead to ground ethics in compassion, pluralism, and the lived realities of human and planetary flourishing.

Dimension	Hedonist Utilitarianism	Parfit (*On What Matters*)	Modern Humanism
Core Aim	Reduce morality to pleasure/pain calculus. Single metric: pleasure	Prove convergence of major moral theories into one unified moral truth	Promote flourishing of individuals and planet via science's learning process, plural values of justice, dignity, sustainability
Promise of Certainty	Mathematical calculation of pleasure and pain offers clear answers	Rational "proof" that morality is objective and universal	Rejects fixed certainty; morality evolves with human understanding dialogue and empathy
Risk	Oversimplifies human flourishing; ignores justice and rights	Collapses pluralism into one "true" morality; Seeks proof where none exists	Less certainty, but richer, inclusive, adaptable ethics. Accepts ambiguity, embraces science, compassion and capabilities

T31.1 Utilitarianism vs. Parfit vs. Modern Humanism

32

Afterword - In Defence of Depth

We live in an age where brevity is mistaken for insight, infotainment passes for analysis, and bite-sized tweets simulate understanding without demanding thought.

This book is an act of resistance against the illusion that depth can be compressed without distortion. It explores life's questions with the patience and nuance they deserve—giving space for ideas to unfold, and for readers to engage in the slow, vital work of thinking, playing, and debating.

As C. P. Cavafy wrote:[1]

"As you set out for Ithaka
hope your road is a long one,
full of adventure, full of discovery."

This book hopes there will be many summer mornings when, with joy, you enter harbours you're seeing for the first time. It is not a sprint toward answers, but a slow unfolding of questions.

"Ithaka gave you the marvellous journey.
Without her you wouldn't have set out."

The adventure of these ideas is our Ithaka. They are a synthesis of ideas from across centuries and cultures. Without them, this journey would not exist. If it offers no final answer, let it not disappoint for that was never the aim.

"Wise as you will have become, so full of experience,
you'll have understood by then what these Ithakas mean."

33

Select Bibliography

"It is not merely that we read them; we converse with them, we quarrel with them, we are instructed by them, we are moved by them. They are not dead; they are immortal companions."[1] (Lytton Strachey)

This select bibliography of my closest immortal humanist companions is not exhaustive, many great thinkers are absent, and for those included, only one or two key works are highlighted. Organized by first name, it offers a curated selection of writings and notable reflections related to the ideas in this book.

Albert Camus, *Neither Victims Nor Executioners*, trans. Dwight Macdonald (New York: Continuum, 1980)

Sean B. Carroll, *Brave Genius: A Scientist, a Philosopher, and Their Daring Adventures from the French Resistance to the Nobel Prize* (New York: Crown, 2013)

Amartya Sen, *Development as Freedom* (New York: Alfred A. Knopf, 1999).

____ *The Idea of Justice* (Cambridge, MA: Belknap Press of Harvard University Press, 2009).

Andrew Dessler, *Introduction to Modern Climate Change*, 3rd ed. (Cambridge: Cambridge University Press, 2021)

Anne Applebaum, *Twilight of Democracy: The Seductive Lure of Authoritarianism* (New York: Doubleday, 2020)

Select Bibliography

Basil Henry Liddell Hart, *Deterrent or Defence: A Fresh Look at the West's Military Position* (New York: Praeger, 1960).

Bertrand Russell, *Why I Am Not a Christian: And Other Essays on Religion and Related Subjects*, ed. Paul Edwards (London: George Allen & Unwin, 1957)

Carl Sagan, *Cosmos*. Originally published 1980. New York: Ballantine Books, 2013

____ *The Demon-Haunted World: Science as a Candle in the Dark* (New York: Ballantine Books, 1995)

Clair Patterson, "Historical Changes in Integrity and Worth of Scientific Knowledge." *Geochimica et Cosmochimica Acta* 58, no. 15 (1994: 3141–3143).

Lydia Denworth, *Toxic Truth: A Scientist, a Doctor, and the Battle over Lead* (Boston: Beacon Press, 2009)

Daniel Dennett, *Elbow Room: The Varieties of Free Will Worth Wanting*. Cambridge, MA: MIT Press, 1984

Daniel Ellsberg, *The Doomsday Machine: Confessions of a Nuclear War Planner* (New York: Bloomsbury, 2017)

Daniel Kahneman (2011). *Thinking, Fast and Slow*. Farrar, Straus and Giroux

David Graeber, *Bullshit Jobs: A Theory* (New York: Simon & Schuster, 2018)

____ and **David Wengrow**, *The Dawn of Everything: A New History of Humanity* (New York: Farrar, Straus and Giroux, 2021)

E. Brian Davies, *Science in the Looking Glass: What Do Scientists Really Know?* (Oxford: Oxford University Press, 2003).

Edward O. Wilson, Half-Earth: Our Planet's Fight for Life (New York: Liveright Publishing, 2016).

____ *Letters to a Young Scientist* (New York: Liveright Publishing, 2013).

Epicurus, *The Epicurus Reader: Selected Writings and Testimonia*, edited and translated by Jeffrey Henderson (Indianapolis: Hackett Publishing Company, 1994)

Frans de Waal, *Our Inner Ape: A Leading Primatologist Explains Why We Are Who We Are* (New York: Riverhead Books, 2005)

Garrett Hardin, *Living Within Limits: Ecology, Economics, and Population Taboos* (New York: Oxford University Press, 1993)

Herbert L. Needleman, "Clair Patterson and Robert Kehoe: Two Views on Lead Toxicity," *Environmental Research* 78, no. 2 (1998)

Jacques Monod, *Chance and Necessity: An Essay on the Natural Philosophy of Modern Biology*, trans. Austryn Wainhouse (New York: Vintage Books, 1972)

Jared Diamond, *Collapse: How Societies Choose to Fail or Succeed* (New York: Viking, 2005).

John Hunt, *The Ascent of Everest* (London: Hodder & Stoughton, 1953)

John Maynard Keynes, *Essays in Persuasion* (London: Macmillan, 1931; repr., New York: W.W. Norton, 1963).

Zachary D. Carter, *The Price of Peace: Money, Democracy, and the Life of John Maynard Keynes* (New York: Random House, 2020)

Joseph Stiglitz, *The Road to Freedom: Economics and the Good Society*. New York: W. W. Norton & Company, 2024

Karl Popper, "Science: Conjectures and Refutations," in *Philosophy of Science: The Central Issues*, ed. Martin Curd, J. A. Cover, and Christopher Pincock (New York: W. W. Norton & Company, 2013)

Kathleen Vohs, Money priming can change people's thoughts, feelings, motivations, and behaviours: An update on 10 years of experiments. *Journal of Experimental Psychology: General, 144*(4), e86–e93, Available at: https://doi.org/10.1037/xge0000091

Keith Stanovich, *What Intelligence Tests Miss: The Psychology of Rational Thought* (New Haven: Yale University Press, 2009)

Martha Nussbaum, *Creating Capabilities: The Human Development Approach* (Cambridge, MA: Belknap Press of Harvard University Press, 2011), 33–34.

____"Who Is the Happy Warrior? Philosophy, Happiness Research, and Public Policy," *International Review of Economics* 59, no. 4 (2012): 351, Available at: https://link.springer.com/article/10.1007/s12232-012-0168-7

Michael Sandel and Thomas Piketty, *Equality: What It Means and Why It Matters* (Ashland, OR: Blackstone Publishing, forthcoming 2025)

Naomi Oreskes and Erik M. Conway, *Merchants of Doubt: How a Handful of Scientists Obscured the Truth on Issues from Tobacco Smoke to Global Warming* (New York: Bloomsbury Press, 2010).

Nick Bullock, *"I Want It and I Want It Now,"* 2011, Available at: https://nickbullock-climber.co.uk/2011/06/02/i-want-it-and-i-want-it-now-i-want-it-and-i-want-it-now-i-want-it-and-i-wa/

Peter Singer, *Ethics in the Real World: 82 Brief Essays on Things That Matter* (Princeton, NJ: Princeton University Press, 2016)

Philip Morrison and Phylis Morrison, *The Ring of Truth: An Inquiry into How We Know What We Know* (New York: Random House, 1987)

Rachel Carson, *Silent Spring* (London: Penguin, 2002)

Richard Feynman, *"Cargo Cult Science,"* commencement address, California Institute of Technology, Pasadena, CA, June 1974. Transcript available at: https://speakola.com/grad/richard-feynman-caltech-1974

Ritchie Robertson, *The Enlightenment: The Pursuit of Happiness, 1680–1790* (New York: Harper, 2021)

Robert Axelrod, *The Evolution of Cooperation* (New York: Basic Books, 1984)

Santiago Ramón y Cajal, Eric A. Newman, Alfonso Araque, and Janet M. Dubinsky, eds., *The Beautiful Brain: The Drawings of Santiago Ramón y Cajal* (New York: Abrams Books, 2017).

Sarah Bakewell, *Humanly Possible: A History of Humanism* (London: Chatto & Windus, 2023)

Shoshana Zuboff, *The Age of Surveillance Capitalism: The Fight for a Human Future at the New Frontier of Power* (New York: PublicAffairs, 2019)

Stanley Milgram, Thomas Blass, ed., *Obedience to Authority: Current Perspectives on the Milgram Paradigm* (Mahwah, NJ: Lawrence Erlbaum Associates, 2000).

Stephen Jay Gould, *Full House: The Spread of Excellence from Plato to Darwin* (New York: Harmony Books, 1996).

____ *Wonderful Life: The Burgess Shale and the Nature of History.* New York: W.W. Norton, 1989

Select Bibliography

Thomas Piketty, *Capital in the Twenty-First Century*, trans. Arthur Goldhammer (Cambridge, MA: Belknap Press of Harvard University Press, 2014).

Tim Berners-Lee, and Mark Fischetti, *Weaving the Web: The Original Design and Ultimate Destiny of the World Wide Web* (New York: HarperCollins, 1999).

Will Durant. *The Story of Philosophy: The Lives and Opinions of the Greater Philosophers.* New York: Simon and Schuster, 1927

Yvon Chouinard, *Let My People Go Surfing: The Education of a Reluctant Businessman*, 10th-anniversary ed. (New York: Penguin, 2016).

Endnotes and Citations

What's in a word – 'Modern Humanism'
1. Johann Wolfgang von Goethe, *Wilhelm Meister's Apprenticeship*, trans. Eric Blackall, ed. Victor Lange, Princeton: Princeton University Press, 1995; originally published 1795–96)

Outline of the book

Objectives and hope
1. Albert Camus, *Neither Victims Nor Executioners*, trans. Dwight Macdonald, New York: Continuum, 1980.
2. Clair Patterson. "Historical Changes in Integrity and Worth of Scientific Knowledge." *Geochimica et Cosmochimica Acta* 58, no. 15 (1994: 3141–3143.

Key to references
1. Bertrand Russell, "Censorship by Progressives," in *Mortals and Others: American Essays 1931–1935*, vol. 2, London: Routledge, 1975, 317–318.
2. Daniel Kahneman, *Thinking, Fast and Slow*, New York: Farrar, Straus. and Giroux, 2011, 28.
3. William Harvey, *The Works of William Harvey*, trans. Robert Willis, London: Sydenham Society, 1847; repr., Philadelphia: University of Pennsylvania Press, 2014)

A libation to the universe
1. Richard Feynman. "The Value of Science." *Engineering and Science* 19, no. 12 (1955: 13–15. Reprinted by permission of Basic Books, an imprint of Hachette Book Group, Inc.

C1 Preface
1. Edward O. Wilson, *Consilience: The Unity of Knowledge* (New York: Alfred A. Knopf, 1998). His vision of unity differs in flavour from mine, yet resonates in spirit.
2. Will Durant. *The Story of Philosophy: The Lives and Opinions of the Greater Philosophers*. New York: Simon and Schuster, 1927. Preface to the Second Edition.
3. Stephen Kotkin, *Stalin: Volume I: Paradoxes of Power, 1878–1928*, New York: Penguin Press, 2014
4. Carl Sagan, *Pale Blue Dot: A Vision of the Human Future in Space*, New York: Random House, 1994, 13.
5. David Graeber and David Wengrow, *The Dawn of Everything: A New History of Humanity* (New York: Farrar, Straus and Giroux, 2021)
6. Ritchie Robertson, *The Enlightenment: The Pursuit of Happiness, 1680–1790*, New York: Harper, 2021, 3.
7. Ibid.
8. Pavel Aksenov. "Stanislav Petrov: The Man Who May Have Saved the World." *BBC News*, September 26, 2013. https://www.bbc.com/news/world-europe-24280831
9. Ibid
10. Charles Darwin. *The Autobiography of Charles Darwin – 1809–1882*. Originally published 1887. New York: W. W. Norton & Company, 1993, 27.
11. Lao Tzu, *Tao Te Ching: A New English Version*, trans. Stephen Mitchell, New York: HarperPerennial Modern Classics, 2006, verse 9
12. Keith E. Stanovich. *The Robot's Rebellion: Finding Meaning in the Age of Darwin*. Chicago: University of Chicago Press, 2004.
13. Richard Dawkins. *The Selfish Gene*. London: Oxford University Press, 1976. Reprint, Oxford: Oxford University Press, 2006, xxix.
14. The title of the book: Jacques Monod. *Chance and Necessity: Essay on the Natural Philosophy of Modern Biology*. New York: Vintage, 1971.
15. Stephen Emmott. *Ten Billion*. New York: Vintage, 2013.

16 Lulu Garcia-Navarro. "Nuclear Secrets, a Compost Heap and the Lost Documents Daniel Ellsberg Never Leaked." *The New York Times*, April 20, 2023.
17 Albert Camus, *Neither Victims Nor Executioners*, trans. Dwight Macdonald, New York: Continuum, 1980.
18 Charles Darwin. 1859. *On the Origin of Species*. Reprint, London: HarperCollins Publishers, 2011, 397.
19 *Unknown: Cosmic Time Machine*. Directed by Shai Gal. Netflix, 2023. Streaming on Netflix.
20 NASA. *Star Basics*. 2024.NASA. https://science.nasa.gov/universe/stars/
21 Carl Sagan, *Pale Blue Dot: A Vision of the Human Future in Space*, New York: Random House, 1994, 7.
22 Note 19
23 Note 19
24 Rachel Carson, Silent Spring, Boston: Houghton Mifflin, 1962)

C2 Narrative Foundations of This Book
1 Daniel Kahneman, *Thinking, Fast and Slow*, New York: Farrar, Straus. and Giroux, 2011.
2 Alfred North Whitehead, *The Aims of Education and Other Essays*, New York: Macmillan, 1929

Part 1 Central questions of life
1 Herbert L. Needleman, "History of Lead Poisoning in the World" 1997, https://www.biologicaldiversity.org/campaigns/get_the_lead_out/pdfs/health/Needleman_1999.pdf
2 Benjamin Franklin. *The Works of Benjamin Franklin: Containing Several Political and Historical Tracts Not Included in Any Former Edition, and Many Letters Official and Private Not Hitherto Published*. Edited by Jared Sparks. Originally published Boston: Hilliard, Gray, and Company, 1882. Reprint, Creative Media Partners, LLC, 2015, vol. 2, 376.

C3 Tetraethyl Lead in Gasoline – History and Politics
1 Classic Car Weekly. "U.S. Auto Production by Year." *Classic Car Weekly*. https://classiccarweekly.wordpress.com/top-u-s-automakers-to-1990/.
2 U.S. Energy Information Administration. *Monthly Energy Review: Table 1.8 – Motor Vehicle Mileage, Fuel Consumption, and Fuel Economy*. Washington, DC: U.S. Department of Energy, https://www.eia.gov/totalenergy/data/browser/?tbl=T01.08.
3 Hale B. Soyster, *Review of the Petroleum Industry in the United States, April 1934*, Circular 11, Washington, D.C.: U.S. Geological Survey, 1934). https://pubs.usgs.gov/publication/cir11
4 U.S. Energy Information Administration, *Annual Energy Review 2011*, 2011, https://www.eia.gov/totalenergy/data/annual/pdf/aer.pdf
5 Jamie Lincoln Kitman, "The Secret History of Lead," *The Nation*, March 20, 2000, https://www.thenation.com/article/archive/secret-history-lead/.
6 William J. Kovarik, *The Ethyl Controversy*, Ph.D. dissertation, University of Maryland, 2013, https://billkovarik.com/wp-content/uploads/2012/08/Ethyl.Controversy.Kovarik.dissertation.pdf
7 "Correspondence," *Scientific American* 123, no. 24 (1920, https://www.scientificamerican.com/issue/sa/1920/12-11/)
8 Peter F. Drucker, *Concept of the Corporation*, originally published 1946, New Brunswick, NJ: Transaction Publishers, 1993, 157.
9 Note 5
10 Brian Foster, "How One Man Accidentally Poisoned the Entire Planet," Glass Almanac, April 11, 2025, https://glassalmanac.com/how-one-man-accidentally-poisoned-the-entire-planet/
11 David Rosner and Gerald Markowitz. *Lead Wars: The Politics of Science and the Fate of America's Children*. Berkeley: University of California Press, 2013
12 Ibid

13 Note 5
14 Christian Warren, *Brush with Death: A Social History of Lead Poisoning*, Baltimore: Johns Hopkins University Press, 2000)
15 David Rosner and Gerald Markowitz, "A 'Gift of God'? The Public Health Controversy over Leaded Gasoline during the 1920s," *American Journal of Public Health* 75, no. 4 (1985)
16 Lydia Denworth, *Toxic Truth: A Scientist, a Doctor, and the Battle over Lead*, Boston: Beacon Press, 2009)
17 Note 6
18 Jerome Nriagu, "Clair Patterson and Robert Kehoe's Paradigm of 'Show Me the Data' on Environmental Lead Poisoning," *Environmental Research* 78, no. 1 (1998: 71–78
19 Ibid
20 Alice Hamilton, *Exploring the Dangerous. Trades*, Boston: Little, Brown and Company, 1943)
21 U.S. Public Health Service, *Proceedings of a Conference to Determine Whether or Not There Is a Public Health Question in the Manufacture, Distribution, or Use of Tetraethyl Lead Gasoline*, Public Health Bulletin no. 158, Washington, D.C.: Government Printing Office, 1925).
22 Note 6
23 Note 6
24 U.S. Public Health Service, *Annual Report of the Surgeon General of the Public Health Service of the United States for the Fiscal Year 1926*, Washington, D.C.: Government Printing Office, 1926, https://commons.wikimedia.org/wiki/File:1925_Proceedings_of_a_Conference_to_Determine_Whether_or_not_there_is_a_Public_Health_Question_in_the_Manufacture,_Distribution,_or_Use_of_Tetraethyl_Lead_Gasoline.pdf
25 U.S. Department of the Interior, Bureau of Mines, "Lead," in *Minerals Yearbook*, 1920–1989, Washington, D.C.: Government Printing Office)
26 Jamie Lincoln Kitman, "The Secret History of Lead," *The Nation*, March 20, 2000, https://www.thenation.com/article/archive/secret-history-lead/
27 Note 16
28 Note 18
29 Note 18
30 Clair C. Patterson, "Contaminated and Natural Lead Environments of Man," *Archives of Environmental Health* 11, no. 3, September 1965: 344–360.
31 Note 14
32 Centers for Disease Control and Prevention, *Guidelines and Recommendations: Childhood Lead Poisoning Prevention* https://www.cdc.gov/lead-prevention/php/guidelines/index.html
33 Centers for Disease Control and Prevention, Data and Statistics: Childhood Lead Poisoning Prevention https://www.cdc.gov/lead-prevention/php/data/index.html
34 UNICEF and Pure Earth, *The Toxic Truth: Children's Exposure to Lead Pollution Undermines a Generation of Future Potential*, 2020,
35 Tata Consultancy Services, *TCS Future-Ready eMobility Study 2025*, https://www.tcs.com/who-we-are/newsroom/press-release/2025-the-year-electric-vehicles-64-percent-consumers-likely-choose-ev-as-their-next-vehicle-reveals-tcs-global-study
36 Wei Peng et al., *Electric Vehicle Transition and Air Pollution Hotspots: Impacts of Battery Mineral Refining*, Environmental Science & Technology 58, no. 2 (2024: 112–129, https://scitechdaily.com/the-dark-side-of-electric-vehicles-a-hidden-pollution-problem/.
37 World Health Organization, *Galvanizing Global Support for a Lead-Free Future*, WHA78.27, Geneva: WHO, May 2025, https://apps.who.int/gb/ebwha/pdf_files/WHA78/A78_R27-en.pdf
38 Center for Global Development, "Global Lead Chromate Exports Pose Poisoning Risk to 1.45 Million Children," April 9, 2025, https://www.cgdev.org/article/

global-lead-chromate-exports-pose-poisoning-risk-145-million-children

C4 Defining Questions
1. Richard P. Feynman, Robert B. Leighton, and Matthew Sands, *The Feynman Lectures on Physics, Vol. I: The New Millennium Edition: Mainly Mechanics, Radiation, and Heat*, New York: Basic Books, 2015, p. 45
2. Edward O. Wilson, Consilience: The Unity of Knowledge, New York: Alfred A. Knopf, 1998.

Part 2 Self-knowledge
1. Jon Krakauer, *Into Thin Air*, 2nd ed., New York: Villard Books, Division of Random House, Inc., 2000.
2. Rachel Nuwer, "The Tragic Tale of Mt Everest's Most Famous Dead Body," *BBC Future*, October 8, 2015 https://www.bbc.com/future/article/20151008-the-tragic-story-of-mt-everests-most-famous-dead-body.

C5 Two Ascents of Everest
1. Tenzing Norgay, *Tiger of the Snows: The Autobiography of Tenzing of Everest*, told to James Ramsey Ullman, New York: G. P. Putnam's Sons, 1955, 251.
2. Edmund Hillary and Edward Levine, "Climb Other Mountains," The New York Times Magazine, May 25, 2003
3. Jon Krakauer, *Into Thin Air*, 2nd ed., New York: Villard Books, Division of Random House, Inc., 2000, 153.
4. John Hunt, *The Ascent of Everest*, London: Hodder & Stoughton, 1953, 18.
5. Note 3
6. Note 4
7. Note 4
8. Note 3
9. Note 4
10. Note 4
11. Note 4
12. Danielle Preiss, "One-Third of Everest Deaths Are Sherpa Climbers," *NPR Illinois*, April 14, 2018, https://www.npr.org/sections/parallels/2018/04/14/599417489/one-third-of-everest-deaths-are-sherpa-climbers
13. Leger C. J, "The 1996 Everest Disaster – The Whole Story," *Base Camp Magazine*, 2016 https://basecampmagazine.com/2016/12/31/the-1996-everest-disaster-the-whole-story/
14. Note 4
15. David Fickling, "We Knocked the Bastard Off," The Guardian, 2003, https://www.theguardian.com/world/2003/mar/13/everest.nepal
16. Eric Shipton, *That Untravelled World: An Autobiography*, London: Hodder & Stoughton, 1969, 179.
17. Edmund Hillary, *High Adventure*, London: Hodder & Stoughton, 1955, 117.
18. Jim Perrin, "Sir Edmund Hillary - Mountaineer Who Dedicated Life After His Ascent of Everest to the People of Nepal," The Guardian, 2008, https://www.theguardian.com/news/2008/jan/11/mainsection.obituaries1
19. Note 17
20. Richard Pearson, "John Hunt Dies," The Washington Post, 1998, https://www.washingtonpost.com/archive/local/1998/11/10/john-hunt-dies/19cfd1e9-97d1-4619-8b00-80a98fcb5e41/
21. Note 3
22. George Lowe, Director, George Lowe, John Noel, and John Stobart, Cinematography, *The Conquest of Everest*, London: Countryman Films and Group 3, 1953).
23. Note 3
24. Wayne E. Askew, "Food for High-Altitude Expeditions: Pugh Got It Right in 1954—A Commentary on the Report by L.G.C.E. Pugh: 'Himalayan Rations with Special Reference to the 1953 Expedition to Mount Everest'," *Wilderness and Environmental Medicine*, 2004,

25 Note 17
26 Note 17
27 Note 3
28 Note 17
29 Note 22
30 Note 3
31 Note 3
32 Note 3

C6 Origins of Human Judgment and Actions
1   Daniel Kahneman, *Thinking, Fast and Slow*, New York: Farrar, Straus. and Giroux, 2011, 256.
2   Ibid
3   Ibid
4   Eta S. Berner and Mark L. Graber, "Overconfidence as a Cause of Diagnostic Error in Medicine," *American Journal of Medicine* 121, no. 5 Suppl, 2008: S2–S23, https://psnet.ahrq.gov/issue/overconfidence-cause-diagnostic-error-medicine.
5   Jon Krakauer, *Into Thin Air: A Personal Account of the Mt. Everest Disaster* (New York: Villard, 1997), 195.
6   Stanley Milgram, "Behavioral Study of Obedience," Journal of Abnormal and Social Psychology 67, no. 4, 1963: 371–378, doi:10.1037/h0040525.
7   Philip G. Zimbardo, Craig Haney, William C. Banks, and David Jaffe, *"The Mind Is a Formidable Jailer: A Pirandellian Prison," The New York Times Magazine*, April 8, 1973, Section 6, pp. 38, ff., https://www.prisonexp.org/selected-references
8   Ibid
9   Christopher R. Browning, *Ordinary Men: Reserve Police Battalion 101 and the Final Solution in Poland*, New York: HarperCollins, 2017).
10 Note 6
11 Frans de Waal, *Are We Smart Enough to Know How Smart Animals Are?*, New York: W. W. Norton, 2016, 15.
12 Maria V. Suntsova and Anton A. Buzdin, "Differences Between Human and Chimpanzee Genomes and Their Implications in Gene Expression, Protein Functions, and Biochemical Properties of the Two Species," BMC Genomics 21, Suppl. 7, 2020: 535, https://doi.org/10.1186/s12864-020-06962-8
13 Ingo Ebersberger, Dirk Metzler, Carsten Schwarz, and Svante Pääbo, *"Genomewide Comparison of DNA Sequences between Humans and Chimpanzees," American Journal of Human Genetics* 70, no. 6, 2002: 1490–1497, https://www.eva.mpg.de/documents/Elsevier/Ebersberger_Genomewide_AmJHumGen_2002_1555923.pdf
14 Kay Prüfer et al., *"The Bonobo Genome Compared with the Chimpanzee and Human Genomes," Nature* 486, no. 7404, 2012: 527–531, https://doi.org/10.1038/nature11128
15 Judgment Day: Intelligent Design on Trial, "Student Handout: Hominidae Family Tree," *NOVA Teachers*, PBS, https://www.pbs.org/wgbh/nova/teachers/activities/3416_id_02.html.
16 Note 5
17 Frans de Waal, *Our Inner Ape: The Best and Worst of Human Nature*, London: Granta Books, 2005.
18 James Reed, *Rise of the Warrior Apes*, directed by James Reed, Keo Films; Discovery International, 2017, documentary film.
19 Note 17
20 Marco Venniro and Scott A. Golden, *"Taking Action: Empathy and Social Interaction in Rats," Neuropsychopharmacology* 45, no. 7, 2020: 1081–1082, https://orcid.org/0000-0001-7653-957X
21 Frans de Waal, *The Age of Empathy: Nature's Lessons for a Kinder Society*, New York: Harmony Books, 2009)
22 Paul Bloom, *Just Babies: The Origins of Good and Evil*. New York: Crown Publishers, 2013.
23 Note 1

Endnotes and Citations

24 Matthew T. Gailliot et al., *"Self-Control Relies on Glucose as a Limited Energy Source: Willpower Is More Than a Metaphor," Journal of Personality and Social Psychology* 92, no. 2, 2007: 325–336, https://psycnet.apa.org/record/2007-00654-010
25 Daria Knoch *et al,*, "Diminishing Reciprocal Fairness by Disrupting the Right Prefrontal Cortex." *Science* 314, no. 5800, 2006: 829–832.
26 Liane Young *et al.,* "Disruption of the Right Temporoparietal Junction with Transcranial Magnetic Stimulation Reduces the Role of Beliefs in Moral Judgments." *Proceedings of the National Academy of Sciences* 107, no. 15, 2010: 6753–6758.
27 Trafton, Anne. "Moral Judgments Can Be Altered by Magnets." *MIT News*, March 30, 2010
28 Ibid
29 Daniel Kahneman, Olivier Sibony, and Cass R. Sunstein, *Noise: A Flaw in Human Judgment,* New York: Little, Brown, 2021, 204.
30 James W. Prescott, "Body Pleasure and the Origins of Violence," *The Bulletin of the Atomic Scientists* 31, no. 9, 1975: 10–20. http://www.deconnection.org/site/images/publicaties/body%20pleasure%20and%20the%20origins%20of%20violence%20-%20j.w.%20prescott.pdf
31 Keith E. Stanovich, *What Intelligence Tests Miss: The Psychology of Rational Thought,* New Haven: Yale University Press, 2009).
32 Ibid
33 Ibid.
34 Daniel C. Dennett, *From Bacteria to Bach and Back: The Evolution of Minds,* New York: W. W. Norton & Company, 2017)
35 Kathleen Vohs, Money priming can change people's thoughts, feelings, motivations, and behaviors: An update on 10 years of experiments, 2015 *Journal of Experimental Psychology: General,* 144(4, e86–e93, https://doi.org/10.1037/xge0000091
36 Ibid
37 Note 1
38 Nick Bullock, *"I Want It and I Want It Now,"* 2011, https://nickbullock-climber.co.uk/2011/06/02/i-want-it-and-i-want-it-now-i-want-it-and-i-want-it-now-i-want-it-and-i-wa/
39 Ibid
40 Robert Trivers, *The Folly of Fools: The Logic of Deceit and Self-Deception in Human Life,* New York: Basic Books, 2011)
41 Hugo Mercier and Dan Sperber, *"Why Do Humans Reason? Arguments for an Argumentative Theory," Behavioral and Brain Sciences* 34, no. 2, 2011: 57–74, https://doi.org/10.1017/S0140525X10000968.
42 Note 1
43 Atul Gawande, *The Checklist Manifesto,* London, England: Profile Books, 2011.
44 John Hunt, *The Ascent of Everest,* London: Hodder & Stoughton, 1953, 287.
45 Daryl Davis, *Klan-destine Relationships: A Black Man's Odyssey in the Ku Klux Klan,* New York: New Horizon Press, 1998)
46 Edmund Hillary, *High Adventure,* New York: Oxford University Press, 1955, 108.
47 Frans de Waal, *The Age of Empathy: Nature's Lessons for a Kinder Society,* New York: Harmony Books, 2009, 6.
48 Martha C. Nussbaum, *Upheavals of Thought: The Intelligence of Emotions,* Cambridge: Cambridge University Press, 2001)
49 Ibid
50 Bertrand Russell, The Triumph of Stupidity, in Mortals and Others: American Essays 1931–1935, vol. 1, London: Routledge, 1975)
51 Robert Hughes, "Modernism's Patriarch," *Time Magazine,* June 10, 1996.

Part 3 A Test of Knowledge
1 Clair C. Patterson, "Historical Changes in Integrity and Worth of Scientific Knowledge," *Geochimica et Cosmochimica* Acta 58, no. 12, 1994: 3199–3206, https://

authors.library.caltech.edu/records/7grf2-2qr03.
2   John R. Huizenga, *Cold Fusion: The Scientific Fiasco of the Century*, Rochester, NY: University of Rochester Press, 1992, 259.

C7 Tetraethyl Lead in Gasoline – The Science

1   *Nicander of Colophon. Theriaca and Alexipharmaca*. Edited and translated by A. S. F. Gow and A. F. Scholfield. Cambridge: Cambridge University Press, 1953
2   Colum Gilfillan, "Lead Poisoning and the Fall of Rome," *Journal of Occupational Medicine* 7, 1965)
3   William J. Kovarik, *The Ethyl Controversy*, Ph.D. dissertation, University of Maryland, 2013, https://billkovarik.com/wp-content/uploads/2012/08/Ethyl.Controversy.Kovarik.dissertation.pdf
4   Christian Warren, *Brush with Death: A Social History of Lead Poisoning*, Baltimore: Johns Hopkins University Press, 2000)
5   Richard Rabin, "Warnings Unheeded: A History of Child Lead Poisoning," *American Journal of Public Health* 79, 1989)
6   Note 3
7   Note 3
8   David Rosner and Gerald Markowitz, "A 'Gift of God'? The Public Health Controversy over Leaded Gasoline during the 1920s," *American Journal of Public Health* 75, no. 4, 1985, https://academiccommons.columbia.edu/doi/10.7916/D8RF64WX/download
9   Ibid
10 Herbert L. Needleman, "Clair Patterson and Robert Kehoe: Two Views on Lead Toxicity," *Environmental Research* 78, no. 2, 1998: 71–78, https://www.sciencedirect.com/science/article/abs/pii/S001393519793807X?via%3Dihub
11 Herbert L. Needleman, "Lead Poisoning," *Annual Review of Medicine* 51, 2000: 383–400, https://sethroberts.org/wp-content/uploads/2024/05/Needleman_2000.pdf.
12 Jerome O. Nriagu, "Clair Patterson and Robert Kehoe's Paradigm of 'Show Me the Data' on Environmental Lead Poisoning," *Environmental Research* 78, no. 1, 1998: 71–78, https://archive.org/download/wikipedia-scholarly-sources-corpus/10.1006%252Fbbrc.2000.2776.zip/10.1006%252Fenrs.1997.3808.pdf.
13 Ibid
14 Clair C. Patterson, "Historical Changes in Integrity and Worth of Scientific Knowledge," *Proceedings of the National Academy of Sciences of the United States of America* 92, no. 19, 1995: 7843–7847, https://authors.library.caltech.edu/records/7grf2-2qr03.
15 *Clair Patterson, oral history interview* by Shirley K. Cohen, California Institute of Technology, Pasadena, 1995. https://oralhistories.library.caltech.edu/32/1/OH_Patterson.pdf.
16 Tim Lane, "Saint Patterson and His Duck Soup," *Iowa Heritage Illustrated* 92, 2005, https://pubs.lib.uiowa.edu/ihi/article/1159/galley/110173/view.
17 Clair Patterson, quoted in "Getting the Lead Out," *Caltech News*, September 21, 2015, https://www.caltech.edu/about/news/getting-lead-out-47935.
18 Lydia Denworth, *Toxic Truth: A Scientist, a Doctor, and the Battle over Lead*, Boston: Beacon Press, 2009, 157.
19 Environmental Protection Agency, EPA, *Air Quality Criteria for Lead*, 1989, https://nepis.epa.gov/Exe/ZyPURL.cgi?Dockey=2000THEX.txt
20 Note 11
21 Adapted from Clair C. Patterson, "Contaminated and Natural Lead Environments of Man," *Archives of Environmental Health* 11, no. 3, 1965: 344–360; and Herbert L. Needleman, "Clair Patterson and Robert Kehoe: Two Views on Lead Toxicity," *Environmental Research* 78, no. 2, 1998: 71–78.
22 Note 12
23 Clair Patterson, "Concentrations of Common Lead in Some Atlantic and Mediterranean Waters and Snow," *Nature* 199, 1963)

24 Note 11
25 Note 19
26 Note 10
27 RC Angrand *et al.*, "Relation of Blood Lead Levels and Lead in Gasoline: An Updated Systematic Review," *Environmental Health* 21, no. 1, December 27, 2022: 138. doi: 10.1186/s12940-022-00936-x.
28 Herbert L. Needleman *et al.*, "Deficits in Psychologic and Classroom Performance of Children with Elevated Dentine Lead Levels," *The New England Journal of Medicine*, March 29, 1979.
29 Note 4
30 Claire B. Ernhart *et al.*, "On Being a Whistleblower: The Needleman Case," *Ethics & Behavior* 3, no. 1, 1993: 73–93, https://gwern.net/doc/statistics/bias/1993-ernhart.pdf
31 Note 19
32 Herbert Needleman, S.K. Geiger, and R. Frank, "Lead and IQ Scores: A Reanalysis," *Science* 227, no. 4688, February 15, 1985: 701–702.
33 J.L Annest, "Trends in the blood lead levels of the U.S. population: The Second National Health and Nutrition Examination Survey, NHANES II. 1976-1980," in *Lead Versus. Health*, M. Rutter and R.R. Jones, eds., New York: John Wiley & Sons. ©1983, John Wiley & Sons, Ltd.
34 Franco Scinicariello *et al.*, "Lead and Delta-Aminolevulinic Acid Dehydratase Polymorphism: Where Does It Lead? A Meta-Analysis," *Environmental Health Perspectives* 115, 2007, https://europepmc.org/article/PMC/1797830
35 Carla Marchetti, "Molecular Targets of Lead in Brain Neurotoxicity," *Neurotoxicity Research* 5, 2003, https://link.springer.com/article/10.1007/BF03033142
36 Jessica Wolpaw Reyes, "Environmental Policy as Social Policy? The Impact of Childhood Lead Exposure on Crime," *The B.E. Journal of Economic Analysis & Policy* 7, no. 1, 2007: Article 51. https://doi.org/10.2202/1935-1682.1796.
37 Ibid
38 Ibid
39 Ibid
40 Ibid

C8 The Cold Fusion Fiasco

1 C. Cookson, "Scientists Claim Nuclear Fusion Produced in Test Tube," *Financial Times*, March 23, 1989.
2 J. Bishop, "Breakthrough in Fusion May Be Announced," *Wall Street Journal*, March 23, 1989.
3 No longer available on the University of Utah website but reproduced in John R. Huizenga, *Cold Fusion: The Scientific Fiasco of the Century*, Appendix I, Oxford, U.K.: Oxford University Press, 1993).
4 Ibid
5 Martin Fleischmann and Stanley Pons, "Electrochemically Induced Nuclear Fusion of Deuterium," *Journal of Electroanalytical Chemistry and Interfacial Electrochemistry* 261, no. 2, 1989: 301–308, https://www.lenr-canr.org/acrobat/Fleischmanelectroche.pdf
6 John R. Huizenga, *Cold Fusion: The Scientific Fiasco of the Century*, Appendix I, Oxford, U.K.: Oxford University Press, 1993).
7 M. Burbidge *et al.*, "Synthesis of the Elements in Stars," *Reviews of Modern Physics* 29, no. 4, 1957: 547–650.
8 Energy.org, DOE, DOE *Explains...Fusion Reactions*, 2024, https://www.energy.gov/science/doe-explainsfusion-reactions
9 Note 6
10 Ibid
11 Ibid
12 Ibid
13 Edmund Newton, "Claims of Cold Fusion in Utah Tantalize Scientists at Caltech,"

Los Angeles Times, May 28, 1989. https://www.latimes.com/archives/la-xpm-1989-05-28-hd-1275-story.html

14 Nathan S. Lewis et al., "Searches for Low-Temperature Nuclear Fusion of Deuterium in Palladium," *Nature* 340, no. 6234, 1989: 525–530, https://www.nature.com/articles/340525a0

15 D. Petrasso et al., "Problems with the γ-Ray Spectrum in the Fleischmann et al. Experiments," *Nature* 339, no. 6221, 1989: 183–185. https://doi.org/10.1038/339183a0.

16 Note 6

17 Ibid

18 Julian Schwinger, *Cold Fusion—Does It Have a Future?*, 1991). https://www.lenr-canr.org/acrobat/SchwingerJcoldfusiona.pdf

19 Note 6

20 Ibid

21 Ibid

22 D. Lindley, "Utah Faculty Protest Cold Fusion Dealings," *Nature* 345, no. 6276, 1990: 561. https://doi.org/10.1038/345561a0.

23 Douglas R. O. Morrison, *Review of Cold Fusion*, 1990). https://www.lenr-canr.org/acrobat/MorrisonDRreviewofco.pdf

24 D. Voss, "What Ever Happened to Cold Fusion," *Physics World* 12, no. 3, 1999: 12–14. doi:10.1088/2058-7058/12/3/14

C9 Science as a Test of Knowledge

1 Richard P. Feynman, Robert B. Leighton, and Matthew Sands, *The Feynman Lectures on Physics*, vol. 1, Reading, MA: Addison-Wesley, 1963, C1-1.

2 *Wrightbrothers.org*, "Some Aeronautical Experiments,", 2024). https://www.wright-brothers.org/History_Wing/Wright_Story/Inventing_the_Airplane/Kitty_Hawk_in_a_Box/Some-Aeronuatical-Experiments-by-Wilbur-Wright.htm

3 James Tobin, *To Conquer the Air: The Wright Brothers and the Great Race for Flight*, New York: Free Press, 2003).

4 Richard P. Feynman, *"Cargo Cult Science,"* commencement address, California Institute of Technology, Pasadena, CA, June 1974. Transcript https://speakola.com/grad/richard-feynman-caltech-1974

5 Karl Popper, "Science: Conjectures and Refutations," in *Philosophy of Science: The Central Issues*, ed. Martin Curd, J. A. Cover, and Christopher Pincock, New York: W. W. Norton & Company, 2013)

6 *McLean v. Arkansas Board of Education*, 529 F. Supp. 1255, E.D. Ark. 1982). https://law.justia.com/cases/federal/district-courts/FSupp/529/1255/2354824/

7 Larry Laudan, "Demystifying Underdetermination," in *Philosophy of Science: The Central Issues*, ed. Martin Curd, J. A. Cover, and Christopher Pincock, New York: W. W. Norton & Company, 2013,

8 Edmund Newton, "Claims of Cold Fusion in Utah Tantalize Scientists at Caltech," *Los Angeles Times*, May 28, 1989. https://www.latimes.com/archives/la-xpm-1989-05-28-hd-1275-story.html

9 Nathan S. Lewis et al., "Searches for Low-Temperature Nuclear Fusion of Deuterium in Palladium," *Nature* 340, no. 6234, 1989: 525–530, https://www.nature.com/articles/340525a0

10 D. Petrasso et al., "Problems with the γ-Ray Spectrum in the Fleischmann et al. Experiments," *Nature* 339, no. 6221, 1989: 183–185. https://doi.org/10.1038/339183a0.

11 George E. P. Box, William G. Hunter, and J. Stuart Hunter, *Statistics for Experimenters: An Introduction to Design, Data Analysis, and Model Building*, New York: Wiley, 1978).

12 Morris Kline, *Mathematics and the Search for Knowledge*, New York: Oxford University Press, 1985, 2.

13 Ibid

14 Philip Morrison and Phylis Morrison, *The Ring of Truth: An Inquiry into How We*

Know What We Know, New York: Random House, 1987, 47.
15 Note 10
16 David Rosner and Gerald Markowitz, "'A Gift of God'? The Public Health Controversy Over Leaded Gasoline During the 1920s," *American Journal of Public Health* 75, 1985)
17 E. B. Wilson, *An Introduction to Scientific Research,* New York, USA: McGraw-Hill, 1952, 46.
18 Robyn Dawes, *Everyday Irrationality: How Pseudo-Scientists, Lunatics, and the Rest of U.S. Systematically Fail to Think Rationally,* New York, USA: Routledge, 2019).
19 Ibid
20 E. B. Wilson, *An Introduction to Scientific Research,* New York, USA: McGraw-Hill, 1952, 46.
21 Ritchie Robertson, *The Enlightenment: The Pursuit of Happiness, 1680–1790,* London: Penguin Books, 2020).
22 Timothy Gowers, *Mathematics: A Very Short Introduction,* Oxford: Oxford University Press, 2002).
23 Garrett Hardin, *Filters Against Folly: How to Survive Despite Economists, Ecologists, and the Merely Eloquent,* New York: Viking Penguin, 1986).
24 Alan D. Sokal, *Beyond the Hoax: Science, Philosophy and Culture,* Oxford: Oxford University Press, 2008, vii.
25 Note 23
26 E. Brian Davies, *Science in the Looking Glass: What Do Scientists Really Know?,* Oxford: Oxford University Press, 2003, x.
27 Philip Anderson, "More Is Different." *Science* 177, no. 4047, 1972: 393–396, https://doi.org/10.1126/science.177.4047.393
28 E. O. Wilson, *Consilience: The Unity of Knowledge,* New York: Alfred A. Knopf, 1998).
29 Ibid
30 Rachael Bedard, "The Doctor Who Hates Medicine," *New York Times,* October 30, 2025, https://www.nytimes.com/2025/10/30/opinion/anti-expert-expert-surgeon-general.html.
31 Herbert Needleman, *History of Lead Poisoning in the World,* 1999, Available at https://www.biologicaldiversity.org/campaigns/get_the_lead_out/pdfs/health/Needleman_1999.pdf
32 Naomi Oreskes and Erik M. Conway, *Merchants of Doubt: How a Handful of Scientists Obscured the Truth on Issues from Tobacco Smoke to Global Warming,* New York: Bloomsbury Press, 2010).
33 John P. A. Ioannidis, *Why Most Published Research Findings Are False,* PLOS Medicine 2, no. 8, 2005: e124. https://doi.org/10.1371/journal.pmed.0020124.
34 Carl Sagan, The Demon-Haunted World: Science as a Candle in the Dark, New York: Random House, 1995)
35 Ibid

Part 4 Evolution and Chance
1 Charles Darwin, *On the Origin of Species by Means of Natural Selection, or the Preservation of Favoured Races in the Struggle for Life,* facsimile of the first edition, Cambridge: Harvard University Press, 1964; originally published London: John Murray, 1859, 61.
2 Jacques Monod, *Chance and Necessity: An Essay on the Natural Philosophy of Modern Biology,* trans. Austryn Wainhouse, New York: Vintage Books, 1972, 180

C10 A Brief History of Life
1 Erwin Schrödinger, *What Is Life? The Physical Aspect of the Living Cell,,* Cambridge: Cambridge University Press, 1944).
2 Jaime Gómez-Márquez, "What is Life?" *Molecular Biology Reports* 48, no. 7, 2021: 6223–6230, https://link.springer.com/article/10.1007/s11033-021-06594-5.
3 Ibid
4 Brent Dalrymple, "The Age of the Earth in the Twentieth Century: A Problem,

Mostly. Solved." *Geological Society of London Special Publications 190*, no. 1, 2001: 205–221, https://doi.org/10.1144/GSL.SP.2001.190.01.14
5   Ben Pearce et al., "Constraining the Time Interval for the Origin of Life on Earth." *Astrobiology* 18, no. 3, 2018: 343–364, https://doi.org/10.1089/ast.2017.1674.
6   Casey McGrath, "Highlight: Unraveling the Origins of LUCA and LECA on the Tree of Life." *Genome Biology and Evolution* 14, no. 6, June 2022: evac072. https://doi.org/10.1093/gbe/evac072.
7   Anja Spang et al., "Evolving Perspective on the Origin and Diversification of Cellular Life and the Virosphere." *Genome Biology and Evolution* 14, no. 6, 2022: evac034. https://doi.org/10.1093/gbe/evac034.
8   Stanley Miller, "A Production of Amino Acids Under Possible Primitive Earth Conditions." *Science* 117, no. 3046, 1953: 528–529, https://doi.org/10.1126/science.117.3046.528.
9   Jeffrey Bada et al., "Reanalysis of Prebiotic Organic Synthesis in Stanley Miller's 1953 Experiment." *Science* 322, no. 5900, 2008: 404–406, https://doi.org/10.1126/science.1161527.
10 Meng, Yifan et al., "Spraying of Water Microdroplets Forms Luminescence and Causes Chemical Reactions in Surrounding Gas." *Science Advances*, 11, no. 5, 2025: eadt8979, https://doi.org/10.1126/sciadv.adt8979.
11 Christopher Conselice et al., "The Evolution of Galaxy Number Density at $z < 8$ and Its Implications." *The Astrophysical Journal* 830, no. 2, 2016: 83, https://doi.org/10.3847/0004-637X/830/2/83.
12 Note 7
13 Louca Stilianos et al., "A Census-Based Estimate of Earth's Bacterial and Archaeal Diversity." *PLOS Biology* 17, no. 2, 2019: e3000106, https://doi.org/10.1371/journal.pbio.3000106
14 Bar-On, Yinon M., Rob Phillips, and Ron Milo. "The Biomass Distribution on Earth." *Proceedings of the National Academy of Sciences*, 115, no. 25, 2018: 6506–6511. https://doi.org/10.1073/pnas.1711842115
15 Ibid
16 Nick Lane *Oxygen: The Molecule that Made the World*. Oxford University Press, 2002.
17 Stephen Jay Gould, *Wonderful Life: The Burgess Shale and the Nature of History*. New York: W.W. Norton, 1989.
18 Ibid
19 Danielle Gruen et al., "Paleozoic Diversification of Terrestrial Chitin-Degrading Bacterial Lineages." *BMC Evolutionary Biology* 19, no. 34, 2019). https://bmcecolevol.biomedcentral.com/articles/10.1186/s12862-019-1357-8.
20 GOLDSROBI, CC0, via Wikimedia Commons, https://en.wikipedia.org/wiki/Timeline_of_the_evolutionary_history_of_life
21 Garrett Hardin. *Nature and Man's Fate*. New York: New American Library, 1959
22 Ibid
23 Ibid
24 Ibid
25 Leigh Van Valen "A New Evolutionary Law." *Evolutionary Theory* 1, 1973: 1–30, https://www.mn.uio.no/cees/english/services/van-valen/evolutionary-theory/volume-1/vol-1-no-1-pages-1-30-l-van-valen-a-new-evolutionary-law.pdf.
26 Ibid
27 Jens Clausen. *Stages in the Evolution of Plant Species*. Ithaca, NY: Cornell University Press, 1951, 133.
28 Note 21
29 Lisa Urry et al., *Campbell Biology*. 12th ed. New York: Pearson, 2021.
30 Sun, Dongchang et al., "Horizontal Gene Transfer Mediated Bacterial Antibiotic Resistance." *Frontiers in Microbiology* 10, 2019: 1933. https://doi.org/10.3389/fmicb.2019.01933.
31 C Venditti et al., Speciation and Bursts of Evolution. *Evo Edu Outreach* 1, 274–280,

2008). https://doi.org/10.1007/s12052-008-0049-4
32 Melania Serafini *et al.*, "Air Pollution: Possible Interaction between the Immune and Nervous. System?" International Journal of Environmental Research and Public Health 19, no. 23, 2022: 16037. https://doi.org/10.3390/ijerph192316037.
33 Giovanni Adami *et al.*, "Association between Long-Term Exposure to Air Pollution and Immune-Mediated Diseases: A Population-Based Cohort Study." *RMD Open* 8, no. 1, 2022: e002055. https://doi.org/10.1136/rmdopen-2021-002055.
34 D. Jablonski, Background and mass extinctions: the alternation of macroevolutionary regimes. *Science.* 1986 Jan 10;231(4734:129-33. doi: 10.1126/science.231.4734.129. PMID: 17842630.
35 De Vos *et al*, "Estimating the Normal Background Rate of Species Extinction." *Conservation Biology* 29, no. 2 (2015): 452–462, https://conbio.onlinelibrary.wiley.com/doi/epdf/10.1111/cobi.12380
36 Ibid
37 Brian Curtice, "Supersaurus May Have Been the Longest Dinosaur to Have Ever Lived." *Modern Sciences*, December 2, 2021, https://modernsciences.org/supersaurus-may-have-been-the-longest-dinosaur-to-have-ever-lived/
38 Clay R. Tabor *et al.*, "Causes and Climatic Consequences of the Impact Winter at the Cretaceous-Paleogene Boundary," *Geophysical Research Letters* 47, no. 6, 2020: e60121, https://doi.org/10.1029/2019GL085572.
39 L. M. Chiappe, "Downsized Dinosaurs: The Evolutionary Transition to Modern Birds. Evolution: Education and Outreach", 2009. 2(2, 248–256, https://evolution-outreach.biomedcentral.com/articles/10.1007/s12052-009-0133-4
40 Berv, S. Singhal *et al.*, "Genome and life-history evolution link bird diversification to the end-Cretaceous. mass extinction," *Science Advances* 10, 2024: eadp0114, https://doi.org/10.1126/sciadv.adp0114.
41 Ni, Xijun *et al.*, "The Oldest Known Primate Skeleton and Early Haplorhine Evolution." *Nature* 498, no. 7452, 2013: 60–64. https://www.scholars.northwestern.edu/en/publications/the-oldest-known-primate-skeleton-and-early-haplorhine-evolution
42 Michael Rampino *et al.*, "What Causes Mass Extinctions? Large Asteroid/Comet Impacts, Flood-Basalt Volcanism, and Ocean Anoxia—Correlations and Cycles." *Geological Society of America Special Paper* 542, 2019: 271–302. https://doi.org/10.1130/2019.2542(14)
43 Louise Humphrey and Chris Stringer. *Our Human Story.* London: Natural History Museum, 2018
44 Ibid
45 Ibid
46 Hu, Wangjie *et al.*, "Genomic inference of a severe human bottleneck during the Early to Middle Pleistocene transition." *Science* 381, no. 6661, August 31, 2023: 979-984. https://doi.org/10.1126/science.abq7487.
47 Ibid
48 Note 43
49 Dora Koller *et al.*, "Denisovan and Neanderthal Archaic Introgression Differentially Impacted the Genetics of Complex Traits in Modern Populations." *BMC Biology* 20, no. 249, 2022). https://doi.org/10.1186/s12915-022-01449-2.
50 Garrett Hardin, *Nature and Man's Fate.* New York: New American Library, 1959.
51 Note 43
52 Ibid
53 Bret Stetka, *A History of the Human Brain: From the Sea Sponge to CRISPR,* How Our Brain Evolved, Portland, OR: Timber Press, 2021).
54 Jianfeng Feng *et al.*, "Human Brain Computing and Brain-Inspired Intelligence," *National Science Review* 11, no. 5, 2024: nwae144, https://doi.org/10.1093/nsr/nwae144.
55 Tim McMillan, "IARPA Research Achieves the Impossible in New Breakthrough Revealing the Brain's Biggest Mysteries," *The Debrief,* April 14, 2025, https://

thedebrief.org/iarpa-research-achieves-the-impossible-in-new-breakthrough-revealing-the-brains-biggest-mysteries/.
56 Karl Marx and Friedrich Engels, *The Communist Manifesto*, London: Penguin Books, 2015).
57 Steven Pinker, *The Better Angels of Our Nature: Why Violence Has Declined*, New York: Viking, 2011).
58 Francis Fukuyama, *The End of History and the Last Man*, New York: Free Press, 1992).
59 Karl Popper, *The Poverty of Historicism*, London: Routledge & Kegan Paul, 1957).
60 Jared Diamond, *Guns, Germs, and Steel: The Fates of Human Societies*, New York: W.W. Norton & Company, 1997).

C11 Evolution as Chance and Necessity
1 Stephen Jay Gould, *Full House: The Spread of Excellence from Plato to Darwin*, New York: Harmony Books, 1996).
2 WorldAtlas. "How Much Bacteria Is on Earth?" WorldAtlas. https://www.worldatlas.com/how-much-bacteria-is-on-Earth.html.
3 World Animal Foundation. "How Many Animals Are in the World?" World Animal Foundation. https://worldanimalfoundation.org/advocate/how-many-animals-are-in-the-world/
4 Note 1
5 Walter Veit et al., "Evolution, Complexity, and Life History Theory," *Biological Theory* 20, no. 1, 2025: 45–67.
6 Garrett Hardin, *Filters Against Folly: How to Survive Despite Economists, Ecologists, and the Merely Eloquent* (New York: Viking Penguin, 1985).
7 Simon Conway Morris, *The Runes of Evolution: How the Universe Became Self-Aware*, Cambridge: Templeton Press, 2015).
8 Richard Dawkins, *The Blind Watchmaker*, Harlow, England: Penguin Books, 2006).
9 Suzana Herculano-Houzel, "The Remarkable, Yet Not Extraordinary, Human Brain as a Scaled-up Primate Brain and Its Associated Cost," *Proceedings of the National Academy of Sciences* 109, no. 2, 2012: 10661–10668, https://doi.org/10.1073/pnas.1201895109.
10 Richard York and Brett Clark, *The Science and Humanism of Stephen Jay Gould*, New York: Monthly Review Press, 2011).
11 Stephen Jay Gould, *Wonderful Life: The Burgess Shale and the Nature of History*, New York: W. W. Norton, 1989)
12 K Nanglu et al., "The Burgess Shale paleocommunity with new insights from Marble Canyon", *British Columbia. Paleobiology*, 2020, 46(1, 58-81. Cambridge University Press. https://www.cambridge.org/core/journals/paleobiology/article/burgess-shale-paleocommunity-with-new-insights-from-marble-canyon-british-columbia/1F90AF7FFD14AAED92801DAC780C952B.
13 Garrett Hardin, *Nature and Man's Fate*. New York: New American Library, 1959
14 Peter Godfrey-Smith, *Other Minds: The Octopus, the Sea, and the Deep Origins of Consciousness*, New York: Farrar, Straus. and Giroux, 2016)
15 Carl Sagan, *The Varieties of Scientific Experience: A Personal View of the Search for God*, ed. Ann Druyan, New York: Penguin Press, 2006).
16 Mark Cartwright, *The 1944 Plot to Assassinate Hitler*, World History Encyclopedia, December 6, 2024, https://www.worldhistory.org/article/2584/the-1944-plot-to-assassinate-hitler/
17 L. Rastrigin, *This Chancy, Chancy, Chancy World*, trans. R. H. M. Woodhouse, Moscow: Mir Publishers, 1973)
18 Lorenz, Edward N. *Deterministic Nonperiodic Flow*. Journal of the Atmospheric Sciences 20, no. 2, 1963: 130–141.
19 Note 18
20 Sapolsky, Robert M. *Determined: A Science of Life Without Free Will*. New York: Penguin Press, 2023.
21 Daniel Dennett, *Elbow Room: The Varieties of Free Will Worth Wanting*. Cambridge,

MA: MIT Press, 1984.
22. Ivar Ekeland, *The Best of All Possible Worlds: Mathematics and Destiny* (Chicago: University of Chicago Press, 2006).
23. Jacques Monod, *Chance and Necessity: An Essay on the Natural Philosophy of Modern Biology*, trans. Austryn Wainhouse, New York: Alfred A. Knopf, 1971)
24. Sam Harris, *Waking Up: A Guide to Spirituality Without Religion.* New York: Simon & Schuster, 2014
25. Note 14

Part 5 Humanism Principles and Processes
1. Steven M. Opal and J. M. Opal, "The Elimination of Smallpox Showed How Humans Can Work Together to Solve Deadly Global Problems," *British Columbia Humanist Association*, October 4, 2018, https://www.bchumanist.ca/elimination_of_smallpox
2. Tedros Adhanom Ghebreyesus, "Foreword," in *Foresight: Tackling Future Pandemics*, World Health Organization, 2022, https://www.who.int/about/accountability/results/who-results-report-2022-2023/2023/health-emergencies-rapidly-detected-and-responded-to

C12 Eradication of Smallpox and the COVID-19 Crisis
1. Ron Sender *et al.*, *Revised Estimates for the Number of Human and Bacteria Cells in the Body*. PLOS Biology, 2016.
2. Statista. "Deadliest Pandemics and Epidemics," https://www.statista.com/chart/34077/deadliest-pandemics-epidemics/
3. Merle Eisenberg and Lee Mordechai, "The Justinianic Plague: An Interdisciplinary Review," *Byzantine and Modern Greek Studies* 43, no. 2, 2019: 156–80, https://middleagesforeducators.princeton.edu/justinianic-plague
4. Ole J. Benedictow, *The Black Death 1346–1353: The Complete History*, Woodbridge: Boydell Press, 2004)
5. Jeffery K. Taubenberger and David M. Morens, "1918 Influenza: The Mother of All Pandemics," *Emerging Infectious. Diseases* 12, no. 1, 2006: 15–22, https://wwwnc.cdc.gov/eid/article/12/1/05-0979_article
6. Anothony Fauci, 2003. 'HIV and AIDS: 20 years of science', *Nature medicine*, 9(7, pp. 839–843. https://doi.org/10.1038/nm0703-839.
7. Christophe Fraser *et al.*, "Factors That Make an Infectious Disease Outbreak Controllable." *Proceedings of the National Academy of Sciences of the United States of America* 101, no. 16, April 20, 2004: 6146–6151, https://www.pnas.org/doi/full/10.1073/pnas.0307506101
8. Catherine Thèves *et al.*, "History of Smallpox and Its Spread in Human Populations," *Microbiology Spectrum* 4, no. 4, July 2016, https://doi.org/10.1128/microbiolspec.poh-0004-2014.
9. Gareth Williams, *Angel of Death: The Story of Smallpox*, London: Palgrave Macmillan, 2010)
10. Ibid
11. Bonnie L. Ford, "Aztec Empire Smallpox Epidemic," EBSCO *Research Starters*, 2022.
12. Note 9
13. Donald Henderson, *Smallpox: The Death of a Disease*. Amherst, NY: Prometheus. Books, 2009.
14. Note 9
15. Igor V. Babkin and Irina N. Babkina, "The Origin of the Variola Virus," *Viruses* 7, no. 3, 2015: 1100–1112
16. Note 8
17. Note 15
18. Note 9
19. Centers for Disease Control and Prevention, *How Smallpox Spreads*, CDC, 2024. https://www.cdc.gov/smallpox/causes/index.html
20. Note 8

21 Note 13
22 Note 9
23 Mike Bray, "New Data in a 200-Year Investigation," *Clinical Infectious. Diseases* 38, no. 1, January 1, 2004: 90–91, https://academic.oup.com/cid/article-abstract/38/1/90/356548?redirectedFrom=fulltext
24 Ibid
25 Note 13
26 A. W. Downie, 1939. *The Immunological Relationship of the Virus. of Spontaneous. Cowpox to Vaccinia Virus.* British Journal of Experimental Pathology, 20(2, 158–176. PMCID: PMC2065307. https://europepmc.org/articles/PMC2065307/pdf/brjexppathol00194-0051.pdf
27 Note 13
28 Ibid
29 Ibid
30 Ibid
31 "Do We Still Use Bifurcated Needles?" *TimesMojo*, July 7, 2022. https://www.timesmojo.com/do-we-still-use-bifurcated-needles/
32 Note 13
33 Ibid
34 Ibid
35 Ibid
36 Moncef Bouzouaya, "Avian Coronaviruses: Characteristics of Epidemiological Interest, in Comparative Medicine." *Bulletin de L'Académie Nationale de Médecine* 205, no. 7, 2021: 737–741. https://doi.org/10.1016/j.banm.2021.03.004.
37 Dorothy Hamre and John Procknow, "A New Virus Isolated from the Human Respiratory Tract," *Proceedings of the Society for Experimental Biology and Medicine* 121, no. 1, 1966: 190–193, https://doi.org/10.3181/00379727-121-30734.
38 World Health Organization. "Severe Acute Respiratory Syndrome, SARS." *World Health Organization*, https://www.who.int/health-topics/severe-acute-respiratory-syndrome#tab=tab_1
39 World Health Organization. "Summary of Probable SARS Cases with Onset of Illness from 1 November 2002 to 31 July 2003." *World Health Organization*, 2003. https://www.who.int/publications/m/item/summary-of-probable-sars-cases-with-onset-of-illness-from-1-november-2002-to-31-july-2003.
40 World Health Organization. *Middle East Respiratory Syndrome Coronavirus., MERS-CoV).* WHO. Last updated March 2025. https://www.who.int/news-room/fact-sheets/detail/middle-east-respiratory-syndrome-coronavirus-%28mers-cov%29
41 Wu, Wenjuan, *et al.*, "Clinical Features of Patients Infected with 2019 Novel Coronavirus. in Wuhan, China." *The Lancet* 395, no. 10223, January 24, 2020: 497–506. https://doi.org/10.1016/S0140-6736(20)30183-5.
42 *BBC News*, "Li Wenliang: Coronavirus. Whistleblower Doctor Dies from Virus," BBC News, February 7, 2020, https://www.bbc.com/news/world-asia-china-51364382.
43 World Health Organization, "Novel Coronavirus. – Thailand," *WHO Disease Outbreak News*, January 14, 2020, https://www.who.int/emergencies/disease-outbreak-news/item/2020-DON234
44 European Journalism Observatory, "Wuhan Mass Banquet 18 Jan 2020," *EJO*, April 14, 2020, https://en.ejo.ch/ethics-quality/china-coronavirus-and-the-media/attachment/wuhan-mass-banquet-18-jan-2020
45 "WHO Declines to Declare International Emergency Over China Virus," *Reuters*, January 23, 2020, https://www.reuters.com/article/us-china-health-who-idUSKBN1ZM1G9/
46 Cirium, "Nearly 10,000 Chinese Flights Suspended as Coronavirus. Outbreak Escalates," *Cirium*, January 31, 2020, https://www.cirium.com/thoughtcloud/nearly-10000-chinese-flights-suspended-as-coronavirus-outbreak-escalates/
47 World Health Organization, "Statement on the Meeting of the International Health

Regulations, 2005. Emergency Committee Regarding the Outbreak of Novel Coronavirus., 2019-nCoV," *WHO News*, January 23, 2020,
48 World Health Organization, "Statement on the Second Meeting of the International Health Regulations, 2005. Emergency Committee Regarding the Outbreak of Novel Coronavirus., 2019-nCoV," *WHO News*, January 30, 2020, https://www.who.int/news/item/30-01-2020-statement-on-the-second-meeting-of-the-international-health-regulations-%282005%29-emergency-committee-regarding-the-outbreak-of-novel-coronavirus-%282019-ncov%29
49 Ibid
50 Swedish Government Official Reports. *Summary of SOU 2020:80 – The Elderly Care in the Pandemic*. Stockholm: Ministry of Health and Social Affairs, 2020. https://www.government.se/contentassets/2b394e1186714875bf29991b4552b374/summary-of-sou-2020_80-elderly-care-during-the-pandemic.pdf.
51 Ibid
52 Petra Zimmermann and Nigel Curtis, "Why Is COVID-19 Less Severe in Children? A Review of the Proposed Mechanisms Underlying the Age-Related Difference in Severity of SARS-CoV-2 Infections," *Archives of Disease in Childhood* 106, no. 5, 2021: 429–434, https://adc.bmj.com/content/106/5/429
53 The College of Physicians of Philadelphia, "Vaccine Development, Testing, and Regulation," *History of Vaccines*, https://historyofvaccines.org/vaccines-101/how-are-vaccines-made/vaccine-development-testing-and-regulation?q=content/articles/vaccine-development-testing-and-regulation
54 Shiyao Xu *et al.*, "Real-World Effectiveness and Factors Associated with Effectiveness of Inactivated SARS-CoV-2 Vaccines: A Systematic Review and Meta-Regression Analysis," *BMC Medicine* 21, Article 160, 2023, https://bmcmedicine.biomedcentral.com/articles/10.1186/s12916-023-02861-3
55 Sarah Gilbert and Catherine Green, *Vaxxers: The Inside Story of the Oxford Vaccine and the Race Against the Virus*, London: Hodder & Stoughton, 2021)
56 Kai Yuan Leong *et al.*, "Revolutionizing Immunization: A Comprehensive Review of mRNA Vaccine Technology and Applications," *Virology Journal* 22, Article 71, 2025, https://virologyj.biomedcentral.com/articles/10.1186/s12985-025-02645-6
57 Marzieh Soheili *et al.*, "The Efficacy and Effectiveness of COVID-19 Vaccines Around the World: A Mini-Review and Meta-Analysis," *Annals of Clinical Microbiology and Antimicrobials* 22, 2023: Article 42, https://ann-clinmicrob.biomedcentral.com/articles/10.1186/s12941-023-00594-y
58 E. Dolgin, 2021. 'The tangled history of mRNA vaccines', *Nature, London*, 597(7876), pp. 318–324. https://doi.org/10.1038/d41586-021-02483-w.
59 Note 56
60 Spencer Kimball, "Covid Pandemic Drives Pfizer's 2022 Revenue to Record $100 Billion," *CNBC*, January 31, 2023, https://www.cnbc.com/2023/01/31/the-covid-pandemic-drives-pfizers-2022-revenue-to-a-record-100-billion.html
61 Tom Espiner, "AstraZeneca to Take Profits from Covid Vaccine," *BBC News*, November 12, 2021, https://www.bbc.com/news/business-59256223
62 Kevin Dunleavy, "AstraZeneca's COVID Vaccine Sales Reached $4B Last Year, but Don't Expect That Much in 2022," *Fierce Pharma*, February 10, 2022, https://www.fiercepharma.com/pharma/astrazeneca-vaccine-sales-reached-4-billion-last-year-but-don-t-expect-2022-company-says
63 Savannah Behrmann and Jeanine Santucci, "Trump and Fauci: A Timeline of Their Relationship During COVID," *USA Today*, October 28, 2020
64 Rem Rieder, "Trump's Statements About the Coronavirus," *FactCheck.org*, March 18, 2020.
65 Tommy Beer, "All The Times Trump Compared Covid-19 To The Flu, Even After He Knew Covid-19 Was Far More Deadly," *Forbes*, September 10, 2020
66 Daniel Wolfe and Daniel Dale, "All of the Times President Trump Said Covid-19 Will Disappear," *CNN*, October 2020
67 Akshay Syal, "Trump Repeats Inaccurate Claim About Masks, Citing CDC Study,"

NBC News, October 15, 2020
68 John Fritze and Courtney Subramanian, "Trump Announces 'Halt' in U.S. Funding to World Health Organization Amid Coronavirus. Pandemic," *MedicalXpress*, April 15, 2020
69 Joshua P. Cohen, "Trump Promoted Hydroxychloroquine, A Drug Now Linked To 17,000 Deaths," *Forbes*, January 7, 2024

C13 Building Blocks of Modern Humanism
1 Julian Huxley, *The Future of Man*, New York: Harper & Brothers, 1959)
2 Larry Laudan, *Science at the Bar - Causes for Concern*, Science, Technology, & Human Values
3 Clifford, William Kingdon. *The Ethics of Belief. In The Ethics of Belief: Essays by William Kingdon Clifford*, William James, and A.J. Burger. Revised edition. Scotts Valley, CA: CreateSpace, 2008. Originally published 1877.
4 Martha C. Nussbaum, "Emotions as Judgments of Value and Importance," in *Thinking about Feeling: Contemporary Philosophers on Emotions*, ed. Robert C. Solomon, New York: Oxford University Press, 2004)
5 William Firth Wells, *Airborne Contagion and Air Hygiene: An Ecological Study of Droplet Infections*. Cambridge, MA: Harvard University Press, 1955.
6 Ben Bunnell, *How One Digital Book Led to an Important COVID-19 Discovery*, Google Blog, https://blog.google/products/search/how-one-digital-book-led-to-an-important-covid-19-discovery/
7 Donald K. Milton, "What Was the Primary Mode of Smallpox Transmission? Implications for Biodefense," *Frontiers in Cellular and Infection Microbiology*, November 29, 2012
8 Dyani Lewis, "Why the WHO Took Two Years to Say COVID Is Airborne," *Nature*, April 6, 2022, https://www.nature.com/articles/d41586-022-00925-7
9 Kerry Allen and Zhaoyin Feng, "China Covid-19: How State Media and Censorship Took on Coronavirus," *BBC News*, December 28, 2020, https://www.bbc.com/news/world-asia-china-55355401
10 Robert B. Laughlin, *The Crime of Reason: And the Closing of the Scientific Mind*, New York: Basic Books, 2008)
11 Heidi Chial. "DNA Sequencing Technologies Key to the Human Genome Project." *Nature Education* 1, no. 1, 2008: 219, https://www.nature.com/scitable/topicpage/dna-sequencing-technologies-key-to-the-human-828/.
12 Marc Lipsitch, quoted in Helen Branswell, "How to Weigh the Risks and Benefits of Reopening," *STAT News*, May 2020, https://www.statnews.com/2020/03/18/we-know-enough-now-to-act-decisively-against-covid-19/
13 Nisreen A. Alwan et al., "Scientific consensus. on the COVID-19 pandemic: we need to act now," *The Lancet* 396, no. 10260, October 14, 2020: e71–e72, https://www.thelancet.com/journals/lancet/article/PIIS0140-6736(20)32153-X/fulltext
14 Note 26
15 Christina Feldman, *Compassion: Listening to the Cries of the World* (Berkeley, CA: Rodmell Press, 2005)
16 *Life on Our Planet*. Directed by Adam Chapman. Produced by Silverback Films and Amblin Television. Narrated by Morgan Freeman. Netflix, 2023.
17 Molière, quoted in *The Quotable Molière*, ed. and trans. H. Gaston Hall, Chapel Hill: University of North Carolina Press, 1970)

Part 6 A Humanist Worldview
1 Tim Berners-Lee, "Tim Berners-Lee on 30 Years of the Web: 'If We Dream a Little, We Can Get the Web We Want,'" *The Guardian*, March 12, 2019, https://www.theguardian.com/technology/2019/mar/12/tim-berners-lee-on-30-years-of-the-web-if-we-dream-a-little-we-can-get-the-web-we-want
2 Sheera Frenkel and Cecilia Kang, *An Ugly Truth: Inside Facebook's Battle for Domination*, London: Little, Brown Book Group, 2021).
3 Ibid

C14 Tim Berners-Lee and Mark Zuckerberg

Endnotes and Citations

1. Tim Berners-Lee and Mark Fischetti, 1999. *Weaving the Web* New York Harper Collins
2. Tim Berners-Lee, *"Hypertext and Our Collective Destiny,"* presented at the Bush Symposium, MIT, October 12, 1995, https://www.w3.org/Talks/9510_Bush/Talk.html
3. Vannevar Bush, *Science, The Endless Frontier,* Washington, DC: U.S. Government Printing Office, 1945). https://www.pi.infn.it/~giorgio/INFN/3M/SciencetheEndlessFrontier.pdf
4. Ted Nelson, "Complex Information Processing: A File Structure for the Complex, the Changing, and the Indeterminate," ACM '65 Proceedings of the 1965 20th National Conference, 84-100, https://www.historyofinformation.com/detail.php?id=830
5. Douglas Engelbart, "Augmenting Human Intellect: A Conceptual Framework," *SRI International,* 1962, http://dougengelbart.org/pubs/books/augment-133150.pdf
6. John Thornhill, "Boldness in Business Person of the Year: Sir Tim Berners-Lee," *Financial Times,* March 15, 2019, https://www.ft.com/content/9d3205a8-15af-11e9-a168-d45595ad076d
7. Web Foundation, *"ForEveryone.net The Web, Past and Future,"* YouTube video, https://www.youtube.com/watch?v=cCE2EyV_IiY
8. Katrina Brooker, "I Was Devastated: Tim Berners-Lee, the Man Who Created the World Wide Web, Has Some Regrets," *Vanity Fair,* July 2018, https://www.vanityfair.com/news/2018/07/the-man-who-created-the-world-wide-web-has-some-regrets
9. Ibid
10. Ibid
11. Ibid
12. Sheera Frenkel and Cecilia Kang, *An Ugly Truth: Inside Facebook's Battle for Domination,* New York: HarperCollins, 2021).
13. Ibid
14. Apostolos Papapostolou, "A Greek Schoolmate Uncovers Zuckerberg's Face(book) and Its Roots," *Greek Reporter,* May 14, 2009,
15. Nicholas Carlson, "Exclusive: Here's the Email Zuckerberg Sent to Cut His Cofounder Out of Facebook," *Business Insider,* May 15, 2012, https://finance.yahoo.com/news/exclusive-heres-email-zuckerberg-sent-152900533.html
16. Matt Lynley and Jim Edwards, "This Was the First News Story Ever Written About 'TheFacebook.com,' From the Site's Birth 10 Years Ago," *Business Insider India,* February 4, 2014, https://finance.yahoo.com/news/first-news-story-ever-written-024200759.html
17. Miguel Helft, "Win for Zuckerberg, Consolation for Twins," *Business Standard,* April 13, 2011, https://www.business-standard.com/article/technology/win-for-zuckerberg-consolation-for-twins-111041300033_1.html
18. Melia Robinson, "How Sean Parker Bounced Back from Being Fired to Change Facebook's History," *ETCIO,* February 10, 2015, https://cio.economictimes.indiatimes.com/news/corporate-news/how-sean-parker-bounced-back-from-being-fired-to-change-facebooks-history/46185974
19. Jason Abbruzzese, *"Mark Zuckerberg's First Big Investor Meeting Included 'a Lot of Staring at the Desk,'"* Mashable, November 10, 2017, https://mashable.com/article/mark-zuckerberg-peter-thiel-reid-hoffman-facebook-investment
20. Note 12
21. Ibid
22. Ibid
23. Kaur, Harmeet. "Zuckerberg Explains How Facebook Handles User Data." *CNN Business,* April 11, 2018. https://money.cnn.com/2018/04/11/technology/facebook-zuckerberg-data/index.html
24. Srank2000. "How Facebook Handles the 4 Petabyte of Data Generated per Day." *Medium,* March 25, 2022. https://medium.com/@srank2000/how-facebook-handles-

the-4-petabyte-of-data-generated-per-day-ab86877956f4.
25 Lesley Stahl, "Aleksandr Kogan: The Link Between Cambridge Analytica and Facebook," *CBS News*, April 22, 2018,, https://www.cbsnews.com/news/aleksandr-kogan-the-link-between-cambridge-analytica-and-facebook/
26 Allison Fass, "Peter Thiel Talks About the Day Mark Zuckerberg Turned Down Yahoo's $1 Billion." *Inc.*, March 12, 2013, https://www.inc.com/allison-fass/peter-thiel-mark-zuckerberg-luck-day-facebook-turned-down-billion-dollars.html
27 Ryan Mac, "Growth at Any Cost: Top Facebook Executive Defended Data Collection in 2016 Memo." *BuzzFeed News*, March 29, 2018, https://www.buzzfeednews.com/article/ryanmac/growth-at-any-cost-top-facebook-executive-defended-data
28 Ibid
29 Ibid
30 Edward Gibbon, *The Decline and Fall of the Roman Empire*. 1st ed. 1776. Reprint, New York: Everyman's Library, 2010.
31 Kashmir Hill, "Inside Facebook's Secret Rulebook for Global Political Speech," *The New York Times*, December 27, 2018, https://www.nytimes.com/2018/12/27/technology/facebook-rulebook.html
32 Ibid
33 Brandy Zadrozny, "Zuckerberg's Testimony Raises New Questions About Facebook's Lobbying Practices," *NBC News*, April 13, 2018, https://www.nbcnews.com/tech/tech-news/zuckerberg-s-testimony-raises-new-questions-facebook-s-lobbying-practices-n865726
34 Theodore Schleifer, "Working for Mark Zuckerberg's Philanthropy Isn't Always Easy Since It Means Working for Mark Zuckerberg." *Vox*, June 26, 2020, https://www.vox.com/recode/2020/6/26/21303664/mark-zuckerberg-facebook-chan-zuckerberg-initiative-philanthropy-tension
35 Shoshana Zuboff, *The Age of Surveillance Capitalism: The Fight for a Human Future at the New Frontier of Power*. New York: PublicAffairs, 2019.
36 Meta Platforms, Inc. Annual Reports, 2008–2022; Statista, "Facebook: number of monthly active users worldwide 2008–2022"; Business of Apps, "Facebook Revenue and Usage Statistics, 2022)"; The New York Times, "Sheryl Sandberg Steps Down as COO of Meta," June 1, 2022.
37 Holly B Shakya, and Nicholas A. Christakis. "Association of Facebook Use With Compromised Well-Being: A Longitudinal Study." *American Journal of Epidemiology* 185, no. 3, February 1, 2017: 203–211, https://doi.org/10.1093/aje/kww189
38 Sam Levin, "Facebook Admits It Poses Mental Health Risk – but Says Using Site More Can Help." *The Guardian*, December 15, 2017, https://www.theguardian.com/technology/2017/dec/15/facebook-mental-health-psychology-social-media
39 Note 15
40 Federal Trade Commission. "FTC Sues Facebook for Illegal Monopolization." *Federal Trade Commission*, December 9, 2020, https://www.ftc.gov/news-events/press-releases/2020/12/ftc-sues-facebook-illegal-monopolization
41 Chris Hughes, "It's Time to Break Up Facebook," The New York Times, May 9, 2019, https://www.nytimes.com/2019/05/09/opinion/sunday/chris-hughes-facebook-zuckerberg.html
42 Mark Zuckerberg, "Facebook's Future Is Young Adults and the Metaverse." *Axios*, October 26, 2021, https://www.axios.com/2021/10/26/facebook-teen-users-young-adults-zuckerberg
43 Ibid

C15 Needs, Purposes, and a Humanist Worldview
1 David Graeber, *Bullshit Jobs: A Theory* New York: Simon & Schuster, 2018, 10, 13.
2 Mihaly Csikszentmihalyi, *Creativity: Flow and the Psychology of Discovery and Invention*. New York: HarperCollins, 1996
3 Mihaly Csikszentmihalyi, Flow: *The Psychology of Optimal Experience*. New York:

Harper & Row, 1990
4   Simon Blackburn, *Think: A Compelling Introduction to Philosophy*, Oxford: Oxford University Press, 1999, 101.
5   Mark Zuckerberg, interview by Ezra Klein, *The Ezra Klein Show*, Vox Media, April 2018, quoted in Henry Farrell, "Mark Zuckerberg Runs a Nation-State, and He's the King," *Center for Internet and Society*, Stanford Law School, April 10, 2018, https://cyberlaw.stanford.edu/publications/mark-zuckerberg-runs-nation-state-and-hes-king/
6   Shoshana Zuboff, "You Are the Object of Facebook's Secret Extraction Operation," *The New York Times*, November 12, 2021, https://cs111.wellesley.edu/content/lectures/lec_connections/surveillance/zuboff-oped2.pdf
7   Billy Perrigo, "Instagram's Body Image Problem May Be Unfixable, Experts Say," *TIME*, September 16, 2021, https://time.com/6098771/instagram-body-image-teen-girls/
8   Sarah Wynn-Williams, *Careless People*, New York: Flatiron Books, 2025).
9   Murray Shanahan, *The Technological Singularity*. Cambridge, MA: MIT Press, 2015.
10 Clair C. Patterson, "Historical Changes in Integrity and Worth of Scientific Knowledge," *Geochimica et Cosmochimica Acta* 58, no. 15, 1994: 3199–3207, https://authors.library.caltech.edu/records/7grf2-2qr03
11 Pjotr Sauer, "One Million and Counting: Russian Casualties Hit Milestone in Ukraine War," *The Guardian*, June 22, 2025, https://www.theguardian.com/world/ng-interactive/2025/jun/22/one-million-and-counting-russian-casualties-hit-milestone-in-ukraine-war
12 Ministry of Statistics and Programme Implementation, MoSPI, "Women and Men in India 2024: Selected Indicators and Data," New Delhi: Government of India, 2025, https://www.mospi.gov.in/publication/women-men-india-2024-selected-indicators-and-data
13 Ming Gao, "China's Gender Imbalance and Its Implications for Human Trafficking, Social Stability, and Economic Development," *Human Rights @ Lund*, March 20, 2025, https://humanrights.blogg.lu.se/2025/03/20/chinas-gender-imbalance-and-its-implications-for-human-trafficking-social-stability-and-economic-development/
14 Matthew White, *The Great Big Book of Horrible Things: The Definitive Chronicle of History's 100 Worst Atrocities*, New York: W. W. Norton & Company, 2011
15 David Sloan Wilson, *Darwin's Cathedral: Evolution, Religion, and the Nature of Society*, Chicago: University of Chicago Press, 2002).
16 Kathleen Vohs, "Money priming can change people's thoughts, feelings, motivations, and behaviors: An update on 10 years of experiments," 2015 *Journal of Experimental Psychology: General*, 144(4, e86–e93, https://doi.org/10.1037/xge0000091
17 Ibid
18 Abraham Maslow,"A Theory of Human Motivation." *Psychological Review* 50, no. 4, 1943: 370–396, https://psycnet.apa.org/doiLanding?doi=10.1037%2Fh0054346
19 Abraham Maslow, *The Farther Reaches of Human Nature*. New York: Viking Press, 1971.
20 Vladimir Kosonogov *et al.*, "Evolutionary Structure of Human Motivation: A Network Analysis Approach," *Personality and Individual Differences* 211 (2023): 112331. https://doi.org/10.1016/j.paid.2023.112331.
21 Martha C. Nussbaum, *Creating Capabilities: The Human Development Approach*, Cambridge, MA: Belknap Press of Harvard University Press, 2011, 33–34.
22 Amartya Sen, *Development as Freedom*, New York: Alfred A. Knopf, 1999, 35–53.
23 Sudhir Anand, *Recasting Human Development Measures*, UNDP Human Development Report Office Discussion Paper, March 2018, https://hdr.undp.org/system/files/documents/anandrecastinghumandevelopmentmeasures.pdf.
24 Note 21
25 Paul Anand *et al.*, "Capabilities and Well-Being: Evidence Based on the Sen–

Nussbaum Approach to Welfare." *Social Indicators Research* 74, no. 1, 2005: 9–55, https://link.springer.com/article/10.1007/s11205-005-6518-z

26 Jamil, Sobia. "Amartya Sen, Martha Nussbaum, and the Capability Approach." Al-Hikmat: *A Journal of Philosophy* 44, 2024: 73–87. https://pu.edu.pk/images/journal/phill/pdf_files/5_v44_24.pdf.

27 Diederik Coetzee, "A Research Study on Expanding Martha Nussbaum's Core Capabilities List: A Necessary Adaptation for the Technological Era." 2022, https://www.researchgate.net/profile/Diederik-Coetzee/publication/370779390_A_Research_Study_on_Expanding_Martha_Nussbaum's_Core_Capabilities_List_A_Necessary_Adaptation_for_the_Technological_Era/links/64635d8c434e26474fec415c/A-Research-Study-on-Expanding-Martha-Nussbaums-Core-Capabilities-List-A-Necessary-Adaptation-for-the-Technological-Era.pdf.

28 Areez Tanbeen Rahman, "Fighting Water Scarcity in Cox's Bazar Refugee Camps," *UNHCR*, January 4, 2019, https://www.unhcr.org/news/stories/fighting-water-scarcity-coxs-bazar-refugee-camps

29 UNHCR, Gender-Based Violence Information Management System Annual Overview: Lebanon 2021, March 21, 2022, https://data.unhcr.org/en/documents/download/96299.

30 Joe Whittaker, Online Radicalisation: What We Know, *European Commission*, 2022, https://home-affairs.ec.europa.eu/system/files/2023-11/RAN-online-radicalisation_en.pdf

31 International Labour Organization, *ILO*, Global Employment Trends for Youth 2024: Executive Summary, August 2024, https://www.ilo.org/sites/default/files/2024-08/GET_2024_ExecSum_EN_0.pdf

32 Sahar Fetrat, "Taliban's Attack on Girls' Education Harming Afghanistan's Future," *Human Rights Watch*, September 17, 2024, https://www.hrw.org/news/2024/09/17/talibans-attack-girls-education-harming-afghanistans-future

33 Sarah Cook, "China's Information Isolation, New Censorship Rules, Transnational Repression," *China Media Bulletin*, Freedom House, February 2021, https://freedomhouse.org/report/china-media-bulletin/2021/chinas-information-isolation-new-censorship-rules-transnational

34 Devika Deshmukh et al., "Household Vulnerability in the Urban Slums of Mumbai, India: Analysis of a Large Cross-Sectional Survey," *medRxiv preprint*, July 20, 2023, https://www.medrxiv.org/content/10.1101/2023.07.20.23292961v1.full.pdf

35 Caroline Hickman et al., "Climate Anxiety in Children and Young People and Their Beliefs About Government Responses to Climate Change: A Global Survey," *Lancet* Planetary Health 5, no. 12, December 2021: 863–873, https://www.thelancet.com/journals/lanplh/article/PIIS2542-5196(21)00278-3/fulltext

36 C.R. DeHaan *et al.*, "Nussbaum's Capabilities and Self-Determination Theory's Basic Psychological Needs: Relating Some Fundamentals of Human Wellness." *Journal of Happiness* Studies 17, no. 5, 2016: 2037–2049, https://link.springer.com/article/10.1007/s10902-015-9684-y

37 Jocelyn Bélanger, *et al.*, "The Psychology of Ideological Obsession: How Need Frustration Fuels Extremism." *Journal of Personality and Social Psychology* 118, no. 5, 2020: 825–848, https://psycnet.apa.org/doiLanding?doi=10.1037%2Fpspp0000249

C16. Activities in the sciences, arts, and humanism

1 Peter Thiel with Blake Masters, *Zero to One: Notes on Startups, or How to Build the Future* (New York: Crown Business, 2014)

2 Joseph A. Schumpeter, *Capitalism, Socialism and Democracy*, New York: Harper & Brothers, 1942)

3 Joseph A. Schumpeter, *The Theory of Economic Development: An Inquiry into Profits, Capital, Credit, Interest, and the Business Cycle*, trans. Redvers Opie, Cambridge, MA: Harvard University Press, 1934)

4 Giovanni Dosi *et al.*, "Embodied and Disembodied Technological Change: The

Sectoral Patterns of Job-Creation and Job-Destruction," IZA Discussion Paper No. 12408, Bonn: Institute of Labor Economics, 2019, https://www.econstor.eu/bitstream/10419/202754/1/dp12408.pdf

5   Adam Smith, *The Theory of Moral Sentiments*. Originally published in 1759. Edited by Ryan Patrick Hanley. Princeton, NJ: Princeton University Press, 2010
6   Adam Smith, *An Inquiry into the Nature and Causes of the Wealth of Nations*. Originally published in 1776. Edited by Edwin Cannan. New York: Modern Library, 1937.
7   Robert Axelrod, T*he Evolution of Cooperation: Revised Edition*, New York: Basic Books, 2006).
8   David Kraines and Vivian Kraines, "Evolution of Learning among Pavlov Strategies in a Competitive Environment with Noise," *Journal of Conflict Resolution* 39, no. 3, 1995: 439–466, https://www.jstor.org/stable/174576.
9   Adam M. Brandenburger and Barry J. Nalebuff, *Co-opetition* (New York: Currency Doubleday, 1996)
10 Robert Axelrod, *A Passion for Cooperation: Adventures of a Wide-Ranging Scientist*, Ann Arbor: University of Michigan Press, 2023).
11 Ibid
12 Tim Berners-Lee, interview in *Scientific American*, April 2010.
13 Casey Newton, "Zuckerberg's Reset Memo," *Platformer*, March 2023.
14 Isaac Asimov, "Humanism," in *The Roving Mind*, Buffalo: Prometheus Books, 1983, 231.
15 Edward Said, *Humanism and Democratic Criticism*, New York: Columbia University Press, 2004)
16 I am grateful to my friend, Alan Thomson, for his insightful observations on how engineers shape technology and the commercially driven design life embedded in products.
17 Robert Oppenheimer, quoted in Kai Bird and Martin J. Sherwin, *American Prometheus: The Triumph and Tragedy of J. Robert Oppenheimer* (New York: Vintage Books, 2006), 321.
18 Geoffrey Cantor, *Michael Faraday: Sandemanian and Scientist* (New York: St. Martin's Press, 1991)
19 Richard Van Noorden, "The Pacifist Physicist," *Nature* 520, no. 7545 (2015): 24–26.
20 John H. Lienhard, *The Engines of Our Ingenuity: An Engineer Looks at Technology and Culture*, New York: Oxford University Press, 2000)
21 Samuel C. Florman, *The Existential Pleasures of Engineering*, 2nd ed., New York: St. Martin's Press, 1994)
22 Lewis Mumford, *Technics and Civilization*, with a foreword by Langdon Winner, Chicago: University of Chicago Press, 2010).
23 "Big Tech Cozies Up to New Administration After Spending Record Sums on Lobbying Last Year," *Issue One*, https://issueone.org/articles/big-tech-spent-record-sums-on-lobbying-last-year/
24 "Big Tech, Big Cash: Washington's New Power Players," *Public Citizen*, https://www.citizen.org/article/big-tech-lobbying-update/
25 Berit Brogaard, "The Quest for Immortality: What Do Scientists Say?" *Psychology Today*, November 15, 2024, https://www.psychologytoday.com/gb/blog/the-superhuman-mind/202409/the-quest-for-immortality-what-do-scientists-say
26 Epicurus, *The Epicurus Reader: Selected Writings and Testimonia*, ed. and trans. Brad Inwood and Lloyd P. Gerson, Indianapolis: Hackett Publishing Company, 1994)
27 International Atomic Energy Agency, *Spent Fuel and Radioactive Waste Information System*, SRIS, https://sris.iaea.org/home
28 Garrett Hardin, *Living Within Limits: Ecology, Economics, and Population Taboos*, New York: Oxford University Press, 1993)
29 Jacob Bronowski, *Science and Human Values*, New York: Harper & Row, 1956)
30 Ibid
31 George Sylvester Viereck, "What Life Means to Einstein," *The Saturday Evening*

Post, October 26, 1929, https://www.saturdayeveningpost.com/wp-content/uploads/satevepost/what_life_means_to_einstein.pdf
32 E. O. Wilson, *Consilience: The Unity of Knowledge.* New York: Alfred A. Knopf, 1998.
33 For the interested reader, Popper and Penrose have suggested "divisions" of the world into three distinct parts. E Brian Davies summarizes these arguments and suggests they are similar in approach to E. O. Wilson's Consilience. See E. Brian Davies, *Why Beliefs Matter: Reflections on the Nature of Science,,* Oxford: Oxford University Press, 2010). In my view the conjoined framework described in this book is different from Popper and Penrose's platonic approaches.
34 "Mentions of Climate Change Removed from Federal Agencies' Websites," *Sabin Center for Climate Change Law,* https://climate.law.columbia.edu/content/mentions-climate-change-removed-federal-agencies-websites
35 PEN America, *Freedom to Write Index 2024,* April 24, 2025, https://pen.org/report/freedom-to-write-index-2024/
36 Jacob Bronowski, *Science and Human Values,* New York: Harper & Row, 1956)
37 Adam Hochschild, *King Leopold's Ghost: A Story of Greed, Terror, and Heroism in Colonial Africa,* Boston: Mariner Books, 1998).
38 Joseph Conrad, *Heart of Darkness,* in *Youth, Heart of Darkness, The End of the Tether,* London: Blackwood's Magazine, 1902; reprint, Penguin Classics, 2007).
39 Chinua Achebe, *Things Fall Apart,* London: Heinemann, 1958).
40 Sven Lindqvist, *Terra Nullius: A Journey Through No One's Land,* trans. Sarah Death, London: Granta Books, 2007).
41 Harshini S. and Praveenkumar N., "The Devastating Effect of Kurtz's Inhumanity in Joseph Conrad's Heart of Darkness," *International Journal of Creative Research Thoughts* 12, no. 5, 2024: 1–8, https://ijcrt.org/papers/IJCRT2405058.pdf
42 Geetanjali Tewari, "Cultural Collision: Colonialism and the Disintegration of Igbo Society in *Things Fall Apart* by Chinua Achebe," *International Journal of Applied Research* 11, no. 2, 2025: 360–362, https://www.allresearchjournal.com/archives/2025/vol11issue2/PartE/11-4-78-158.pdf
43 Ida Nursoo, "Indigenous Law, Colonial Injustice and the Jurisprudence of Hybridity," *Journal of Legal Pluralism and Unofficial Law* 50, no. 1, 2018: 1–23, https://openresearch-repository.anu.edu.au/items/2bea269d-9dd6-4660-ab31-fdf9e91dedf0
44 Frank Wilczek, *A Beautiful Question: Finding Nature's Deep Design,* New York: Penguin Press, 2015)
45 Anjan Chatterjee. *The Aesthetic Brain: How We Evolved to Desire Beauty and Enjoy Art.* Oxford: Oxford University Press, 2014.
46 Dissanayake, Ellen. *Homo Aestheticus: Where Art Comes From and Why.* Seattle: University of Washington Press, 1995.
47 Anjan Chatterjee and Oshin Vartanian, "Neuroscience of Aesthetics," *Annals of the New York Academy of Sciences* 1369, no. 1, 2016: 172–194. 5, https://nyaspubs.onlinelibrary.wiley.com/doi/abs/10.1111/nyas.13035
48 Zaira Cattaneo, "Neural Correlates of Visual Aesthetic Appreciation: Insights from Non-Invasive Brain Stimulation," *Experimental Brain Research* 238, no. 1–16, 2020, https://link.springer.com/article/10.1007/s00221-019-05685-x.
49 Ibid
50 Lucretius Titus Carus, *The Nature of Things,* translated by A.E. Stallings, London: Penguin Classics, 2007,
51 Note 26
52 John Maynard Keynes, Economic Possibilities for Our Grandchildren, in *Essays in Persuasion* (New York: Harcourt Brace, 1932)
53 *UNESCO Institute for Statistics,* "February 2025 UIS Data Release: Explore the Latest Progress on SDG 9.5 Research and Development through Key Indicators," last updated March 21, 2025 https://www.unesco.org/en/articles/february-2025-uis-data-release-explore-latest-progress-sdg-95-research-and-development-through-key

Endnotes and Citations

54 E. O. Wilson, *Letters to a Young Scientist*, New York: Liveright Publishing, 2013, 10.
55 Clair C. Patterson, "Historical Changes in Integrity and Worth of Scientific Knowledge," *Geochimica et Cosmochimica Acta* 58, no. 15, 1994: 3199–3207, https://authors.library.caltech.edu/records/7grf2-2qr03
56 E.O. Wilson, *Half-Earth: Our Planet's Fight for Life*, New York: Liveright Publishing, 2016).
57 Lily Tomlin, quoted in *The Yale Book of Quotations*, ed. Fred R. Shapiro, New Haven: Yale University Press, 2006)

Part 7 Planetary Flourishing
1 Nelson Mandela, *Notes to the Future: Words of Wisdom*, ed. Sello Hatang and Verne Harris, New York: Atria Books, 2012, 141.
2 Wendell Berry, *The Unforeseen Wilderness: An Essay on Kentucky's Red River Gorge*, Lexington: University Press of Kentucky, 1971, 40.
3 Harriet Jacobs, *Incidents in the Life of a Slave Girl*, ed. Jean Fagan Yellin, Cambridge, MA: Harvard University Press, 1987, 201.

C17 The Shallow Pond
1 Peter Singer. "Famine, Affluence, and Morality." *Philosophy & Public Affairs* 1, no. 3, Spring 1972: 229–243, https://rintintin.colorado.edu/~vancecd/phil308/Singer2.pdf
2 Ibid
3 Peter Singer, *The Life You Can Save: How to Do Your Part to End World Poverty*, New York: Random House, 2009, 13–15.
4 Ibid
5 "The Life You Can Save – Nonprofit – High Impact Charities," *The Life You Can Save*, https://www.thelifeyoucansave.org/.
6 World Bank Group, "June 2025 Update to Global Poverty Lines," World Bank Factsheet, June 5, 2025, https://www.worldbank.org/en/news/factsheet/2025/06/05/june-2025-update-to-global-poverty-lines
7 Joe Hasell, Bertha Rohenkohl, and Pablo Arriagada, "$3 a Day: A New Poverty Line Has Shifted the World Bank's Data on Extreme Poverty," Our World in Data, August 10, 2025, https://ourworldindata.org/new-international-poverty-line-3-dollars-per-day
8 Akanksha Likhar and Manoj S. Patil, "Importance of Maternal Nutrition in the First 1,000 Days of Life and Its Effects on Child Development: A Narrative Review," *Cureus* 14, no. 10 (2022): e30083 https://assets.cureus.com/uploads/review_article/pdf/114933/20240724-319105-j0vxwx.pdf

C18 Collapse of the Aral Sea
1 Philip Micklin, "The Aral Sea Disaster," *Annual Review of Earth and Planetary Sciences* 35, 2007: 47–72, https://doi.org/10.1146/annurev.earth.35.031306.140120.
2 Ibid
3 Ibid
4 Brain, Stephen. "The Great Stalin Plan for the Transformation of Nature." *Environmental History* 15, no. 4, 2010: 670–700, https://doi.org/10.1093/envhis/emq091.
5 Makuch, Ben. "Ghost Lake: The Aral Sea Disaster." *Vice*, 2012, https://www.vice.com/en/article/avnjm5/ghost-lake-the-aral-sea-disaster.
6 Jonathan B. Tucker and Raymond A. Zilinskas, eds., "The 1971 Smallpox Epidemic in Aralsk, Kazakhstan, and the Soviet Biological Warfare Program," Occasional Paper No. 9, Monterey, CA: *Center for Nonproliferation Studies*, Monterey Institute of International Studies, 2002, https://nonproliferation.org/the-1971-smallpox-epidemic-in-aralsk-kazakhstan-and-the-soviet-biological-warfare-program-commentaries/.
7 Ibid
8 Anonymous proverb, commonly attributed to Cree Nation, quoted in Alan AtKisson, Believing Cassandra: *How to Be an Optimist in a Pessimist's World*, London: Earthscan, 2010)

C19 The UN Durban Conference 2001

1. United Nations General Assembly, *Decade for Action to Combat Racism and Racial Discrimination*, A/RES/3057, XXVIII, adopted 2 November 1973, https://www.refworld.org/legal/resolution/unga/1973/en/9348.
2. Ibid
3. United Nations, *Report of the World Conference to Combat Racism and Racial Discrimination*, Geneva, 14–25 August 1978, A/CONF.92/40, New York: United Nations, 1978, https://digitallibrary.un.org/record/2019.
4. United Nations, *Report of the Second World Conference to Combat Racism and Racial Discrimination*, Geneva, 1–12 August 1983, A/CONF.119/26, New York: United Nations, 1983, https://documents.un.org/doc/undoc/gen/n83/233/38/pdf/n8323338.pdf.
5. United Nations, *World Conference on Human Rights*, Vienna, 14–25 June 1993, *Vienna Declaration and Programme of Action*, New York: United Nations, 1993, https://www.un.org/en/conferences/human-rights/vienna1993.
6. United Nations General Assembly, *World Conference Against Racism, Racial Discrimination, Xenophobia and Related Intolerance*, Resolution 52/111, adopted 12 December 1997, A/RES/52/111, https://documents.un.org/access.nsf/get?OpenAgent&DS=A/RES/52/111&Lang=E.
7. United Nations Secretariat, *Compilation of the Secretariat's Draft Declaration and Programme of Action and Final Documents of the Regional Inter-Governmental Meetings Held in Strasbourg, Santiago, Dakar and Tehran*, A/CONF.189/PC.2/29, Geneva: United Nations, 2 May 2001, https://digitallibrary.un.org/record/439461.
8. United Nations, *Report of the World Conference Against Racism, Racial Discrimination, Xenophobia and Related Intolerance*, Durban, South Africa, 31 August–8 September 2001, A/CONF.189/12, New York: United Nations, 2002, https://digitallibrary.un.org/record/456677; see also United Nations, *Durban Declaration and Programme of Action*, https://www.ohchr.org/en/publications/reference-publications/durban-declaration-and-programme-action;
9. Colin L. Powell, "Remarks on the U.S. Withdrawal from the World Conference Against Racism," *U.S. Department of State*, 3 September 2001, https://2001-2009.state.gov/secretary/former/powell/remarks/2001/4789.htm.
10. Note 8
11. NGO Forum, *Declaration of the NGO Forum of the World Conference Against Racism*, Durban, South Africa, 1 September 2001, as cited in "NGO Forum at Durban Conference 2001," NGO Monitor, https://ngo-monitor.org/reports/ngo_forum_at_durban_conference_/;
12. Mary Robinson, statement at the World Conference Against Racism, Durban, South Africa, 1 September 2001, as cited in "Mary Robinson Denies Equal Rights to Palestinians," *Islamic Human Rights Commission*, 8 September 2001, https://www.ihrc.org.uk/press-releases/mary-robinson-denies-equal-rights-to-palestinians/.

C20 Poverty, Conservation, and Historical Responsibilities
1. Epictetus, Enchiridion, trans. Sharon Lebell, in *The Art of Living: The Classical Manual on Virtue, Happiness, and Effectiveness* (New York: HarperOne, 2007)
2. Jeremy Bentham, *The Classical Utilitarians: Bentham and Mill*, ed. John Troyer, Indianapolis: Hackett Publishing Company, 2003; originally published 1789)
3. John Stuart Mill, *Utilitarianism*, ed. Roger Crisp, Oxford: Oxford University Press, 1998; originally published 1863)
4. Singer, Peter. *The Life You Can Save: How to Do Your Part to End World Poverty*. New York: Random House, 2009
5. Peter Singer, *Famine, Affluence, and Morality*, rev. ed., Oxford: Oxford University Press, 2016).
6. Note 4
7. Note 5
8. "The Life You Can Save – Nonprofit – High Impact Charities," *The Life You Can Save*, https://www.thelifeyoucansave.org/.

Endnotes and Citations

9 Ibid
10 E.O. Wilson, *Half-Earth: Our Planet's Fight for Life*, New York: Liveright Publishing, 2016).
11 Ibid
12 Ibid
13 António Guterres, "Aral Sea 'Probably Biggest Ecological Catastrophe of Our Time', Secretary-General Says, Stressing Need to Act Forcefully in Preventing Tragedy from Multiplying," *United Nations Press Release*, June 10, 2017, https://press.un.org/en/2017/sgsm18565.doc.htm
14 Claudia Wieners et al., "Global Tipping Points Report 2025" (*Exeter: University of Exeter*, 2025), https://global-tipping-points.org/resources-gtp/
15 Note 10
16 Dennis Liu, "E.O. Wilson Biodiversity Foundation: Implementing the Half-Earth Project," *Rewilding*, May 31, 2021, https://rewilding.org/e-o-wilson-biodiversity-foundation-implementing-the-half-earth-project/
17 Yvon Chouinard, *Let My People Go Surfing: The Education of a Reluctant Businessman*, 10th-anniversary ed., New York: Penguin, 2016).
18 Charlie King, "Yvon Chouinard: The Founder of Patagonia," *Sustainability Magazine*, October 4, 2024, https://sustainabilitymag.com/articles/yvon-chouinard-the-founder-of-patagonia.
19 Savanteum, "How I Saved Patagonia From Bankruptcy" Yvon Chouinard, YouTube video, 11:12, March 15, 2019, https://www.youtube.com/watch?v=Z1Ja8YEYaTI.
20 Ibid
21 Garrett Hardin, "The Tragedy of the Commons," *Science* 162, no. 3859, December 13, 1968: 1243–1248, https://www.garretthardinsociety.org/articles_pdf/tragedy_of_the_commons.pdf
22 Joseph E. Stiglitz, *The Road to Freedom: Economics and the Good Society*, London: Penguin Books, 2023)
23 World Bank, One Health for Central Asia (Washington, DC: World Bank Group, 2022), https://www.worldbank.org/content/dam/infographics/780xany/2022/apr/Presentations/CA-One-Health-Infographic-en.pdf
24 Whitaker, R. and Lenin, J., 2024. *Snakes, Drugs and Rock 'n' Roll: My Early Years.* Mysuru: HarperCollins India.
25 S. R. Rai, *Silent Valley: Whispers of Reason*, Thiruvananthapuram: Kerala Sastra Sahitya Parishad, 1986).
26 Ibid
27 ibid
28 Government of India, "Notification of Silent Valley National Park," *Kerala Gazette Extraordinary*, no. 1005, Trivandrum: Agriculture [Forest Special-A] Department, 15 November 1984)
29 Sven Lindqvist, *Exterminate All the Brutes*, trans. Joan Tate, New York: The New Press, 1996)
30 Desiderius. Erasmus, *Adages*, trans. William Barker, Toronto: University of Toronto Press, 2001)
31 Alfred North Whitehead, *Adventures of Ideas*, New York: The Macmillan Company, 1933; reprint, New York: The Free Press, 1967,
32 Ibid
33 Buckminster Fuller, *Operating Manual for Spaceship Earth*, New York: E.P. Dutton, 1969)
34 Nikul Joshi, "Caste System in Ancient India," *World History Encyclopedia*, last modified November 20, 2017, https://www.worldhistory.org/article/1152/caste-system-in-ancient-india/.
35 Sonalde Desai, Reeve Vanneman, and National Council of Applied Economic Research, *India Human Development Survey-II, IHDS-II, 2011–12*, Inter-university Consortium for Political and Social Research [distributor], 2018-08-08. https://doi.

org/10.3886/ICPSR36151.v6
36 United Nations Development Programme, UNDP. and Oxford Poverty and Human Development Initiative, OPHI). *Global Multidimensional Poverty Index 2021: Unmasking Disparities by Ethnicity, Caste and Gender.* UNDP and OPHI, 2021, https://hdr.undp.org/content/2021-global-multidimensional-poverty-index-mpi.
37 Kaivan Munshi, "Caste and the Indian Economy," *Journal of Economic Literature* 57, no. 4 (December 2019): 781–834, https://www.aeaweb.org/articles?id=10.1257/jel.20171307.
38 John Parker and Richard Rathbone, *African History: A Very Short Introduction*, Oxford University Press, 2007
39 Adam Hochschild, *King Leopold's Ghost: A Story of Greed, Terror, and Heroism in Colonial Africa*, Boston: Houghton Mifflin, 1998).
40 Stuart A. Reid, *The Lumumba Plot: The Secret History of the CIA and a Cold War Assassination*, New York: Alfred A. Knopf, 2023)
41 Joana de Deus Pereira and Anne-Marie Weeden, "Congo's Fragile Truce? Foreign Interference and Conflict Minerals in the DRC," *Royal United Services Institute*, May 7, 2025, https://www.rusi.org/explore-our-research/publications/commentary/congos-fragile-truce-foreign-interference-and-conflict-minerals-drc.
42 Leander Heldring and James A. Robinson, *Colonialism and Development in Africa*, African Economic History Network Working Paper No. 5/2012, https://www.econstor.eu/bitstream/10419/242634/1/aehn-wp-05.pdf.
43 "An Act for the better ordering and governing of Negroes" Laws of Enslavement and Freedom in the Anglo-Atlantic World, https://slaveryandfreedomlaws.lib.unb.ca/laws/17
44 Angelina Clapp, Kyel Towler, and Ryan Ritter, "'Slaves of the State': 13th Amendment, Mass Incarceration and the Prison Industrial Complex," *Civic Learning and Democratic Engagement*, James Madison University, September 17, 2020, https://sites.lib.jmu.edu/civic/2020/09/17/slaves-of-the-state-13th-amendment-mass-incarceration-and-the-prison-industrial-complex.
45 Brennan Center for Justice, *Voting Laws Roundup: July 2021*, https://www.brennancenter.org/our-work/research-reports/voting-laws-roundup-july-2021.
46 Thomas Piketty, *A Brief History of Equality*, Cambridge, MA: Belknap Press, 2022, esp. chs. 1–3. and Thomas Piketty, *Capital in the Twenty-First Century*, Cambridge, MA: Belknap Press, 2014, pp. 25–27, 237–245.
47 Jared Diamond, *Collapse: How Societies Choose to Fail or Succeed*, New York: Viking, 2005).
48 Joan C. Williams, *Outclassed: How the Left Lost the Working Class and How to Win Them Back*, New York: St. Martin's Press, 2025).

Part 8
1 Buckminster Fuller, *Critical Path*, New York: St. Martin's Press, 1981)

C21 Keynes' Vision for the Future
1 John Maynard Keynes, Economic Possibilities for our Grandchildren, 1930, in *Essays in Persuasion*, London: Macmillan, 1931.
2 Ibid

C22 Diagnosing the Crisis and Evaluating Possibilities
1 John Maynard Keynes, *The General Theory of Employment, Interest and Money*, originally published 1936, Cham: Palgrave Macmillan, 2018).
2 Zachary D. Carter, *The Price of Peace: Money, Democracy, and the Life of John Maynard Keynes*, New York: Random House, 2020).
3 Robert Skidelsky, John Maynard *Keynes: 1883–1946: Economist, Philosopher, Statesman* (New York: Penguin Press, 2009)
4 Stanley Loomis, *Paris in the Terror: June 1793–July 1794*, Philadelphia: J. B. Lippincott, 1964)
5 Keynes, John Maynard. *The End of Laissez-Faire*. Originally published 1926. Amherst, NY: Prometheus Books, 2004.
6 U.S. Bureau of Economic Analysis, "National Income and Product Accounts Tables,"

https://www.bea.gov/data/gdp/gross-domestic-product
7. International Monetary Fund, *World Economic Outlook Database*, April 2025, https://www.imf.org/en/Publications/WEO/weo-database/2025/April
8. World Bank, "Fact Sheet: An Adjustment to Global Poverty Lines," last modified May 2, 2022, https://www.worldbank.org/en/news/factsheet/2022/05/02/fact-sheet-an-adjustment-to-global-poverty-lines
9. Hannah Ritchie, "How Much of the World's Land Would We Need to Feed the Global Population with the Average Diet of a Given Country?" *Our World in Data*, October 3, 2017, https://ourworldindata.org/agricultural-land-by-global-diets
10. REN21, *Renewables 2025 Global Status Report*, Paris: REN21 Secretariat, 2025, https://www.ren21.net/gsr-2025/
11. World Economic Forum, *Fostering Effective Energy Transition 2025*, Geneva: World Economic Forum, 2025, https://reports.weforum.org/docs/WEF_Fostering_Effective_Energy_Transition_2025.pdf
12. *World Population Review*, "Average Workweek by Country 2025," https://worldpopulationreview.com/country-rankings/average-work-week-by-country
13. International Labour Organization, *Working Time Statistics, ILO Modelled Estimates*, 2025 ed., https://www.ilo.org/sites/default/files/2025-01/WESO25_Trends_Report_EN.pdf
14. *Demand Sage*, "Average Screen Time Statistics 2025, Global Data," May 29, 2025, https://www.demandsage.com/screen-time-statistics/
15. Datareportal, *Digital 2025: India Overview*, https://datareportal.com/reports/digital-2025-india
16. Datareportal, *Digital 2025: China Overview*, https://datareportal.com/reports/digital-2025-china
17. Datareportal, *Digital 2025: Nigeria Overview*, https://datareportal.com/reports/digital-2025-nigeria
18. José Ortega y Gasset, *The Revolt of the Masses*, trans. Anthony Kerrigan, ed. Kenneth Moore, Notre Dame, IN: University of Notre Dame Press, 1985)
19. Thorstein Veblen, *The Theory of the Leisure Class: An Economic Study in the Evolution of Institutions*, 1899; repr., Oxford: Oxford University Press, 2007, ed. Robert Lekachman.
20. Xi Luan and Jialu You, "Rat Race: Work Hours and Economic Inequality," *Journal of the Knowledge Economy*, published online July 2025, https://doi.org/10.1007/s13132-024-02369-y.
21. John Kenneth Galbraith, *The Affluent Society*, 1958; repr., London: Penguin Books, 1999, with new introduction by the author.
22. Kaitlin Woolley and Marissa A. Sharif, "The Psychology of Your Scrolling Addiction," *Harvard Business Review*, January 31, 2022, https://hbr.org/2022/01/the-psychology-of-your-scrolling-addiction.
23. Jessica Hughes-Nind *et al.*, "The Association Between Motivations for Social Media Use, Stress and Academic Attainment," *Current Psychology* 43, 2024: 28025–28037, https://doi.org/10.1007/s12144-024-06392-9.
24. Stephen Jay Gould, *The Mismeasure of Man*, revised and expanded ed., New York: W. W. Norton & Company, 1996)
25. James W. Prescott, "Body Pleasure and the Origins of Violence," *The Bulletin of the Atomic Scientists* 31, no. 9, 1975: 10–20. http://www.deconnection.org/site/images/publicaties/body%20pleasure%20and%20the%20origins%20of%20violence%20-%20j.w.%20prescott.pdf
26. M. E. Street, and S. Bernasconi, (2021). Microplastics, environment and child health. *Italian Journal of Pediatrics*, 47, 75. https://ijponline.biomedcentral.com/articles/10.1186/s13052-021-01034-3

C23 Approaches to Economic and Social Structures
1. Abhijit V. Banerjee and Esther Duflo, *Good Economics for Hard Times: Better Answers to Our Biggest Problems* (New York: PublicAffairs, 2019).
2. Amartya Sen, *Development as Freedom* (New York: Alfred A. Knopf, 1999).

3   Amartya Sen, *The Idea of Justice* (Cambridge, MA: Belknap Press of Harvard University Press, 2009).
4   Thomas Piketty, *Capital in the Twenty-First Century*, trans. Arthur Goldhammer, Cambridge, MA: Harvard University Press, 2014)
5   Joseph Stiglitz, *The Price of Inequality: How Today's Divided Society Endangers Our Future*, New York: W. W. Norton & Company, 2012)
6   Ibid
7   Michael J. Sandel, *What Money Can't Buy: The Moral Limits of Markets* (New York: Farrar, Straus and Giroux, 2012).
8   Michael J. Sandel, *The Tyranny of Merit: What's Become of the Common Good?* (New York: Farrar, Straus and Giroux, 2020).
9   David Graeber, Debt: *The First 5,000 Years*, Brooklyn, NY: Melville House, 2011
10 Ibid
11 Daron Acemoglu and James A. Robinson, *Why Nations Fail: The Origins of Power, Prosperity, and Poverty*, New York: Crown Business, 2012).
12 Michael Sandel, interview by Madhavdas Gopalakrishnan, *Bennett University Thought Leaders Series*, Times Now, February 24, 2024,
13 Frank Dikötter, Mao's Great Famine: The History of China's Most Devastating Catastrophe (New York: Walker & Company, 2010), estimates up to 45 million deaths during the Great Leap Forward. Combined with Cultural Revolution casualties, total deaths under Mao's regime are often cited as exceeding 70 million. See also R.J. Rummel, China's Bloody Century (New Brunswick: Transaction Publishers, 1991), who estimates the toll at approximately 77 million.
14 Banerjee, Abhijit V., and Esther Duflo. *Poor Economics: A Radical Rethinking of the Way to Fight Global Poverty*. New York: PublicAffairs, 2011.
15 Ludwig von Mises, *Human Action: A Treatise on Economics*, 1st ed. (New Haven: Yale University Press, 1949); reprint, Scholar's Edition (Auburn, AL: Ludwig von Mises Institute, 1998).
16 Murray N. Rothbard, *Man, Economy, and State: A Treatise on Economic Principles*, 1st ed. (Princeton: D. Van Nostrand, 1962); reprint, Scholar's Edition (Auburn, AL: Ludwig von Mises Institute, 2004).
17 Kenneth J. Arrow, *Social Choice and Individual Values*, 1st ed., New York: John Wiley & Sons, 1951); 3rd ed., New Haven: Yale University Press, 2012)
18 Eric Maskin and Amartya Sen, *The Arrow Impossibility Theorem*, New York: Columbia University Press, 2014)
19 Note 3
20 John Maynard Keynes, *Essays in Persuasion*, ed. Donald Moggridge, London: Palgrave Macmillan, 2010, Originally published 1932.

C24 Approaches to Democracy – Power and Politics
1   Ritchie Robertson, *The Enlightenment: The Pursuit of Happiness 1680–1790* (London: Penguin Books, 2020).
2   Note 1
3   Note 1
4   Anne Applebaum, *Twilight of Democracy: The Seductive Lure of Authoritarianism* (New York: Doubleday, 2020).
5   Daron Acemoglu and James A. Robinson, *The Narrow Corridor: States, Societies, and the Fate of Liberty*, New York: Penguin Press, 2019)
6   Note 1
7   Daron Acemoglu and Simon Johnson, *Power and Progress: Our Thousand-Year Struggle Over Technology and Prosperity*, New York: PublicAffairs, 2023).
8   Yanis Varoufakis, *Technofeudalism: What Killed Capitalism* (London: The Bodley Head, 2023).
9   Joseph E. Stiglitz, *The Road to Freedom: Economics and the Good Society* (New York: W. W. Norton & Company, 2023)
10 Note 4

Endnotes and Citations

11 Amartya Sen, *Development as Freedom*, New York: Alfred A. Knopf, 1999)
12 Thomas Piketty, *Time for Socialism: Dispatches from a World on Fire*, 2016–2021, New Haven: Yale University Press, 2021)
13 Marc Lipsitch, "We Know Enough Now to Act Decisively Against Covid-19. Social Distancing Is a Good Place to Start," *STAT*, March 18, 2020, https://www.statnews.com/2020/03/18/we-know-enough-now-to-act-decisively-against-covid-19/
14 Donald Kagan, *The Great Dialogue: History of Greek Political Thought from Homer to Polybius*, New York: Free Press, 1965)
15 Note 1
16 David Graeber and David Wengrow, *The Dawn of Everything: A New History of Humanity*, New York: Farrar, Straus and Giroux, 2021).
17 Ibid
18 Ibid
19 David Graeber, *Pirate Enlightenment, or the Real Libertalia*, New York: Farrar, Straus and Giroux, 2023)
20 James C. Scott, *The Art of Not Being Governed: An Anarchist History of Upland Southeast Asia*, New Haven: Yale University Press, 2009)
21 Polycarp Ikuenobe, "African Communal Ethics," in *The Palgrave Handbook of African Social Ethics*, ed. Nimi Wariboko and Toyin Falola (Cham: Palgrave Macmillan, 2020), 129–148, https://doi.org/10.1007/978-3-030-36490-8_8.
22 Mingsheng Wang, "Confucian Ethics of Governance and Its Values in Modern Era," in *The History and Logic of Modern Chinese Politics* (Singapore: Springer, 2021), 3–15, https://doi.org/10.1007/978-981-16-3716-2_1.
23 Roy, Radical Humanist: Selected Writings, ed. Innaiah N. (Amherst, NY: Prometheus Books, 2004)
24 Ibid
25 David Graeber, *Fragments of an Anarchist Anthropology*, Chicago: Prickly Paradigm Press, 2004).
26 Michael J. Sandel, *Democracy's Discontent: America in Search of a Public Philosophy* (Cambridge, MA: Harvard University Press, 1996)
27 Karl Popper, *The Open Society and Its Enemies*, vol. 1–2 (London: Routledge, 1945).
28 M. N. Roy, *M. N. Roy Radical Humanist: Selected Writings*. Edited by Innaiah Narisetti. Amherst, NY: Prometheus Books, 2010

C25 A Framework for Humanist Education

1 Sarah Bakewell, *Humanly Possible: the Great humanist experiment in living* (London: Penguin Press, 2023).
2 John Dewey, *Democracy and Education:* With a Critical Introduction by Patricia H. Hinchey, New York: Myers Education Press, 2018; orig. pub. 1916)
3 Wilhelm von Humboldt, The Limits of State Action, ed. and trans. J.W. Burrow (Cambridge: Cambridge University Press, 1969)
4 Thomas Piketty, *Capital in the Twenty-First Century*, Cambridge, MA: Harvard University Press, 2014)
5 Alfred North Whitehead, *The Aims of Education and Other Essays*, New York: Macmillan, 1929)
6 R. Buckminster Fuller, *Education Automation: Comprehensive Learning for Emergent Humanity*, Carbondale: Southern Illinois University Press, 1962)
7 Alfred North Whitehead, *The Aims of Education and Other Essays*, New York: Macmillan, 1929)
8 Ibid
9 Note 2
10 John Allen Paulos, *Innumeracy: Mathematical Illiteracy and Its Consequences*, New York: Hill and Wang, 1988)
11 Ibid

12 Jessica Wolpaw Reyes, "Environmental Policy as Social Policy? The Impact of Childhood Lead Exposure on Crime," *The B.E. Journal of Economic Analysis & Policy* 7, no. 1, 2007: Article 51. https://doi.org/10.2202/1935-1682.1796.
13 World Health Organization. "Health Community Calls for Urgent Action for Clean Air Ahead of WHO Conference." *WHO News Release*, January 27, 2025, https://www.who.int/news/item/27-01-2025-health-community-calls-for-urgent-action-for-clean-air-ahead-of-who-conference
14 Barbara A. Oakley, *A Mind for Numbers: How to Excel at Math and Science, Even If You Flunked Algebra*, New York: Jeremy P. Tarcher/Penguin, 2014)
15 Denis Diderot. *Selected Writings*. Translated by Leonard W. Tancock, Macmillan, 1966. ISBN: 9780020845803.
16 Rachel Carson, *The Sense of Wonder*, New York: Harper & Row, 1965)
17 Lucy Jones, *Losing Eden: Why Our Minds Need the Wild* (London: Allen Lane, 2020).
18 Sven Lindqvist, *Dig Where You Stand: How to Research Your Job*, edited by Andrew Flinn and Astrid von Rosen, translated by Ann Henning Jocelyn, London: Repeater Books, 2023)
19 Ibid
20 Richard P. Feynman, T*he Quotable Feynman*, ed. Michelle Feynman, Princeton, NJ: Princeton University Press, 2015)
21 John H. Lienhard, *The Engines of Our Ingenuity: An Engineer Looks at Technology and Culture*, Oxford University Press, 2003)
22 Salman Khan, *The One World Schoolhouse: Education Reimagined* (New York: Twelve, 2012).
23 Juvenal, Satires, trans. Niall Rudd, Oxford: Oxford University Press, 1991, Satire 10, line 81.

Part 9 Directions Towards a Better World
1 Carl Sagan, *Pale Blue Dot: A Vision of the Human Future in Space*, New York: Random House, 1994).

C26 An Ape Looks at a Pale Blue Dot
1 Carl Sagan, *The Varieties of Scientific Experience: A Personal View of the Search for God*, ed. Ann Druyan, New York: Penguin Press, 2006
2 Tony Reichhardt, "The First Photo from Space," *Air & Space Magazine*, Smithsonian Institution, March 2012, https://www.smithsonianmag.com/air-space-magazine/the-first-photo-from-space-13721411/
3 E. O. Wilson *Half-Earth: Our Planet's Fight for Life*. New York: Liveright Publishing Corporation, 2016.
4 "View of Earth from V-2 #13, Launched October 24, 1946," White Sands Missile Range and *Johns Hopkins University Applied Physics Laboratory*, https://www.smithsonianmag.com/air-space-magazine/the-first-photo-from-space-13721411/
5 Robert Williams et al., "The Hubble Deep Field: Observations, Data Reduction, and Galaxy Photometry," *The Astronomical Journal* 112, no. 4, 1996: 1335–1389, https://ui.adsabs.harvard.edu/abs/1996AJ....112.1335W/abstract
6 NASA, "27 Key Images in Hubble History," https://www.nasa.gov/feature/27-key-images-in-hubble-history/
7 Carl Sagan, *Pale Blue Dot: A Vision of the Human Future in Space*, New York: Random House, 1994.
8 NASA/JPL-Caltech, *Pale Blue Dot Revisited*, image processed by Kevin M. Gill with input from Candy Hansen and William Kosmann, February 12, 2020, The Pale Blue Dot - Download - NASA Science
9 Ibid
10 Ibid

C27 Augmenting Reasoned Discourse
1 Joseph Luft and Harrington Ingham, "The Johari Window, a Graphic Model of Interpersonal Awareness," *Proceedings of the Western Training Laboratory in Group Development*, University of California, Los Angeles, 1955.

Endnotes and Citations

2 Ibid
3 E. O. Wilson, *Consilience: The Unity of Knowledge*, New York: Alfred A. Knopf, 1998
4 Douglas Wilson, "Seven Theses on the Age of the Earth," *Blog & Mablog*, March 7, 2014, https://dougwils.com/books-and-culture/s21-atheism-and-apologetics/seven-theses-on-the-age-of-the-earth.html
5 Philip Delaquis and Barbara Miller, directors. *Wisdom of Happiness*. Das Kollektiv für audiovisuelle Werke and Mons Veneris Films, 2024. Documentary film. Available via Gere Foundation, https://gerefoundation.org/en/wisdom-of-happiness/
6 Thupten Jinpa, *Illuminating the Mind: Exploring Buddhism and Science with the Dalai Lama*, Mind & Life Institute and Wisdom Academy, transcript of Lesson 1, https://wisdomexperience.org/wp-content/uploads/2021/10/Lesson-01_Transcript_Thupten-Jinpa.pdf
7 Sam Harris, *Waking Up: A Guide to Spirituality Without Religion*, New York: Simon & Schuster, 2014
8 Francis Bacon, *Novum Organum: With Other Parts of the Great Instauration*, ed. Peter Urbach and John Gibson, Chicago: Open Court Publishing, 1994, originally published 1620.
9 Martin Rees, *On the Future: Prospects for Humanity*, Princeton, NJ: Princeton University Press, 2018.
10 Ritchie Robertson. The *Enlightenment: The Pursuit of Happiness, 1680–1790*. London: Penguin Books, 2020.
11 Amartya Sen, *The Idea of Justice*, London: Penguin Books, 2010.
12 Bertrand Russell, *The Practice and Theory of Bolshevism* Originally published 1920. Scholar's Choice Edition, Creative Media Partners, LLC, 2015.
13 Bertrand Russell, *Why I Am Not a Christian: And Other Essays on Religion and Related Subjects*, ed. Paul Edwards, London: George Allen & Unwin, 1957
14 Bertrand Russell, *The Practice and Theory of Bolshevism*. Originally published 1920. Scholar's Choice Edition, Creative Media Partners, LLC, 2015.
15 Note 16
16 Keith E. Stanovich, *What Intelligence Tests Miss: The Psychology of Rational Thought*, New Haven: Yale University Press, 2009
17 Ibid
18 Ibid
19 Ibid
20 Ibid
21 David Hume, *A Treatise of Human Nature*, ed. David Fate Norton and Mary J. Norton (Oxford: Oxford University Press, 2000; originally published 1739–40) Book III, Part I, Section I
22 Andrew E. Dessler, *Introduction to Modern Climate Change*, 3rd ed., Cambridge: Cambridge University Press, 2021
23 Calvin Beisner, "A Special Letter from Cornwall Alliance President E. Calvin Beisner," *Cornwall Alliance*, July 31, 2025, https://cornwallalliance.org/a-special-letter-from-cornwall-alliance-president-e-calvin-beisner/
24 Darren Woods, quoted in Kevin Crowley, "Exxon CEO Slams EU Climate and Human Rights Directive as Threat to Oil and Gas Competitiveness," *World Oil*, September 18, 2025, https://www.worldoil.com/news/2025/9/18/exxon-ceo-slams-eu-climate-and-human-rights-directive-as-threat-to-oil-and-gas-competitiveness/
25 Donald J. Trump, "Statement by President Trump on the Paris Climate Accord," *The White House*, June 1, 2017, https://trumpwhitehouse.archives.gov/briefings-statements/statement-president-trump-paris-climate-accord/
26 Note 14
27 Sharon Street, "A Darwinian Dilemma for Realist Theories of Value," *Philosophical Studies* 127, no. 1 (2006): 109–166

Chapter 28 From Cleverness to Wisdom
1 Buckminster Fuller, *Critical Path*, New York: St. Martin's Press, 1981

2   Garrett Hardin, *Living Within Limits: Ecology, Economics, and Population Taboos*, New York: Oxford University Press, 1993
3   Ibid
4   Rachel L. Carson, *Silent Spring* (1962; London: Penguin, 2002).
5   Curtice, Brian. "Supersaurus May Have Been the Longest Dinosaur to Have Ever Lived." *Modern Sciences*, December 2, 2021. https://modernsciences.org/supersaurus-may-have-been-the-longest-dinosaur-to-have-ever-lived/
6   Robert H. Gray, "The Extended Kardashev Scale," *The Astronomical Journal* 159, no. 3 (March 2020): 112, https://doi.org/10.3847/1538-3881/ab792b.
7   Richard Feynman, "The Value of Science." *Engineering and Science* 19, no. 12, 1955: 13–15.
8   Carl Sagan, *Pale Blue Dot: A Vision of the Human Future in Space*, New York: Random House, 1994.
9   Ibid
10 Eric M. Jones, "'Where Is Everybody?' An Account of Fermi's Question," *Los Alamos National Laboratory Report* LA-10311-MS, March 1985, https://www.osti.gov/biblio/5746675
11 Robin Hanson, "The Great Filter – Are We Almost Past It?" last modified September 15, 1998, http://mason.gmu.edu/~rhanson/greatfilter.html.
12 Carl Sagan, *A Path Where No Man Thought: Nuclear Winter and the End of the Arms Race* (New York: Random House, 1990), 26.
13 Rachel Carson, *Man's War Against Nature*, London: Penguin UK, 2021.
14 Annie Jacobsen, *Nuclear War: A Scenario*, New York: Dutton, 2024.
15 Amartya Sen, *The Idea of Justice*, London: Belknap Press, 2011.
16 Richard York and Brett Clark, *The Science and Humanism of Stephen Jay Gould*, New York: Monthly Review Press, 2011.
17 James Baldwin, *No Name in the Street*, New York: Dial Press, 1972, 47.
18 Albert Einstein, "My Credo," in *Living Philosophies*, ed. Clifton Fadiman, New York: Simon and Schuster, 1931, https://einstein-website.de/en/credo/
19 Albert Camus, *Return to Tipasa*, in *Lyrical and Critical Essays*, trans. Ellen Conroy Kennedy, New York: Vintage, 1970)
20 Leon C. Megginson, "Lessons from Europe," *Southwestern Social Science Quarterly* 44, no. 1, 1963)
21 Marcus Aurelius, *Meditations*, trans. Gregory Hays, New York: Modern Library, 2002, Book 6, section 36.

Appendices
1   Stephen Hawking, *Black Holes and Baby Universes and Other Essays*, New York: Bantam Books, 1993, 14.
2   Pierre-Simon Laplace, *The Analytic Theory of Probabilities*, trans. Richard J. Pulskamp, Cincinnati: Xavier University, 2021). Originally published as *Théorie Analytique des Probabilités*, 2nd ed., Paris: Courcier, 1814).
3   W.S. Anglin, *Mathematics: A Concise History and Philosophy*, New York: Springer-Verlag, 1994, 5
4   *Understanding Humanism: War and Peace – A Humanist Perspective*, Humanists UK, 2021.

C29 Aspects from the Philosophy of Science
1   Martin Curd, J. A. Cover, and Chris Pincock, *Philosophy of Science: The Central Issues*, 2nd ed. (New York: W. W. Norton & Company, 2013)

C30 Plausible reasoning and some probability calculations
1   Federation of American Scientists, *Home*, https://fas.org
2   L. Rastrigin, *This Chancy, Chancy, Chancy World*, trans. R. H. M. Woodhouse, Moscow: Mir Publishers, 1984)
3   Ibid
4   Harold Jeffreys, *Scientific Inference*, 3rd ed. (Cambridge: Cambridge University Press, 1973).

5 Jordan Ellenberg, *How Not to Be Wrong: The Hidden Maths of Everyday Life*, London: Allen Lane, 2014).
6 Colin Howson and Peter Urbach, Scientific Reasoning: The Bayesian Approach, 2nd ed. (Chicago: Open Court, 2006).
7 Jessica Wolpaw Reyes, "Environmental Policy as Social Policy? The Impact of Childhood Lead Exposure on Crime," *The B.E. Journal of Economic Analysis & Policy* 7, no. 1, 2007: Article 51. https://Doi.org/10.2202/1935-1682.1796.
8 Colin Howson and Peter Urbach, Scientific Reasoning: The Bayesian Approach, 2nd ed. (Chicago: Open Court, 2006).
9 Henri Poincaré, *The Calculus of Probabilities*, trans. E.T. Bell, New York: Dover Publications, 1960).
10 Robert H. Gray, "The Extended Kardashev Scale," *The Astronomical Journal* 159, no. 3 (March 2020): 112, https://doi.org/10.3847/1538-3881/ab792b.

C31 The Allure of Mathematics and Moral Certainty
1 Derek Parfit, *On What Matters*, vol. 1 (Oxford: Oxford University Press, 2011)
2 Roger Penrose, *Shadows of the Mind: A Search for the Missing Science of Consciousness*, Oxford: Oxford University Press, 1994)
3 Eugene P. Wigner, "The Unreasonable Effectiveness of Mathematics in the Natural Sciences," *Communications on Pure and Applied Mathematics* 13, no. 1, 1960: 1–14, https://personal.lse.ac.uk/ROBERT49/teaching/ph201/Week15_xtra_Wigner.pdf
4 Brian Davies, *Science in the Looking Glass: What Do Scientists Really Know?*, Oxford: Oxford University Press, 2003).
5 Timothy Gowers, *Mathematics: A Very Short Introduction*, Oxford: Oxford University Press, 2002).

C32 Afterword - In Defence of Depth
1 C.P. Cavafy, Collected Poems, trans. Edmund Keeley and Philip Sherrard (Princeton, NJ: Princeton University Press, 1992), 36.

C33 Select Bibliography
1 Lytton Strachey, *Eminent Victorians*, originally published 1918 (London: Penguin Books, 1989)

Index

A

Acemoglu and Johnson
  Technology and power, 330

Acemoglu and Robinson
  Inclusive vs extractive institutions, 316
  Narrow Corridor, 329

Adventure Consultants, 48

Aesthetics
  Part of modern humanism, 249

Age of the Earth. *See* Patterson, Clair

AI and automation
  Humanist education of displaced workers, 349

Axelrod, Robert. See:Competition and cooperation – Axelrod's simulations

B

Bacon, Francis
  Idols of the mind, 363

Banerjee and Duflo
  Economic experiments, 320

Berners-Lee, Tim, 198

Brains
  Humans, 149
  Octopuses, 157

C

Cailliau, Robert, 201

Cambrian explosion, 132

Camus, Albert
  Formidable gamble - path of persuasion, vii

Carlson, Rachel
  Education and nature, 346
  Elixirs of death, 377

Chance. *See* Evoution by natural selection - Shaped by contingency

Types, 158

Cognitive Biases, 63

Cognitive neuroscience, 71

Cold Fusion Fiasco, 98

Competition and cooperation
  Axelrod's simulations, 238
  Co-opetition, 239
  Tech Industry view, 237

Consequences of IS and OUGHT
  Part 6 - Needs, purposes, and activities, 196
  Part 7 - Planetary flourishing, 261
  Part 8 - Structures and frameworks, 289

Conservation
  Half-Earth project, 277

COVID-19, 171
  Excess deaths, 174
  Political and social pressures, 178
  Vaccine development and deployment, 177

Csikszentmihalyi, Mihaly
  Creative flow, 210

D

Darwin, Charles.
  Keeping mind free, 16

Dennett, Daniel
  Free will, 159
  Memes, 74

Dertouzos, Michael, 201

Diamond, Jared. *See* Historical Responsibility - Five lessons from History
  Geography as a key determinant of the success of societies, 150

E

Earth
  Age — Why it matters, 361
  Big five extinction events, 145

Index

First photograph from space, 353
Geological ages, 150
Pale Blue Dot, 354

Economic frameworks
Aligned with humanism, 293
China - cautionary tales, 319
India - cautionary tales, 316
Universal Basic Income, 321

Educations
Humanist education, 341

Ellsberg, Daniel
Is it possible to change the course of history, 18

Emotions
As forms of judgment, 97
Literature and philosophy, 247
Tempered by science, 247

Engineering
As a creative pursuit, 243
Different from science, 241

Environmental damage
Aral Sea, 265
Aral Sea – Lessons, 275
Other examples, 276

Epochs of thought
Across cultures and centuries, 12
The European Enlightenment, 13

ESG
Misuse of framework, 272

Ethyl Corporation
Formation, role of GM, Du Pont family, and SONJ, 28

Everest
1953 - First successful ascent, 59
First ascent in perspective, 61
Sherpas - vital role, 53
Tourism, 54
Two ascents - Case study, 45

Evolution by natural selection, 135
Bell-curve - Change and adaptation, 137
Bell-curve 3-D Multitude of traits, 138
Bell-curve of a trait in a population, 136
Local optimization, 139
Mendel's experiments and findings, 142
Modern synthesis, 143
Net like, 141
Perspectives on human behaviour, 76
Red Queen hypothesis, 140
Shaped by contingency, 155

F

Feynman, Richard P.
Universe of atoms, 6

Fleischmann, Martin. *See* Cold Fusion Fiasco

Flourishing
Aristotle's explanation, 251
As humanist activities, 255
Planetary flourishing, 288
Practical measures to encourage physical affection, 300
Practical measures to encourage physical pleasures, 300
Practical measures to enhance kindness and compassion, 301
Pre-requisites and pleasures, 253
Relational and Cognitive, 254

Free will, 159

Fuller, Buckminster
Critical Path - S curve, 373

G

Genetic drivers of cognition, 67

Gould, Stephen
Adaptations, 154
Evolution and chance, 155
Evolution without purpose, 151

Graeber, David
Bullshit jobs, 210
Debt and mutual aid, 314

H

Hall, Rob
Tolerate no dissension, 52

Hardin, Garrett
  Critique of continuous compounding, 374
  Tragedy of the commons, 278

Henderson, D. A.. *See* Smalpox - Eradication Role of

Historical responsibility
  Challenges of effective dialogues – An example of failure, 286
  Five lessons from History – Jared Diamond, 284
  Humanist approach, 280
  UN Durban Conference 2001, 269

Homo sapiens, 146
  A clever animal – but question over wisdom, 373
  Civilizational rite of passage, 381
  Four supreme challenges, 382
  Near extinction, 147

How the world IS
  Part 2 - Self-knowledge, 44
  Part 3 – Science as a test of knowledge, 83
  Part 4 - Evolution and Chance, 126

How the world OUGHT to be
  Part 5 - Humanist principles and processes, 161

Hunt, John
  Background and appointment to lead in 1953, 56

Hutton, James
  Deep time, 88

I

I, Robot
  Brief description of my personal change, 16

J

Johari framework. *See* Science - Landscape of knowledge

JWST
  James Webb Space Telescope -Example of human achievement, 19

K

Kahneman, Daniel
  Not the same person all the time, 72
  Overcoming biases, 77
  Overconfidence bias, 64
  Priming and Vohs experiments, 75
  System 1 and 2, 23

Kardesheve scale, 378

Kehoe, Robert
  Background and funding by GM, Oil companies and Ethyl Corporation, 33
  Kehoe Rule, 86
  Mexico village test, 87
  Model of biological systems, 118
  Normal level of lead in humans, 87

Kettering, Charles
  Background and pressures to perform for GM, 26

Keynes, John Maynard
  Vision for the future, 291

L

Leaded gasoline. *See* Tetraethyl Lead

Life
  Classification, 130
  Definition, 127
  Origins, 128
  Taxonomy, 131

Literature. *See* Emotion - With reflection through literature and philosophy

M

Maslow, Abraham. *See* Needs and capabilities

Memes, 72

Mendel, Gregor. *See* Evolution by natural selection

Midgley, Thomas
  Search for anti-knock additive to gasoline, 26

Milgram, Stanley
  Experiments on authority bias, 65

Modern Humanism
  Approach to individual differences, 299
  Basics of knowledge and belief, 181
  Building blocks, 180
  Conjoined triplet with science and the creative arts, 8
  Consequences of a break with science and the creative arts, 246
  Dialogic justice, 294
  Different from other meanings of humanism, v
  Discourse as a layered process, 367
  Ethical processes and decision-making, 186
  Expanding circles of concern and circles of influence, 272
  Flourishing without fixed moral truths, 272
  Foundational building bocks, 161
  From persecution, exclusion, tolerance, to humanist discourse, 363
  Human-centred values and system, 192
  Philosophical boundaries and moral humility, 189
  Worldview, 233

Monod, Jacques
  Chance and necessity, 18

Mountain Madness, 48

N

Narrative foundations of this book, 37

Needleman, Herbert, Dr.
  Method of measuring lead in children, 94

Needs and capabilities
  Maslow's model of needs, 217
  Nussbaum's capabilities framework, 219
  Placing Maslow's needs alongside Nussbaum capabilities framework, 222
  Status in 2025, 230

Normal distribution
  Explanation, 396

Nuclear war
  Binomial distribution calculation of probability, 396

Nussbaum, Martha. *See* Needs and capabilities

O

Ortega y Gasset, Jose
  Mass-Man, 297

P

Pandemics
  Modelling the spread, 164
  Most harmful in history, 163

Parfit, Derek
  Critique of his universal morality, 399

Patterson, Clair
  Age of the Earth, 88
  Experiments on lead in the environment, 89
  Lead levels in humans - pre-historic, modern, lead poisoned, 92
  Model of biological systems, 93, 118
  Opposition from oil companies, 92
  True scientific discovery is a community effort, 385

Petrov, Stanislav
  50-50 chance of nuclear catastrophe, 15

Piketty, Thomas
  Equality and participatory socialism, 310

Pons, Stanley. *See* Cold Fusion Fiasco

Popper, Karl
  Against historicism, 150
  Falsification, 105

Open society, 339

Poverty and precarity
  Global extreme poverty 2025, 264
  Shallow Pond - Peter Singer's allegory for moral action to alleviate poverty, 263

Power and politics
  Approaches to democracy, 327
  Egalitarian examples from history, 332, 333
  Non-hierarchical and rotational systems, 335

Priming, 74
  Everest tourist climbers - I want it and I want it now, 76
  Kathleen Vohs experiments, 74

Q

Questions of life
  Part 1 – What really matters, 24
  Roadmap through the book, 37

R

Rationality. *See* Modern Humanism – Discourse as a layered process
  Different from intelligence, 365

Reyes, Jessica
  Violent crime and environmental lead levels 22 year lag, 96

Robertson, Ritchie
  Book on the Enlightenment, 14

Roy, M. N.. *See* Power and politics

Russell, Bertrand
  Religions as harmful dogmas, 365

S

Sagan, Carl
  Existential question, 10
  Pale Blue Dot image, 355
  Sagan scale of civilizations, 379
  Visions we offer our children, 351

Sandel, Michael
  Criticism of market morality and meritocratic justice, 313
  Democracy-Moral foundations, 339

Science
  A test of knowledge, 83, 104
  Cannot address questions of what OUGHT to be, 123
  Design of experiments, 112
  Differences with the creative arts, 245
  Empirical basis, 104
  Feedback loops, 109
  Instruments, 111
  Landscape of knowledge, 359
  Learning process - conceptual framework, 110
  Living with uncertainties, 109
  Not dependent on authorities, 116
  Philosophy of science, 388
  Plausible reasoning, 106
  Reductionism and Emergence, 120
  Replication, 116
  Suggestions for non-specialists, 123
  Targeted observations, 114
  Theories, models, and stories, 117
  Two types of errors, 393
  Use of controls, 112

Scientists. *See* Wilson E. O. Two types of scientists
  Professional - Number worldwide, 258

Sen, Amartya
  Capabilities, justice, and ethics, 309
  Epistemic breakdown of his conception of rationality, 368
  Five instrumental freedoms, 290

Sherpas. *See* Everest - Sherpas - vital role

Singer, Peter. *See* Poverty and precarity - Shallow Pond

Smallpox, 165
  Eradication - Role of D. A. Henderson, 169
  Vaccine – from Vaccina virus, 167
  Vaccine delivery - bifurcated needle - Dr. Benjamin Rubin, 168

Social Choice
  Arrow's Impossibility Theorem,

Stanovich, Keith. *See* Rationality - Different from intelligence

Stiglitz, Joseph
  Reforming capitalism, 312

Strategies for cognitive integrity, 77

T

Technology
  Definition, 243
  S-curve / logistics function, 376

Tetraethy Lead
  Case study Leaded gasoline History and Politics, 25
  Case study - Leaded gasoline - The Science, 84
  CDC guidance on maximum lead levels in children, 35
  Effects of lead at the molecular, 95
  Lead - Effects on children, 94
  Lead - No natural level of lead in humans, 95
  Lead - Status of lead poisoning 2025, 36
  Violent crime and environmental lead levels 22 year lag, 96

Tragedy of the Commons, 278

V

Visions of life
  Part 9 - Directions towards a better world, 351

Vohs, Kathleen. *See* Priming

Voyager missions, 356

W

de Waal, Frans
  Bonobos, 69
  Chimpanzees, 68
  Human behaviour, 67

Whitehead, Alfred North
  Education - three stages, 23

Wilson, E. O. *See* Conservation - Half-Earth project
  Two types of scientists, 258

Work and Leisure
  Bullshit jobs, 210
  Creative projects, 210

Worldviews
  Affected by needs and capability deprivations, 226
  Analysis using Shanahan's AI model, 213
  Grounded in humanist principles, 232, 233
  Historic and contemporary examples from needs and capability deprivations, 228
  Interplay with needs, capabilities, and environments, 211
  Origins, 216

Z

Zuckerberg, Mark, 202

About the Author

Kai Taraporevala, born and based in Mumbai, is a writer, caregiver and physics student whose journey spans diverse cultures and fields of thought. Having lived in India, the U.K., France, the Middle East, and Singapore, Kai reflects on the deep commonalities of human experience across communities, alongside the biases and ideologies that divide us.

He holds master's degrees in physics from IIT Delhi and the University of Dundee, and an MBA from INSEAD. After decades in finance, Kai emerged from Plato's allegorical cave to embrace a life of caregiving and pursue a third master's in physics with a focus on astrophysics at the Open University, U.K., which he will complete by mid-2026. Now 60, this return was driven by lifelong curiosity, philosophical inquiry, personal reflection, and a deep sense of gratitude for the many kindnesses and privileges he's been grateful to receive.

This book is the result of a decade-long creative journey, enriched by workshops, dialogue, and the enduring insights of great humanists. It blends case studies with commentary to explore how we might live wisely in an age of immense power and deep fragility.

Kai finds joy in the outdoors and in conversations with friends. He writes to share, to connect, and to spark meaningful dialogue, inviting readers to reflect, question, imagine — and respond with compassion toward a better world.